Physics and Partial Differential Equations
Volume I

Physics and Partial Differential Equations
Volume I

Tatsien Li
Fudan University
Shanghai, People's Republic of China

Tiehu Qin
Fudan University
Shanghai, People's Republic of China

Translated by Yachun Li
Shanghai Jiao Tong University
Shanghai, People's Republic of China

Society for Industrial and Applied Mathematics
Philadelphia

Beijing, People's Republic of China

Co-published by Higher Education Press, People's Republic of China and the Society for Industrial and Applied Mathematics, Philadelphia, Pennsylvania. All rights reserved.

Copyright © 2012 Higher Education Press.

10 9 8 7 6 5 4 3 2 1

Printed in the United States of America. No part of this book may be reproduced, stored, or transmitted in any manner without the written permission of the publisher. For information, write to the Society for Industrial and Applied Mathematics, 3600 Market Street, 6th Floor, Philadelphia, PA 19104-2688 USA.

Library of Congress Cataloging-in-Publication Data
880-01 Li, Daqian.
 [880-01 Wu li xue yu pian wei fen fang cheng. English]
 Physics and partial differential equations / Tatsien Li, Tiehu Qin ; translated by Yachun Li.
    p. cm. -- (Other titles in applied mathematics)
 Includes bibliographical references and index.
 ISBN 978-1-611972-26-9
1. Mathematical physics. 2. Differential equations, Partial. I. 880-01 Qin, Tiehu. II. Title.
 QC20.7.D5L5313 2012
 530.15'5353--dc23   2012010830

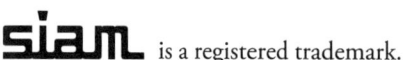

Contents

Preface to the English Edition vii

Preface to the Chinese Edition ix

1   Electrodynamics 1
   1.1   Introduction 1
   1.2   Preliminaries 2
   1.3   Maxwell's Equations in a Vacuum; Lorentz Force 13
   1.4   Electromagnetic Energy and Momentum; Conservation and Transformation Laws of Energy and Momentum 17
   1.5   Mathematical Structure of Maxwell's Equations; Wave Effect of Electromagnetic Fields 22
   1.6   Scalar Potential and Vector Potential of an Electromagnetic Field . . . 31
   1.7   Maxwell's Equations in a Medium 40
   1.8   Electrostatic Fields and Magnetostatic Fields 48
   1.9   Darwin Model 57
   Exercises 65
   Bibliography 67

2   Fluid Dynamics 69
   2.1   System of Ideal Fluid Dynamics 69
   2.2   System of Viscous Fluid Dynamics 90
   2.3   Navier–Stokes Equations 105
   2.4   Shock Waves 109
   2.5   System of One-Dimensional Fluid Dynamics in Lagrangian Representation 120
   Exercises 126
   Bibliography 128

3   Magnetohydrodynamics 131
   3.1   Plasma 131
   3.2   System of Magnetohydrodynamics 133
   3.3   System of Magnetohydrodynamics When the Conductivity σ Is Infinite 144

3.4 Mathematical Structure of Magnetohydrodynamics System 148
3.5 System of One-Dimensional Magnetohydrodynamics 152
Exercises 158
Bibliography 159

4 Reacting Fluid Dynamics **161**
4.1 Introduction 161
4.2 System of Reacting Fluid Dynamics 162
4.3 System of One-Dimensional Reacting Fluid Dynamics 169
Exercises 172
Bibliography 173

5 Elastic Mechanics **175**
5.1 Introduction 175
5.2 Description of Deformation; Strain Tensor 177
5.3 Conservation Laws; Stress Tensor 182
5.4 Constitutive Equation: Relationship Between Stress and
    Deformation 194
5.5 System of Elastodynamics and Its Mathematical Structure 207
5.6 Well-Posed Problems of the System of Elastostatics 224
Exercises 234
Bibliography 235

Appendix A   Cartesian Tensor **237**
A.1 Definition of Tensor 237
A.2 Operations of Tensor 239
A.3 Invariants of the Second-Order Symmetric Tensor 242
A.4 Isotropic Tensor 243
A.5 Differentiation of Tensor 247

Appendix B   Overview of Thermodynamics **249**
B.1 Objective of the Study of Thermodynamics 249
B.2 The First Law of Thermodynamics; Internal Energy 250
B.3 The Second Law of Thermodynamics; Entropy 250
B.4 Legendre Transform 253
B.5 Thermodynamic Functions 256
B.6 Expressions of Internal Energy and Entropy 258

Index **261**

Preface to the English Edition

The first volume of the Chinese edition of this book was published in July 1997, and the second volume was published in June 2000. In July 2000, upon the readers' request, we corrected several typographical errors and republished the first volume.

In this edition, minor typographical errors are corrected, and a small paragraph has been added to section 5.5.4 in Chapter 5, while the remaining text is unchanged.

We would like to take this opportunity to express our sincere thanks to our teachers, friends, and readers for their encouragement and support.

Tatsien Li
Tiehu Qin
Shanghai

Preface to the Chinese Edition

The fundamental equations in many important physical and mechanical disciplines are partial differential equations. Although the names of these equations are well known, and although a considerable amount of research has been done on these equations, it is not an easy task to comprehensively and profoundly understand the related physical and mechanical background. The purpose of this book is to offer some help not only to teachers, graduate students, and senior undergraduate students engaged in studying, researching, and teaching applied partial differential equations, but also to scholars and researchers in other disciplines and application areas, so that they can become proficient in the use of important fundamental equations in modern physics, gain familiarity with the whys, wherefores, and derivation of these equations, understand some commonly used mathematical models more easily, and thus not only study and use partial differential equations more consciously, but also learn to grasp some significant problems in order to properly carry out their research. Therefore, our purpose in writing this book is to build a bridge between physics and partial differential equations.

In this book, starting with the most basic concepts of physics we focus on the whole process of establishing the fundamental equations for physical and mechanical disciplines such as electrodynamics, fluid dynamics, magnetohydrodynamics, reacting fluid dynamics, elastic mechanics, thermoelastic mechanics, viscoelastic mechanics, kinetic theory of gases, special relativity, and quantum mechanics. At the same time, we give a brief description of the mathematical structures and features of these equations, including their types and basic characteristics, the behavior of solutions, and some approaches commonly used to solve these equations. We also selectively introduce some worldwide research results from recent years, including those of the authors and their research group. We hope that readers who are unfamiliar with the related physical and mechanical disciplines can gain access to the core of these disciplines in an easy-to-digest way in a short time so as to complete as soon as possible their transition from physics to mathematics and from related physical and mechanical fields to their mathematical models described by partial differential equations. On the other hand, for readers who are more familiar with the related physical and mechanical disciplines, we hope that, through their in-depth understanding of the mathematical structures and features of the fundamental equations, they will ultimately see the advantage of effective mathematical tools and expressions in more clearly presenting the basic contents of physics, and consequently use modern mathematical concepts, methods, and tools more consciously to solve related physical and mechanical problems.

This book is divided into two volumes, each consisting of five chapters. The contents of each chapter are relatively independent; however, all of the chapters echo and relate with each other to a certain extent. Exercises and a bibliography are included in each chapter.

The vast majority of chapters are not meant to be difficult for those readers who have taken basic undergraduate courses in mathematics and physics. This book can be used as a textbook for graduate courses or elective senior undergraduate courses, as well as a reference book or extracurricular reading material.

Since the second half of 1987, the contents of this book have been continuously and successfully taught in Fudan University as both an elective senior undergraduate course and a required graduate degree course. The lecture notes have been constantly supplemented and revised, and it is on these that the final version of this book is based.

The authors would like to thank Higher Education Press for its enthusiastic support of the publication of this book, and to thank Professor Sixu Guo for careful and meticulous typesetting. Thanks also go to Dr. Zhijie Cai for his responsible and proficient typing of the entire manuscript, and to Dr. Yingqiu Gu for his assistance in conforming of all the physical units in this book to the international system of units (SI units). In particular, we are grateful to Minyou Qi, Professor of Mathematics in the Department of Mathematics at Wuhan University, and Guangjiong Ni, Professor of Physics in the Department of Physics at Fudan University. They have carefully reviewed the manuscript and supplied many helpful comments and suggestions. Their hard work enriched this book.

As mathematicians, the authors may have a superficial understanding of physics to a certain extent, so errors and omissions are inevitable. We hope that readers will not spare their comments and corrections.

Tatsien Li
Tiehu Qin
November 10, 1996

Chapter 1
Electrodynamics

1.1 Introduction

Electrodynamics is aimed at the study of fundamental properties of *electromagnetic fields*, that is, the laws of motion of electromagnetic fields and their interaction with charged substances.

Electric and magnetic fields were first introduced as description tools. Gradually it was realized that they also exist as substances which have their own laws of motion and can interact with charged substances. Moreover, they are also equipped with basic features of motion such as energy and momentum, which can be converted into each other in the interaction with charged substances.

In order to understand electromagnetic fields, we need to know how *electric field intensity* $\boldsymbol{E} = (E_1, E_2, E_3)$ and *magnetic induction intensity* $\boldsymbol{B} = (B_1, B_2, B_3)$ change with the spatial location (x_1, x_2, x_3) and time t; i.e., we need to grasp the functions $\boldsymbol{E}(t, x_1, x_2, x_3)$ and $\boldsymbol{B}(t, x_1, x_2, x_3)$, from which we can figure out the distribution and change laws of the electromagnetic fields, calculate the forces acting on charged substances by electromagnetic fields (given by the *Lorentz force formula*), and calculate energy and momentum of electromagnetic fields.

The laws of motion of electromagnetic fields are described by a system of partial differential equations in terms of $\boldsymbol{E}$ and $\boldsymbol{B}$; these are called *Maxwell's equations*. Under prescribed conditions for specific problems, we can solve $\boldsymbol{E}(t, x_1, x_2, x_3)$ and $\boldsymbol{B}(t, x_1, x_2, x_3)$ from Maxwell's equations so as to determine the electromagnetic field under consideration and all of its features. Just like $F = ma$ in Newtonian mechanics, Maxwell's equations, the fundamental equations in electrodynamics, are the basis and starting point for any discussion related to electromagnetic fields.

Therefore, the study of electrodynamics can be roughly summarized as follows:

(a) To establish Maxwell's equations according to the basic features of electromagnetic fields, which are the basic equations and general framework of electrodynamics.

(b) To simplify Maxwell's equations according to various specific situations, so as to obtain their solutions and carry on the corresponding physical interpretation.

We will focus on (a) while giving some explanations for selective points of (b).

1.2 Preliminaries

First we recall some necessary preliminaries to be used in the establishment of Maxwell's equations, including electric field, magnetic field, and electromagnetic induction, etc., and we give their corresponding mathematical forms. In the next section, we will combine them organically into the desired Maxwell's equations.

1.2.1 Coulomb's Law, Divergence, and Curl of Electrostatic Field

1.2.1.1 Coulomb's Law; Electric Field Intensity

Coulomb's law resulted from experiments and is the foundation of the theory of electrostatics. It is stated as follows: Let q and q_1 be two point electric charges at rest in a vacuum, and let $\boldsymbol{r}_1$ be the position vector in the direction from charge q_1 to charge q, with distance $r_1 = |\boldsymbol{r}_1|$. Then charge q_1 acts upon charge q with a force

$$\boldsymbol{F} = k \frac{q q_1 \boldsymbol{r}_1}{r_1^3}, \tag{1.1}$$

where $k > 0$ is a constant depending on the system of units. In the international standard system of units (SI units), which we employ here, force, distance, and charge are measured by Newton (N), meter (m), and Coulomb (C), respectively, and

$$k = \frac{1}{4\pi \varepsilon_0}, \tag{1.2}$$

where $\varepsilon_0 = 8.85419 \times 10^{-2}\, \mathrm{C}^2/(\mathrm{N} \cdot \mathrm{m}^2)$ is called the *dielectric constant* in a vacuum.

One should note that Coulomb's law is valid only for a point charge at rest in a vacuum.

For the case of several point charges q, q_i ($i = 1, \ldots, n$), according to the principle of superposition of vectors, the force upon q due to other charges q_i ($i = 1, \ldots, n$) is given by

$$\boldsymbol{F} = \frac{1}{4\pi \varepsilon_0} \sum_{i=1}^{n} \frac{q q_i \boldsymbol{r}_i}{r_i^3},$$

where $\boldsymbol{r}_i$ is the position vector pointing from q_i to q, $r_i = |\boldsymbol{r}_i|$.

Similarly, the force acting on a point charge q due to continuous distribution of charges in the spatial domain Ω with charge density $\rho(x_1, x_2, x_3)$, from the principle of superposition, can be given by

$$\boldsymbol{F} = \frac{1}{4\pi \varepsilon_0} \int_{\Omega} \frac{q \rho \boldsymbol{r}}{r^3} \mathrm{d}V,$$

where $\boldsymbol{r}$ is the position vector pointing from volume element $\mathrm{d}V$ to point charge q, $r = |\boldsymbol{r}|$.

According to Coulomb's law, the space around the charge has a special physical property: any charge in this space has force upon it. Any space with this property is known as an *electric field*, which can be described mathematically by the vector field indicating the forces upon charges. Here the electric field is due to the existence of charges. However, we will see later that an electric field is a form of existence of substances, which can exist independently without charges (for example, changing a magnetic field can generate an electric field). The electric field generated by the static charges is called the *electrostatic field*.

1.2 Preliminaries

The force on a charge is different from point to point in an electric field. In order to describe the force in the electric field, we take the force on a static unit positive charge (test charge) at a particular point to measure the intensity of the electric field at this point, which is called the *electric field intensity*, denoted by $\boldsymbol{E} = (E_1, E_2, E_3)$. In the case of the electrostatic field, it is a function with respect to only (x_1, x_2, x_3). For an electric field evolving with time t, electric field intensity can be defined in a similar way, but the test charge must be static, and in this case, it is also a function with respect to t.

From Coulomb's law, a static charge q in the electric field with intensity $\boldsymbol{E}$ experiences a force

$$\boldsymbol{F} = q\boldsymbol{E}. \tag{1.3}$$

As to the actual measurement of electric field intensity, we should be careful that the original electric field is not changed much by introducing the test charge. Therefore, it is not necessary to take a unit positive charge as the test charge. Instead, we can take a smaller point charge q and determine the electric field intensity $\boldsymbol{E}$ from $\boldsymbol{E} = \boldsymbol{F}/q$.

From (1.1), (1.2), the intensity of an electric field caused by a single point charge q_1 is

$$\boldsymbol{E} = \frac{1}{4\pi\varepsilon_0} \frac{q_1 \boldsymbol{r}_1}{r_1^3}, \tag{1.4}$$

where $\boldsymbol{r}_1$ is the vector centered at q_1; similarly, the intensity of an electric field caused by several point charges q_i ($i = 1, \ldots, n$) can be given by

$$\boldsymbol{E} = \frac{1}{4\pi\varepsilon_0} \sum_{i=1}^{n} \frac{q_i \boldsymbol{r}_i}{r_i^3}; \tag{1.5}$$

while for an electric field generated by continuous distribution of charges in Ω with density ρ, the intensity can be obtained as

$$\boldsymbol{E}(P) = \frac{1}{4\pi\varepsilon_0} \int_{\Omega} \frac{\rho(P')\boldsymbol{r}_{P'P}}{r_{P'P}^3} \mathrm{d}V_{P'}, \tag{1.6}$$

where $P : (x_1, x_2, x_3)$, $P' : (x_1', x_2', x_3')$, $\boldsymbol{r}_{P'P} = \overrightarrow{P'P} = (x_1 - x_1', x_2 - x_2', x_3 - x_3')$, $\mathrm{d}V_{P'} = \mathrm{d}x_1' \mathrm{d}x_2' \mathrm{d}x_3'$.

The above formulas are valid only for the electric field generated by static charges.

1.2.1.2 Gauss's Law

First, we introduce the concept of *electric flux*.

We can define electric field lines after determining electric field intensity $\boldsymbol{E}$. The *electric field line* is the integral curve of vector field $\boldsymbol{E}$, i.e., the curve tangential to $\boldsymbol{E}$ everywhere, which satisfies

$$\frac{\mathrm{d}x_1}{E_1} = \frac{\mathrm{d}x_2}{E_2} = \frac{\mathrm{d}x_3}{E_3}, \tag{1.7}$$

whose direction is given by that of $\boldsymbol{E}$. Electric field lines should fill the whole space. However, in order to reflect the strength of distribution of the electric field, we normally follow this rule: the electric field lines are dense where the electric field intensity is large, and the electric field lines are sparse where the electric field intensity is small. We formulate, at a point in the electric field, that the number of electric field lines passing through a unit surface area is $\pm|\boldsymbol{E}| = \boldsymbol{E} \cdot \boldsymbol{n}$, where $\boldsymbol{n}$ is the outward unit normal to the surface, and $\boldsymbol{E} \cdot \boldsymbol{n}$ is the inner product (scalar product) of $\boldsymbol{E}$ and $\boldsymbol{n}$. If the direction of $\boldsymbol{n}$ is in accordance with $\boldsymbol{E}$, it takes a plus sign; otherwise it takes a minus sign. Then for a general element of surface area dS with outward unit normal $\boldsymbol{n}$, the number of electric field lines passing through dS along $\boldsymbol{n}$ is $\boldsymbol{E} \cdot \boldsymbol{n} dS$, which is called the *electric flux* passing through dS along $\boldsymbol{n}$. Thus the electric flux passing through an arbitrary surface S along $\boldsymbol{n}$ is $\int_S \boldsymbol{E} \cdot \boldsymbol{n} dS$, where dS is the surface element of S. From the above definition we can see that the electric flux is actually the flux of vector field $\boldsymbol{E}$ passing through the corresponding surface.

Gauss's Law. *In an electrostatic field, the outward electric flux passing through an arbitrary closed surface Γ is equal to the algebraic sum of charges enclosed in the surface Γ divided by ε_0.*

Based on Gauss's law, if Γ contains only point charges with algebraic sum Q, then

$$\int_\Gamma \boldsymbol{E} \cdot \boldsymbol{n} dS = \frac{1}{\varepsilon_0} Q; \tag{1.8}$$

for continuously distributed charges with density ρ, it holds that

$$\int_\Gamma \boldsymbol{E} \cdot \boldsymbol{n} dS = \frac{1}{\varepsilon_0} \int_\Omega \rho dV, \tag{1.9}$$

where Ω is the domain enclosed by Γ, and $\boldsymbol{n}$ is the outward unit normal to Γ.

Proof of Gauss's Law. From the principle of superposition, it is sufficient to prove the case when the inside point charge is a unit positive charge, and, without loss of generality, we assume that the charge is at the origin. Then from (1.5), the electric field intensity is

$$\boldsymbol{E} = \frac{1}{4\pi\varepsilon_0} \frac{\boldsymbol{r}}{r^3}, \tag{1.10}$$

where $\boldsymbol{r} = (x_1, x_2, x_3)$. Therefore

$$\boldsymbol{E} \cdot \boldsymbol{n} dS = \frac{1}{4\pi\varepsilon_0} \frac{1}{r^2} \cos\theta dS,$$

where θ is the angle between the outward unit normal $\boldsymbol{n}$ to dS and $\boldsymbol{r}$. It is easy to see that

$$d\omega = \frac{\cos\theta}{r^2} dS$$

is the solid angle formed by dS with respect to the origin, whose sign depends on whether θ is an acute or an obtuse angle. Since any closed surface forms a solid angle of 4π with respect to any interior point, we have

$$\int_\Gamma \boldsymbol{E} \cdot \boldsymbol{n} dS = \frac{1}{4\pi\varepsilon_0} \int_\Gamma d\omega = \frac{1}{\varepsilon_0}.$$

The proof is completed. □

1.2 Preliminaries

As a result, the electric flux passing through a closed surface depends only on the total charges inside the surface and has nothing to do with either the distribution of charges or the outside charges!

Formula (1.9) is the integral form of Gauss's law. Now we come to its differential form. By Green's formula

$$\int_\Gamma \boldsymbol{E} \cdot \boldsymbol{n} \, \mathrm{d}S = \int_\Omega \mathrm{div}\, \boldsymbol{E} \, \mathrm{d}V, \tag{1.11}$$

we obtain from (1.9) that

$$\int_\Omega \mathrm{div}\, \boldsymbol{E} \, \mathrm{d}V = \frac{1}{\varepsilon_0} \int_\Omega \rho \, \mathrm{d}V,$$

which holds for arbitrary domain Ω in the electric field, and thus we obtain the differential form of Gauss's law

$$\mathrm{div}\, \boldsymbol{E} = \frac{\rho}{\varepsilon_0}, \tag{1.12}$$

which is Gauss's law for the case of continuously distributed charges.

For point charges, we can also rewrite (1.8) as (1.12) by using the δ-function. For simplicity, we suppose that there is only one charge Q located at the origin, with density $Q\delta(x_1,x_2,x_3)$. According to (1.8) and (1.11), we still have

$$\mathrm{div}\, \boldsymbol{E} = \frac{1}{\varepsilon_0} Q\delta(x_1,x_2,x_3), \tag{1.13}$$

which is the differential form of Gauss's law for the case of point charges.

From (1.12), (1.13), we know that **the electrostatic field is a field with source**, whose source is the charge. Every unit positive charge issues an electric flux of $\frac{1}{\varepsilon_0}$, while every unit negative charge absorbs an electric flux of $\frac{1}{\varepsilon_0}$. Later on, we will see that this is an important difference between electrostatic field and magnetostatic field, and between general electric field and magnetic field as well.

Next we will show that electrostatic field is irrotational. First we prove that, for an arbitrary closed curve l, it holds that

$$\int_l \boldsymbol{E} \cdot \mathrm{d}\boldsymbol{l} = 0. \tag{1.14}$$

This implies that the circulation of electric field intensity $\boldsymbol{E}$ along an arbitrary closed curve l is zero, i.e. the work done by the electrostatic field along any closed curve is zero. As before, we prove (1.14) only for the case when the electrostatic field is generated by a single positive unit charge at the origin. The electric field intensity $\boldsymbol{E}$ is then given by (1.10), and hence

$$\oint_l \boldsymbol{E} \cdot \mathrm{d}\boldsymbol{l} = \frac{1}{4\pi\varepsilon_0} \oint_l \frac{\boldsymbol{r}}{r^3} \cdot \mathrm{d}\boldsymbol{r}.$$

where $\boldsymbol{r} \cdot \mathrm{d}\boldsymbol{r} = \frac{1}{2}\mathrm{d}(\boldsymbol{r} \cdot \boldsymbol{r}) = \frac{1}{2}\mathrm{d}(r^2) = r\,\mathrm{d}r$, and therefore

$$\oint_l \boldsymbol{E} \cdot \mathrm{d}\boldsymbol{l} = \frac{1}{4\pi\varepsilon_0} \oint_l \frac{1}{r^2} \mathrm{d}r = -\frac{1}{4\pi\varepsilon_0} \oint_l \mathrm{d}\left(\frac{1}{r}\right) = 0,$$

so (1.14) is true.

We now turn to the differential form of (1.14). From the Stokes formula, (1.14) can be rewritten as
$$\int_S \text{rot}\, \boldsymbol{E} \cdot \boldsymbol{n} \, \mathrm{d}S = 0,$$
where S is an arbitrary surface enclosed by l. Due to the arbitrariness of l and S, we immediately arrive at
$$\text{rot}\, \boldsymbol{E} = \boldsymbol{0}, \tag{1.15}$$
which implies that **the electrostatic field is an irrotational field**.

Now we introduce the concept of potential of an electrostatic field by using its irrotationality. Since the vector field $\boldsymbol{E}$ is irrotational, there exists a scalar function $\phi(x_1, x_2, x_3)$ such that
$$\boldsymbol{E} = -\text{grad}\, \phi; \tag{1.16}$$
here ϕ (which can differ up to an arbitrary additive constant) is called the *electrostatic field potential* (see Lemma 1.2 in section 1.6 of this chapter for reference). The minus sign on the right-hand side of (1.16) is used to indicate that the electric field intensity points toward the direction in which the electric potential decreases.

In the SI units, the unit of electric potential is Volt (V), and the unit of electric field intensity is Volt/meter (V/m) accordingly.

Actually, the potential function ϕ of the electrostatic field can be given by
$$\phi(x_1, x_2, x_3) = -\int_{(x_1^0, x_2^0, x_3^0)}^{(x_1, x_2, x_3)} \boldsymbol{E} \cdot \mathrm{d}\boldsymbol{l} + \phi_0, \tag{1.17}$$
where (x_1^0, x_2^0, x_3^0) is any given point inside the electric field, and ϕ_0 is an arbitrary constant. From (1.15) we know that the integral on the right-hand side of the above formula is independent of the integral path.

It is easy to verify directly that for the electrostatic field generated by the point charge Q at the origin, if it is assumed that the potential ϕ vanishes at infinity, then
$$\phi(x_1, x_2, x_3) = \frac{1}{4\pi\varepsilon_0} \frac{Q}{r},$$
where $r = \sqrt{x_1^2 + x_2^2 + x_3^2}$.

Then, for an electrostatic field generated by continuously distributed charges with density ρ in Ω, the potential ϕ is
$$\phi(P) = \frac{1}{4\pi\varepsilon_0} \int_\Omega \frac{\rho(P')}{r_{P'P}} \mathrm{d}V_{P'}.$$

We can summarize in the following: **the electrostatic field is an irrotational field with source, whose divergence is $\frac{\rho}{\varepsilon_0}$ and curl is zero**. This is a corollary of Coulomb's law. We will see later on that, although Coulomb's law is valid only for the electrostatic field and cannot apply to general fields, the conclusion that the divergence of the electric field intensity is $\frac{\rho}{\varepsilon_0}$ is still true for general fields, while the irrotationality is true only for the electrostatic field.

1.2.2 Ampère–Biot–Savart Law; Divergence and Curl of Magnetostatic Field

1.2.2.1 Electric Current Density,; Conversation Law of Charges

If the charge is not at rest, then its directional flow forms an electric current. The electric current density vector j is introduced to describe the state of the current, which is a function of t, x_1, x_2, x_3, measuring the current flow of a point in the conductor at a particular time. Its definition is the following: The *current density* j is a vector in the direction of the current flow with magnitude equal to the amount of the charge across a unit area perpendicular to the direction of current flow in unit time. Thus, for the surface element dS with unit normal vector n, the charge across dS in the direction of n in unit time (i.e., current) is

$$dI = j \cdot n\,dS.$$

In SI units, the unit of current is Ampère (A), and 1 Ampère= 1 Coulomb/second (C/s).

The *conservation law of electric charges* tells us the following: The amount of charges is conserved. Therefore, the increased amount of charges in any closed surface Γ in unit time is equal to the amount of charges flowing into the surface Γ during the same time period, i.e.,

$$\frac{d}{dt}\int_\Omega \rho\, dV = -\int_\Gamma j \cdot n\, dS, \tag{1.18}$$

where Ω is the domain enclosed by Γ, and n is the outward unit normal to Γ. Then, by Green's formula (1.11), it follows that

$$\int_\Omega \frac{\partial}{\partial t}\rho\, dV = -\int_\Omega \operatorname{div} j\, dV. \tag{1.19}$$

Due to the arbitrariness of domain Ω, (1.19) implies that

$$\frac{\partial \rho}{\partial t} + \operatorname{div} j = 0. \tag{1.20}$$

This is the differential form of the conservation law of electric charges, which is known as the *continuity equation of electric current*. If we denote the macroscopic velocity of current by v, it is easy to see that $j = \rho v$. Inserting it into (1.20), we can see that it has the same form as the continuity equation in continuum mechanics (see Chapters 2 and 5).

In the case of stationary current, although there are moving charges, the distribution of charge and current is time independent everywhere, and then we have

$$\operatorname{div} j = 0, \tag{1.21}$$

which implies that the stationary current has no source. The integral form of (1.21) is

$$\int_\Gamma j \cdot n\, dS = 0, \tag{1.22}$$

where Γ denotes any closed surface in the electric field.

In the case of stationary current, the charges in the conductor form a time-independent steady distribution. Experiments show that the electric field generated by a steady current still obeys Coulomb's law, and then can be treated as an electrostatic field, whose electric field intensity is

$$E(P) = \frac{1}{4\pi\varepsilon_0} \int_\Omega \frac{\rho(P')\mathbf{r}_{P'P}}{r_{P'P}^3} \mathrm{d}V_{P'}. \tag{1.23}$$

At this time, E is still an irrotational field with source, satisfying

$$\mathrm{div}\, E = \frac{\rho}{\varepsilon_0},$$
$$\mathrm{rot}\, E = \mathbf{0}.$$

In other words, the electric field generated by a steady current still obeys the law of electrostatic field.

1.2.2.2 Ampère–Biot–Savart Law; Magnetic Induction Intensity

It has been found that there is force upon an electrified conducting wire when it is near another conducting wire with electric current flow. If the two currents are moving in the same direction, the two wires attract each other; otherwise, they repel each other. Generally speaking, there is force upon a current element jdV when it is close to another current. Any physical object with this property is called a *magnetic field*, which can be mathematically described by the vector field reflecting the force upon current elements. Later on, we will see that the introduction of magnetic field is not only as a tool for dealing with problems, but also as a reflection of objective physical phenomena. A magnetic field is not necessarily generated by electric current; changing an electric field can also generate a magnetic field.

We consider a steady electric current distribution $j(x_1, x_2, x_3)$. Experimental results show that the force upon the current element $j(P)\mathrm{d}V_P$ at point P in this distribution due to current element $j(P')\mathrm{d}V_{P'}$ at point P' is given by

$$\frac{\mu_0}{4\pi} j(P)\mathrm{d}V_P \times \left(\frac{j(P')\mathrm{d}V_{P'} \times \mathbf{r}_{P'P}}{r_{P'P}^3} \right), \tag{1.24}$$

where $\mu_0 = 4\pi \times 10^{-7}$ Volt · second/(Amp · meter) (V·s/(A·m)) is the *magnetic conductivity* in a vacuum, and the symbol "×" between vectors represents the outer product (vector product). Thus the total force upon current element $j(P)\mathrm{d}V_P$ at point P is

$$\mathrm{d}F(P) = \frac{\mu_0}{4\pi} j(P)\mathrm{d}V_P \times \int_\Omega \frac{j(P')\mathrm{d}V_{P'} \times \mathbf{r}_{P'P}}{r_{P'P}^3}, \tag{1.25}$$

where Ω is the domain of the current distribution. This is the *Ampère–Biot–Savart law*. Let

$$B(P) = \frac{\mu_0}{4\pi} \int_\Omega \frac{j(P')\mathrm{d}V_{P'} \times \mathbf{r}_{P'P}}{r_{P'P}^3}. \tag{1.26}$$

Equation (1.25) can be rewritten as

$$\mathrm{d}F(P) = j(P)\mathrm{d}V_P \times B(P), \tag{1.27}$$

where $B(P)$ is called the *magnetic induction intensity* at point P, with unit Tesla (T), 1 Tesla = 1 N/(A·m). This relation shows that the force upon the current element depends

1.2 Preliminaries

not only on the magnitude of B but also on its direction. Moreover, the magnitude of B at one point equals the maximum force upon the unit current element at this point, and its direction is either in the same or the opposite direction of the current element when the force is zero.

Formally, formula (1.27) is similar to the the following formula of the force upon the charge element $\rho(P)\mathrm{d}V_P$ in an electrostatic field:

$$\mathrm{d}\boldsymbol{F}(P) = \rho(P)\mathrm{d}V_P \boldsymbol{E}(P), \tag{1.28}$$

while the definition (1.26) of magnetic induction intensity B is similar to the following definition of the electric field intensity:

$$\boldsymbol{E}(P) = \frac{1}{4\pi\varepsilon_0} \int_\Omega \frac{\rho(P')\boldsymbol{r}_{P'P}}{r_{P'P}^3} \mathrm{d}V_{P'}. \tag{1.29}$$

The magnetic field B generated by steady electric current is a function with respect to only (x_1, x_2, x_3), which is called the *magnetostatic field*.

1.2.2.3 Ampère Theorem

For the magnetostatic field, its magnetic induction intensity B satisfies the following.

Ampère Theorem. *For any closed curve l, it holds that*

$$\oint_l \boldsymbol{B} \cdot \mathrm{d}\boldsymbol{l} = \mu_0 \int_S \boldsymbol{j} \cdot \boldsymbol{n} \mathrm{d}S, \tag{1.30}$$

where S is any surface surrounded by l, whose direction of the outward unit normal $\boldsymbol{n}$ forms a right-hand coordinate system with respect to the rotary direction of l.

Actually, the surface integral on the right-hand side of (1.30) is independent of the choice of surface S, which can be deduced from the continuity equation (1.22) of steady currents.

Formula (1.30) is the integral form of the Ampère theorem. From the Stokes formula

$$\oint_l \boldsymbol{B} \cdot \mathrm{d}\boldsymbol{l} = \int_S \mathrm{rot}\,\boldsymbol{B} \cdot \boldsymbol{n} \mathrm{d}S, \tag{1.31}$$

using formula (1.30), and noticing the arbitrariness of S, we have

$$\mathrm{rot}\,\boldsymbol{B} = \mu_0 \boldsymbol{j}. \tag{1.32}$$

which is the differential form of the Ampère theorem. It is equivalent to the integral form (1.30) if all the quantities under consideration are smooth. This relation shows that **the magnetostatic field is a rotational field, whose curl is $\mu_0 \boldsymbol{j}$**.

Proof of the Ampère Theorem. We prove the differential form (1.32). We denote the gradient, curl, and divergence at (x_1, x_2, x_3) and (x'_1, x'_2, x'_3) by grad, rot, div and grad', rot', div', respectively.

Noticing that

$$\frac{\boldsymbol{r}_{P'P}}{r_{P'P}^3} = -\,\mathrm{grad}\,\frac{1}{r_{P'P}},$$

and using the following formula in vector analysis:

$$\text{rot}(\phi A) = \text{grad}\,\phi \times A + \phi\,\text{rot}\,A,$$

B given by (1.26) can be rewritten as

$$\begin{aligned} \boldsymbol{B}(P) &= \frac{\mu_0}{4\pi} \int_\Omega \text{grad}\,\frac{1}{r_{P'P}} \times \boldsymbol{j}(P') \mathrm{d}V_{P'} \\ &= \frac{\mu_0}{4\pi} \int_\Omega \left(\text{rot}\,\frac{\boldsymbol{j}(P')}{r_{P'P}} - \frac{1}{r_{P'P}} \text{rot}\,\boldsymbol{j}(P') \right) \mathrm{d}V_{P'} \\ &= \text{rot}\,\boldsymbol{A}(P), \end{aligned} \quad (1.33)$$

where

$$\boldsymbol{A}(P) = \frac{\mu_0}{4\pi} \int_\Omega \frac{\boldsymbol{j}(P')}{r_{P'P}} \mathrm{d}V_{P'}. \quad (1.34)$$

Thus, from formula (1.33) and the formula in vector analysis, we have

$$\text{rot}\,\boldsymbol{B}(P) = \text{rot}(\text{rot}\,\boldsymbol{A}(P)) = \text{grad}\,\text{div}\,\boldsymbol{A}(P) - \Delta\boldsymbol{A}(P), \quad (1.35)$$

where Δ is the *Laplace operator* with respect to (x_1, x_2, x_3):

$$\Delta = \frac{\partial^2}{\partial x_1^2} + \frac{\partial^2}{\partial x_2^2} + \frac{\partial^2}{\partial x_3^2}.$$

First, we calculate the first term on the right-hand side of (1.35). From (1.34), we have

$$\begin{aligned} &\text{div}\,\boldsymbol{A}(P) \\ &= \frac{\mu_0}{4\pi} \int_\Omega \text{div}\left(\frac{\boldsymbol{j}(P')}{r_{P'P}}\right) \mathrm{d}V_{P'} \\ &= \frac{\mu_0}{4\pi} \int_\Omega \text{grad}\,\frac{1}{r_{P'P}} \cdot \boldsymbol{j}(P') \mathrm{d}V_{P'} \\ &= -\frac{\mu_0}{4\pi} \int_\Omega \text{grad}'\,\frac{1}{r_{P'P}} \cdot \boldsymbol{j}(P') \mathrm{d}V_{P'} \\ &= -\frac{\mu_0}{4\pi} \int_\Omega \text{div}'\left(\frac{\boldsymbol{j}(P')}{r_{P'P}}\right) \mathrm{d}V_{P'} \\ &\quad + \frac{\mu_0}{4\pi} \int_\Omega \frac{\text{div}'\,\boldsymbol{j}(P')}{r_{P'P}} \mathrm{d}V_{P'}, \end{aligned} \quad (1.36)$$

which, by the continuity equation (1.21) of steady current and Green's formula, can be rewritten as,

$$\text{div}\,\boldsymbol{A}(P) = -\frac{\mu_0}{4\pi} \int_{\partial\Omega} \frac{\boldsymbol{j}(P')}{r_{P'P}} \cdot \boldsymbol{n}\,\mathrm{d}S_{P'}, \quad (1.37)$$

where $\partial\Omega$ is the boundary of Ω. Since Ω contains the domain occupied by the entire current, $\boldsymbol{j} \cdot \boldsymbol{n} = 0$ should hold on its boundary, then $\text{div}\,\boldsymbol{A}(P) = 0$, and thus

$$\text{grad}\,\text{div}\,\boldsymbol{A}(P) = \boldsymbol{0}. \quad (1.38)$$

1.2 Preliminaries

Now let us look at the second term on the right-hand side of (1.35). From (1.34), every component of $A(P)$ is a volume potential. We already know that the volume potential,

$$u(P) = -\frac{1}{4\pi} \int_\Omega \frac{f(P')}{r_{P'P}} \mathrm{d}V_{P'},$$

is a solution of the *Poisson equation* $\Delta u(P) = f(P)$. Then from (1.34) we have

$$\Delta A(P) = -\mu_0 j(P). \tag{1.39}$$

Equation (1.32) immediately follows from (1.35), (1.38), and (1.39). □

From the above deduction, we can also see that **the magnetostatic field is a solenoidal (source-free) field**, i.e.,

$$\mathrm{div}\, B = 0. \tag{1.40}$$

In fact, from (1.33) we have

$$\mathrm{div}\, B = \mathrm{div}\, \mathrm{rot}\, A = 0.$$

The integral form of (1.40) is

$$\int_S B \cdot n \mathrm{d}S = 0, \tag{1.41}$$

where S is any closed surface in the domain under consideration. It shows that for any closed surface, the outward magnetic induction flux is always zero; in other words, the magnetic induction flux flowing out of the closed surface always equals those flowing into the surface, so their algebraic sum is zero.

The source-free property of the magnetic field is distinct from the electric field. This shows that the magnetic field line is always a closed curve, and positive and negative magnetic charges always exist in pairs. From the symmetry of the electric and magnetic fields, it is possible to speculate on the existence of a magnetic monopole, which has been analyzed much theoretically. In spite of the report that the magnetic monopole has been found in cosmic rays, it has not yet been confirmed. This is an interesting research topic for the theoretical physical community. If the magnetic monopole exists, the magnetic field will also become a field with source, and the entire theory would be greatly modified accordingly.

Thus **the magnetic field generated by a steady electric current (magnetostatic field) is a rotational source-free field**, which can be deduced from the Ampère–Biot–Savart law. Later on, we will see that, although the Ampère–Biot–Savart law is valid only for the magnetostatic field, the source-free property ($\mathrm{div}\, B = 0$) of the magnetic field is also true for general cases. But the conclusion of $\mathrm{rot}\, B = \mu_0 j$ will contradict the conservation law of electric charges in general cases, which has to be modified.

1.2.3 Faraday's Law of Electromagnetic Induction

For the electrostatic field (electric field generated by static charges) or the electric field generated by a steady current, we already know that, for an arbitrary closed curve l, it holds that

$$\oint_l E \cdot \mathrm{d}l = 0,$$

which implies that the work done by the electric field along an arbitrary closed curve is zero. But the electric field is not necessarily generated by charges; changing the magnetic field can also generate an electric field, i.e., induction electromotive force. At this time the above conclusion is no longer valid. From experiments, Faraday summarized the following.

Faraday's law of electromagnetic induction. *The electric field circulation $\oint_l \boldsymbol{E} \cdot d\boldsymbol{l}$ along any closed curve l is proportional to the decreasing rate of magnetic induction flux $\int_S \boldsymbol{B} \cdot \boldsymbol{n} dS$ across any surface S enclosed by this curve, that is,*

$$\oint_l \boldsymbol{E} \cdot d\boldsymbol{l} = -\int_S \frac{\partial \boldsymbol{B}}{\partial t} \cdot \boldsymbol{n} dS, \tag{1.42}$$

where the direction of the outward unit normal $\boldsymbol{n}$ and the rotary direction of l constitute a right-handed coordinate system.

The minus sign in (1.42) means the following: When the magnetic induction flux increases in the direction of $\boldsymbol{n}$, i.e.,

$$\int_S \frac{\partial \boldsymbol{B}}{\partial t} \cdot \boldsymbol{n} dS > 0,$$

the induction electromotive force should inhibit this increasing trend, and thus the magnetic field generated by the induced current is always along the direction that hinders changes to the magnetic flux which produces the induced current. Therefore, as to the induction electromotive force, $\oint_l \boldsymbol{E} \cdot d\boldsymbol{l} < 0$. This is exactly *Lenz's law*.

The $\boldsymbol{E}$ in the integral on the left-hand side of (1.42) contains the electric field generated not only by changing the magnetic field, but also by static charges and steady current, since $\oint_l \boldsymbol{E} \cdot d\boldsymbol{l} = 0$ always holds for the latter.

From the Stokes formula, and noticing the arbitrariness of l and S, we obtain immediately from (1.42) the differential form of Faraday's law of electromagnetic induction,

$$\operatorname{rot} \boldsymbol{E} = -\frac{\partial \boldsymbol{B}}{\partial t}, \tag{1.43}$$

which implies that changing the magnetic field can also generate the electric field.

From Faraday's law of electromagnetic induction we can see that the irrotationality $\operatorname{rot} \boldsymbol{E} = \boldsymbol{0}$, which holds in both the electrostatic field and the electric field generated by steady current, is no longer valid in general cases, which should be modified according to (1.43). Equation (1.43) applies to the general situation, that is, the electric field generated by static charges or steady current, as well as by changing the magnetic field. This formula will be part of Maxwell's equations describing the general rules of the electromagnetic field.

Besides, to make (1.42) reasonable, the integral on its right-hand side should be independent of the choice of the surface S; that is, for two arbitrary surfaces S_1 and S_2 both enclosed by l, it holds that

$$\int_{S_1} \frac{\partial \boldsymbol{B}}{\partial t} \cdot \boldsymbol{n} dS = \int_{S_2} \frac{\partial \boldsymbol{B}}{\partial t} \cdot \boldsymbol{n} dS.$$

Thus we can deduce immediately that, for any closed surface S,

$$\frac{d}{dt} \int_S \boldsymbol{B} \cdot \boldsymbol{n} dS = 0;$$

that is, for any closed surface, the outward magnetic flux does not change in time. This can be reduced to the differential form

$$\frac{\partial}{\partial t} \operatorname{div} \boldsymbol{B} = 0. \tag{1.44}$$

Therefore if the magnetic field has no source at the beginning, for instance, there is no magnetic field ($\boldsymbol{B} = \boldsymbol{0}$) or there is only a magnetic field generated by a steady current (div $\boldsymbol{B} = 0$), then div $\boldsymbol{B} = 0$ always holds in later time, although the magnetic field may change; that is, the outward magnetic induction flux across any closed surface remains zero. This shows that, the magnetic field can still be assumed to be solenoidal, i.e.,

$$\operatorname{div} \boldsymbol{B} = 0. \tag{1.45}$$

From the preceding subsection, we know that the electric current can generate the magnetic field, and now we know that changing the magnetic field can also generate the electric field, which implies that the electric and magnetic fields are closely linked. In the following we will see that changing the electric field will also generate the magnetic field. Thus, the electric and magnetic fields are more closely connected, and a complete form of Maxwell's equations will be given accordingly.

1.3 Maxwell's Equations in a Vacuum; Lorentz Force

1.3.1 Maxwell's Equations in a Vacuum

In the previous section, we outlined some experimental laws for electromagnetic phenomena, which apply to different situations. Our goal is to establish a universal law for the electromagnetic phenomena that can completely determine the motion of electromagnetic fields and be applicable to all possible situations. This requires summarization and extension from special to general cases starting from known facts. The process relies on some assumptions. We need to verify through practice the validity of these assumptions and their corresponding conclusions which, in turn will be used as a guide to practice and play a significant role once verified by experiments.

The general law concerning the electromagnetic field was first developed by Maxwell and is called *Maxwell's equations*.

In order to get Maxwell's equations, we first list the known results in the following and then generalize them:

$$\operatorname{div} \boldsymbol{E} = \frac{\rho}{\varepsilon_0} \quad \text{or} \quad \int_\Gamma \boldsymbol{E} \cdot \boldsymbol{n} \mathrm{d}S = \frac{1}{\varepsilon_0} \int_\Omega \rho \mathrm{d}V, \tag{1.46}$$

$$\operatorname{rot} \boldsymbol{E} = -\frac{\partial \boldsymbol{B}}{\partial t} \quad \text{or} \quad \oint_l \boldsymbol{E} \cdot \mathrm{d}\boldsymbol{l} = -\int_S \frac{\partial \boldsymbol{B}}{\partial t} \cdot \boldsymbol{n} \mathrm{d}S, \tag{1.47}$$

$$\operatorname{div} \boldsymbol{B} = 0 \quad \text{or} \quad \int_S \boldsymbol{B} \cdot \boldsymbol{n} \mathrm{d}S = 0, \tag{1.48}$$

$$\operatorname{rot} \boldsymbol{B} = \mu_0 \boldsymbol{j} \quad \text{or} \quad \oint_l \boldsymbol{B} \cdot \mathrm{d}\boldsymbol{l} = \mu_0 \int_S \boldsymbol{j} \cdot \boldsymbol{n} \mathrm{d}S, \tag{1.49}$$

and

$$\frac{\partial \rho}{\partial t} + \operatorname{div} \boldsymbol{j} = 0 \quad \text{or} \quad \frac{\mathrm{d}}{\mathrm{d}t} \int_\Omega \rho \mathrm{d}V = -\int_\Gamma \boldsymbol{j} \cdot \boldsymbol{n} \mathrm{d}S. \qquad (1.50)$$

Each of the above five formulas is limited to different situations. Formula (1.46) comes from Coulomb's law, which is applicable to the electrostatic field or the electric field generated by steady current; (1.49) comes from the Ampère–Biot–Savart law, which is applicable to the magnetostatic field; (1.47) is generalized from unsteady cases, while experiments were restricted to the scope of slow changes, and whether it is valid for fast-change cases is still to be explored; as for (1.48), it implies that the magnetic field has no source, that is, there is no magnetic monopole, which is still regarded as a reasonable assumption so far; and (1.50) holds for general cases.

Now we consider how to extend the above results to general cases so as to describe the general law for the electromagnetic field.

Although (1.46) is deduced from Coulomb's law for the electrostatic field, it can be regarded as applicable to general situation since it does not conflict with experiments; in other words, although Coulomb's law is not valid for general cases, it is still true that every unit positive charge sends out an electric flux of $\frac{1}{\varepsilon_0}$. We suppose that Faraday's law of electromagnetic induction still holds for fast-change cases; that is, (1.47) holds for the general cases, and so does (1.48).

All of the above need not to be modified. But (1.49) cannot apply to unsteady cases; otherwise, it will contradict the conservation law of charges. In fact, from (1.49) we have

$$\operatorname{div} \boldsymbol{j} = \frac{1}{\mu_0} \operatorname{div} \operatorname{rot} \boldsymbol{B} = 0.$$

Then $\frac{\partial \rho}{\partial t} = 0$ follows immediately from (1.50); that is, ρ as well as $\boldsymbol{j}$ are both independent of t, which is reduced to steady cases. Since the conservation law of charges is a general law verified by experiments, (1.49) has to be modified. Note that differentiating (1.46) with respect to t gives

$$\frac{\partial \rho}{\partial t} = \varepsilon_0 \operatorname{div} \frac{\partial \boldsymbol{E}}{\partial t}. \qquad (1.51)$$

Therefore, the contradiction will be solved if we rewrite (1.49) as

$$\operatorname{rot} \boldsymbol{B} = \mu_0 \varepsilon_0 \frac{\partial \boldsymbol{E}}{\partial t} + \mu_0 \boldsymbol{j}. \qquad (1.52)$$

In fact, acting div on both sides of the above formula, we have

$$\operatorname{div}\left(\varepsilon_0 \frac{\partial \boldsymbol{E}}{\partial t} + \boldsymbol{j}\right) = 0. \qquad (1.53)$$

Equations (1.51) and (1.53) lead to equation (1.50), the conservation law of charge.

Equation (1.52) tells us that not only the current but also changing the electric field can generate the magnetic field, which is just like that changing the magnetic field can generate the electric field. Comparing (1.47) with (1.52) we find that $\frac{\partial \boldsymbol{B}}{\partial t}$ and $\frac{\partial \boldsymbol{E}}{\partial t}$ have opposite signs, which is because the generated magnetic field is determined by the right-hand spiral rule when the electric field increases, while the induced current is produced to prevent the magnetic field from increasing and so is determined by the left-hand rule when the magnetic field increases.

1.3 Maxwell's Equations in a Vacuum; Lorentz Force

Combining (1.46)–(1.48) and (1.52), we obtain *Maxwell's equations in a vacuum*, whose differential form is

$$\operatorname{div} \boldsymbol{E} = \frac{\rho}{\varepsilon_0}, \tag{1.54}$$

$$\operatorname{rot} \boldsymbol{E} = -\frac{\partial \boldsymbol{B}}{\partial t}, \tag{1.55}$$

$$\operatorname{div} \boldsymbol{B} = 0, \tag{1.56}$$

$$\operatorname{rot} \boldsymbol{B} = \mu_0 \left(\varepsilon_0 \frac{\partial \boldsymbol{E}}{\partial t} + \boldsymbol{j} \right), \tag{1.57}$$

accompanied by the equation of conservation law of charges

$$\frac{\partial \rho}{\partial t} + \operatorname{div} \boldsymbol{j} = 0. \tag{1.58}$$

We always assume ρ and $\boldsymbol{j}$ satisfy (1.58). The first two formulas of Maxwell's equations determine the divergence and curl of the electric field, while the last two formulas determine the divergence and curl of the magnetic field, and the electric field and the magnetic field are related through the second and fourth formulas. This relation is the basis on which the electromagnetic field moves in the form of wave motion.

The system of Maxwell's equations obtained above was of course only a hypothesis at first, whose correctness was verified through practice. In 1862, Maxwell derived the wave nature of the electromagnetic field from this system, predicted the existence of the *electromagnetic wave*, pointed out that its propagation speed is the speed of light, and accordingly developed the electromagnetic theory of light—light is a kind of electromagnetic wave. Twenty years later, Hertz confirmed the existence of electromagnetic waves by experimental methods, which verified the theory of Maxwell. Now we know that not only a variety of light waves, but also many other waves such as radio waves, heat rays, X-rays, γ-rays, etc. are all electromagnetic waves with wave length in different scopes, which provide sufficient evidence for the correctness of Maxwell's theory.

Here we can see that a fundamental system of equations in physical science is required not only to be a noncontradictory system (compatibility) in mathematics, but also to be consistent with the results from experiments and practice, and the latter is of fundamental importance. Maxwell's equations provide us with an important example.

If the distribution of charge and current is known in the electromagnetic field, Maxwell's equations will include six unknown functions $\boldsymbol{E} = (E_1, E_2, E_3)$ and $\boldsymbol{B} = (B_1, B_2, B_3)$, while eight equations are included formally—two more than the number of unknown functions. But we can point out that the six equations of fundamental importance in the system are

$$\varepsilon_0 \frac{\partial \boldsymbol{E}}{\partial t} - \frac{1}{\mu_0} \operatorname{rot} \boldsymbol{B} = -\boldsymbol{j}, \tag{1.59}$$

$$\frac{\partial \boldsymbol{B}}{\partial t} + \operatorname{rot} \boldsymbol{E} = \boldsymbol{0}. \tag{1.60}$$

In fact, just as we stated before, $\operatorname{div} \boldsymbol{B} = 0$ is satisfied automatically for all t as long as it holds at $t = 0$, which can be proved immediately by acting div on both sides of (1.60). Similarly, under the conservation law (1.58) of charges, (1.54) holds for all t as long as it holds at $t = 0$. To demonstrate this, acting div on both sides of (1.59) and noticing (1.58), we have

$$\varepsilon_0 \frac{\partial}{\partial t} \operatorname{div} \boldsymbol{E} = \frac{\partial \rho}{\partial t},$$

i.e.,

$$\frac{\partial}{\partial t}\left(\operatorname{div} \boldsymbol{E} - \frac{\rho}{\varepsilon_0}\right) = 0.$$

So (1.54) holds for all t as long as it holds at $t = 0$. Then the two equations without t-derivatives, (1.54) and (1.56), are reduced to additional requirements for initial values, and we need only concentrate on the remaining six equations (1.59), (1.60), whose mathematical structure will be discussed specifically later.

1.3.2 Lorentz Force

If the electric charge and its movement situation are not given, in order to determine the electromagnetic field, we have to solve Maxwell's equations together with the movement equations of the charged body. For this purpose, we are going to discuss the force acting on charge and current by the electromagnetic field.

We first summarize the known facts as follows. First, the static charge in the electrostatic field will receive force. At this time, the force upon the charge with unit volume, i.e., the volume density of force (referred to as force density for short) is

$$\boldsymbol{f} = \rho \boldsymbol{E}. \tag{1.61}$$

Besides, the steady current in the magnetostatic field will receive force. At this time, the force upon the current with unit volume, i.e., the force density, is

$$\boldsymbol{f} = \boldsymbol{j} \times \boldsymbol{B}. \tag{1.62}$$

Lorentz generalized the above formulas into the general case for the force acting upon a moving charged body in a vacuum by the electromagnetic field. The moving charged body will receive forces simultaneously from the electric field and the magnetic field because of the existence of both charge and current. Lorentz assumed, regardless of the movement state of the charged body, that the force density (the force per unit volume) for the general electromagnetic field can be determined by the following:

$$\boldsymbol{f} = \rho \boldsymbol{E} + \boldsymbol{j} \times \boldsymbol{B} \tag{1.63}$$

or

$$\boldsymbol{f} = \rho \boldsymbol{E} + \rho \boldsymbol{v} \times \boldsymbol{B}, \tag{1.64}$$

where $\boldsymbol{v}$ is the macro-velocity of electric charge flow. The above formula is called the *Lorentz force formula*. Here, the first term represents the force which this unit volume charge received from the electric field, and the second term represents the force which the moving-charge-generated current received from the magnetic field. What should be noticed is that $\boldsymbol{E}$ and $\boldsymbol{B}$ in the above formula represent the total electric field and magnetic field at this unit volume, including the electric field and the magnetic field generated by the charged body itself. The Lorentz force formula has also been verified in practice.

Maxwell's equations and the Lorentz force formula, together with the conservation law of charges, constitute the foundation of electrodynamics. By combining them with

1.4 Electromagnetic Energy and Momentum; Conservation and Transformation Laws of Energy and Momentum

the mechanical equations of motion of a charged body, we can completely determine the electromagnetic field and the movement of charge. This will be specified in the discussion about magnetohydrodynamics equations (see Chapter 3) when we study the motion of charged fluids in an electromagnetic field.

Now, based on the known fundamental laws of electromagnetic phenomena—Maxwell's equations and the Lorentz force formula—we are going to reveal the existence of electromagnetic energy and electromagnetic momentum and establish their quantitative expressions as well as the conservation and transformation laws of energy and momentum. The awareness of a new form of energy or momentum is always obtained through such a relation with well-known forms of energy and momentum that they can transform with each other yet remain conserved as a whole. Next we are going to understand electromagnetic energy and momentum by studying the change of the total mechanical energy and momentum of a charged body in an electromagnetic field.

1.4.1 Electromagnetic Energy, Conservation, and Transformation Laws of Energy

We now study the rate of change $\frac{dU_m}{dt}$ of the total mechanical energy U_m of a moving charged body in a vacuum caused by the electromagnetic field. When no any other form of energy enters, from the Lorentz force formula (1.64), we have

$$\frac{dU_m}{dt} = \int_\Omega \boldsymbol{f} \cdot \boldsymbol{v} dV = \int_\Omega (\rho \boldsymbol{E} + \rho \boldsymbol{v} \times \boldsymbol{B}) \cdot \boldsymbol{v} dV$$
$$= \int_\Omega \rho \boldsymbol{E} \cdot \boldsymbol{v} dV = \int_\Omega \boldsymbol{j} \cdot \boldsymbol{E} dV. \qquad (1.65)$$

From Maxwell's equation (1.59), we have

$$\boldsymbol{j} \cdot \boldsymbol{E} = \frac{1}{\mu_0} \boldsymbol{E} \cdot \operatorname{rot} \boldsymbol{B} - \frac{\varepsilon_0}{2} \frac{\partial (E^2)}{\partial t}, \qquad (1.66)$$

where $|\boldsymbol{E}|$ is denoted by E. Noticing

$$\operatorname{div}(\boldsymbol{E} \times \boldsymbol{B}) = \operatorname{rot} \boldsymbol{E} \cdot \boldsymbol{B} - \boldsymbol{E} \cdot \operatorname{rot} \boldsymbol{B}$$

and using Maxwell's equation (1.60), (1.66) can be rewritten as

$$\boldsymbol{j} \cdot \boldsymbol{E} = \frac{1}{\mu_0}(\operatorname{rot} \boldsymbol{E} \cdot \boldsymbol{B} - \operatorname{div}(\boldsymbol{E} \times \boldsymbol{B})) - \frac{\varepsilon_0}{2} \frac{\partial (E^2)}{\partial t}$$
$$= -\frac{1}{2} \frac{\partial}{\partial t} \left(\varepsilon_0 E^2 + \frac{1}{\mu_0} B^2 \right) - \operatorname{div} \boldsymbol{S}, \qquad (1.67)$$

where $B = |\boldsymbol{B}|$, and

$$\boldsymbol{S} = \frac{1}{\mu_0} \boldsymbol{E} \times \boldsymbol{B} \qquad (1.68)$$

is called the *Poynting vector*.

Thus, from (1.65) and Green's formula, we get

$$\begin{aligned}\frac{\mathrm{d}U_m}{\mathrm{d}t} &= -\frac{1}{2}\frac{\mathrm{d}}{\mathrm{d}t}\int_\Omega \left(\varepsilon_0 E^2 + \frac{1}{\mu_0}B^2\right)\mathrm{d}V - \int_\Omega \mathrm{div}\,\boldsymbol{S}\,\mathrm{d}V \\ &= -\frac{1}{2}\frac{\mathrm{d}}{\mathrm{d}t}\int_\Omega \left(\varepsilon_0 E^2 + \frac{1}{\mu_0}B^2\right)\mathrm{d}V - \int_{\partial\Omega} \boldsymbol{S}\cdot\boldsymbol{n}\,\mathrm{d}S.\end{aligned}$$

Since the integral domain contains the whole space of the electromagnetic field, the integral on the boundary on the right-hand side is zero. Then

$$\frac{\mathrm{d}U_m}{\mathrm{d}t} = -\frac{1}{2}\frac{\mathrm{d}}{\mathrm{d}t}\int_\Omega \left(\varepsilon_0 E^2 + \frac{1}{\mu_0}B^2\right)\mathrm{d}V. \qquad (1.69)$$

It shows that, when the electromagnetic field changes, the total mechanical energy of the charged body changes accordingly, while the total mechanical energy recovers when the electromagnetic field does. This indicates that the change of the energy U_m does not imply its elimination or creation; rather, it implies the transformation into another form of energy, that is, *electromagnetic energy*. From (1.69), the increment of U_m is the decrement of

$$U_{e,m} = \frac{1}{2}\int_\Omega \left(\varepsilon_0 E^2 + \frac{1}{\mu_0}B^2\right)\mathrm{d}V. \qquad (1.70)$$

Thus $U_{e,m}$ is the quantitative expression of electromagnetic energy, and (1.69) represents the *conservation and transformation law of electromagnetic energy*.

1.4.2 Electromagnetic Momentum; Conservation and Transformation Law of Momentum

When there is no any other external force, according to Lorentz force formula (1.64), the rate of change of the total mechanical momentum $\boldsymbol{G}_m$ of the charged body is given by

$$\frac{\mathrm{d}\boldsymbol{G}_m}{\mathrm{d}t} = \int_\Omega \boldsymbol{f}\,\mathrm{d}V = \int_\Omega (\rho\boldsymbol{E} + \rho\boldsymbol{v}\times\boldsymbol{B})\mathrm{d}V. \qquad (1.71)$$

Next we will prove

$$\frac{\mathrm{d}\boldsymbol{G}_m}{\mathrm{d}t} = -\frac{\mathrm{d}}{\mathrm{d}t}\int_\Omega \frac{1}{c^2}\boldsymbol{S}\,\mathrm{d}V, \qquad (1.72)$$

where $\boldsymbol{S}$ is the Poynting vector defined in (1.68), and $c \stackrel{\mathrm{d}}{=} \frac{1}{\sqrt{\varepsilon_0\mu_0}} = 2.997\,925 \times 10^8$ m/s is the *light speed* in a vacuum. Equation (1.72) shows that the change of the electromagnetic field will cause the change of the total mechanical momentum of the charged body, which will recover when the electromagnetic field does. Thus the change of the momentum does not

1.4 Electromagnetic Energy and Momentum

mean that it is eliminated or created, but that it is changed into another form of momentum, that is, *electromagnetic momentum*, whose quantitative expression is

$$G_{e,m} = \int_\Omega \frac{1}{c^2} S \, dV. \tag{1.73}$$

Equation (1.72) represents the *conservation and transformation law of (electromagnetic) momentum*.

***Proof of* (1.72).** From Maxwell's equations (1.54) and (1.57), we have

$$\rho = \varepsilon_0 \operatorname{div} E,$$
$$\rho v = j = \frac{1}{\mu_0} \operatorname{rot} B - \varepsilon_0 \dot{E},$$

where $\dot{E} = \frac{\partial E}{\partial t}$. Inserting it into the Lorentz force formula (1.64), we get

$$f = \varepsilon_0 (\operatorname{div} E) E + \left(\frac{1}{\mu_0} \operatorname{rot} B - \varepsilon_0 \dot{E} \right) \times B. \tag{1.74}$$

From Maxwell's equations (1.55) and (1.56), we have

$$\frac{1}{\mu_0} (\operatorname{div} B) B + \varepsilon_0 (\operatorname{rot} E + \dot{B}) \times E = 0, \tag{1.75}$$

where $\dot{B} = \frac{\partial B}{\partial t}$. Adding (1.74) and (1.75) gives

$$f = \varepsilon_0 (\operatorname{div} E) E + \frac{1}{\mu_0} (\operatorname{div} B) B$$
$$+ \left(\frac{1}{\mu_0} \operatorname{rot} B - \varepsilon_0 \dot{E} \right) \times B + \varepsilon_0 (\operatorname{rot} E + \dot{B}) \times E. \tag{1.76}$$

To simplify the right-hand side of (1.76), we first need to prove a formula in vector analysis:

$$(\operatorname{div} E) E + \operatorname{rot} E \times E = \operatorname{div}(E \otimes E) - \frac{1}{2} \operatorname{grad}(E^2), \tag{1.77}$$

where $E \otimes E$ is the *tensor product* of E and E (see Appendix A), i.e., the *dyadic vector*, whose matrix form is

$$E \otimes E = \begin{pmatrix} E_1^2 & E_1 E_2 & E_1 E_3 \\ E_2 E_1 & E_2^2 & E_2 E_3 \\ E_3 E_1 & E_3 E_2 & E_3^2 \end{pmatrix},$$

and $\operatorname{div}(E \otimes E)$ is the divergence of this second-order tensor, that is, the row vector formed by the divergence of the column vector of the above matrix.

In fact, it is clear that the component of $(\operatorname{rot} E \times E)$ in the x_1 direction is

$$(\operatorname{rot} E \times E)_1 = \left(\frac{\partial E_1}{\partial x_3} - \frac{\partial E_3}{\partial x_1} \right) E_3 - \left(\frac{\partial E_2}{\partial x_1} - \frac{\partial E_1}{\partial x_2} \right) E_2.$$

Thus, the component on the left-hand side of (1.77) in the x_1 direction is

$$\left(\frac{\partial E_1}{\partial x_1} + \frac{\partial E_2}{\partial x_2} + \frac{\partial E_3}{\partial x_3}\right) E_1 + \left(\frac{\partial E_1}{\partial x_3} - \frac{\partial E_3}{\partial x_1}\right) E_3$$
$$- \left(\frac{\partial E_2}{\partial x_1} - \frac{\partial E_1}{\partial x_2}\right) E_2,$$

while the component on the right-hand side of (1.77) in the x_1 direction is

$$\frac{\partial (E_1^2)}{\partial x_1} + \frac{\partial}{\partial x_2}(E_2 E_1) + \frac{\partial}{\partial x_3}(E_3 E_1) - E_1 \frac{\partial E_1}{\partial x_1}$$
$$- E_2 \frac{\partial E_2}{\partial x_1} - E_3 \frac{\partial E_3}{\partial x_1}$$
$$= E_1 \left(\frac{\partial E_1}{\partial x_1} + \frac{\partial E_2}{\partial x_2} + \frac{\partial E_3}{\partial x_3}\right) + \left(\frac{\partial E_1}{\partial x_3} - \frac{\partial E_3}{\partial x_1}\right) E_3$$
$$- \left(\frac{\partial E_2}{\partial x_1} - \frac{\partial E_1}{\partial x_2}\right) E_2,$$

which is equal to the component on the left-hand side of (1.77) in the x_1 direction. The same can be verified for the components in the x_2 and x_3 directions. Thus (1.77) is proved.

Similarly, we have

$$(\text{div } \boldsymbol{B})\boldsymbol{B} + \text{rot } \boldsymbol{B} \times \boldsymbol{B} = \text{div}(\boldsymbol{B} \otimes \boldsymbol{B}) - \frac{1}{2}\text{grad}(B^2). \quad (1.78)$$

By using (1.77) and (1.78), (1.76) can be rewritten as

$$\boldsymbol{f} = -\text{div}\left(\frac{1}{2}\left(\varepsilon_0 E^2 + \frac{1}{\mu_0}B^2\right)\boldsymbol{I} - \varepsilon_0 \boldsymbol{E} \otimes \boldsymbol{E} - \frac{1}{\mu_0}\boldsymbol{B} \otimes \boldsymbol{B}\right)$$
$$- \varepsilon_0(\dot{\boldsymbol{E}} \times \boldsymbol{B} + \boldsymbol{E} \times \dot{\boldsymbol{B}}),$$

where $\boldsymbol{I}$ is the unit second-order tensor, i.e., unit matrix. Here we also used the obvious relations

$$\text{grad}(E^2) = \text{div}(E^2 \boldsymbol{I}), \qquad \text{grad}(B^2) = \text{div}(B^2 \boldsymbol{I}).$$

From this and noticing definition (1.68) of the Poynting vector, we can obtain immediately

$$\boldsymbol{f} = -\text{div } \boldsymbol{\Phi} - \frac{1}{c^2}\frac{\partial \boldsymbol{S}}{\partial t}, \quad (1.79)$$

where c is the speed of light in a vacuum, and

$$\boldsymbol{\Phi} = \frac{1}{2}\left(\varepsilon_0 E^2 + \frac{1}{\mu_0}B^2\right)\boldsymbol{I} - \varepsilon_0 \boldsymbol{E} \otimes \boldsymbol{E} - \frac{1}{\mu_0}\boldsymbol{B} \otimes \boldsymbol{B}. \quad (1.80)$$

From (1.71) and (1.79), and using Green's formula, we have

$$\frac{dG_m}{dt} = -\frac{d}{dt}\int_\Omega \frac{1}{c^2}\boldsymbol{S}dV - \int_{\partial \Omega} \boldsymbol{\Phi} \boldsymbol{n} dS; \quad (1.81)$$

here and hereafter, $\boldsymbol{\Phi} \boldsymbol{n}$ represents the product between the matrix and the vector in the usual sense. Similarly to the argument about electromagnetic energy, the second term on the right-hand side of the above equation is zero. Thus (1.72) is proved. □

1.4.3 Electromagnetic Energy (Momentum) Density; Electromagnetic Energy Flux (Momentum Flux) Density

Now let us consider an arbitrarily given area Ω in the electromagnetic field. Integrating (1.67) over Ω and using Green's formula, we can get

$$\int_\Omega \boldsymbol{j} \cdot \boldsymbol{E} \mathrm{d}V + \frac{1}{2}\frac{\mathrm{d}}{\mathrm{d}t}\int_\Omega \left(\varepsilon_0 E^2 + \frac{1}{\mu_0}B^2\right)\mathrm{d}V = -\int_{\partial\Omega} \boldsymbol{S}\cdot\boldsymbol{n}\mathrm{d}S, \qquad (1.82)$$

where $\boldsymbol{n}$ is the outward unit normal vector. By (1.65), the first term on the left-hand side of the above formula is the rate of change $\frac{\mathrm{d}}{\mathrm{d}t}U_m(\Omega)$ of the total mechanical energy of the charged body. Thus, (1.82) indicates that the increment of the total energy (the sum of the total mechanical energy and electromagnetic energy) in Ω during unit time is equal to the in-flow flux of vector $\boldsymbol{S}$ across $\partial\Omega$, i.e., the in-flow electromagnetic energy. So

$$\frac{1}{2}\left(\varepsilon_0 E^2 + \frac{1}{\mu_0}B^2\right)$$

is called the *electromagnetic energy density* (i.e., the electromagnetic energy per unit volume), and the Poynting vector

$$\boldsymbol{S} = \frac{1}{\mu_0}\boldsymbol{E}\times\boldsymbol{B} \qquad (1.83)$$

is also called the *electromagnetic energy flux density vector*. Equation (1.82) is the conservation and transformation law of energy in any given finite domain, and (1.67) is its differential form.

Similarly we can discuss momentum. Integrating (1.79) over Ω, and from Green's formula, we can get

$$\int_\Omega \boldsymbol{f}\mathrm{d}V + \frac{\mathrm{d}}{\mathrm{d}t}\int_\Omega \frac{1}{c^2}\boldsymbol{S}\mathrm{d}V = -\int_{\partial\Omega}\boldsymbol{\Phi}\boldsymbol{n}\mathrm{d}S,$$

that is,

$$\frac{\mathrm{d}\boldsymbol{G}_m(\Omega)}{\mathrm{d}t} + \frac{\mathrm{d}}{\mathrm{d}t}\int_\Omega \frac{1}{c^2}\boldsymbol{S}\mathrm{d}V = -\int_{\partial\Omega}\boldsymbol{\Phi}\boldsymbol{n}\mathrm{d}S, \qquad (1.84)$$

where $\boldsymbol{G}_m(\Omega)$ represents the total mechanical momentum of the charged body in Ω. Equation (1.84) indicates that the increment of the total momentum (the sum of the total mechanical momentum and electromagnetic momentum) in Ω during unit time is equal to the in-flow flux of vector $\boldsymbol{\Phi}$ across $\partial\Omega$. Then

$$\frac{1}{c^2}\boldsymbol{S}$$

is called the *electromagnetic momentum density vector*, and $\boldsymbol{\Phi}$ is called the *electromagnetic momentum flux density tensor*, which means the following: $\boldsymbol{\Phi}\boldsymbol{n}\mathrm{d}S$ is the electromagnetic momentum across area element $\mathrm{d}S$ in the direction of normal $\boldsymbol{n}$ during unit time. Equation (1.84) is the conservation and transformation law of momentum in any given finite domain, and (1.79) is its differential form.

1.5 Mathematical Structure of Maxwell's Equations; Wave Effect of Electromagnetic Fields

1.5.1 Mathematical Structure of Maxwell's Equations

We pointed out in section 1.3 that the equations of fundamental importance in Maxwell's equations are the following six equations concerning six unknown functions $\boldsymbol{E}$ and $\boldsymbol{B}$ if the distribution of charge and current in the electromagnetic field is given:

$$\varepsilon_0\mu_0 \frac{\partial \boldsymbol{E}}{\partial t} - \operatorname{rot} \boldsymbol{B} = -\mu_0 \boldsymbol{j}, \tag{1.85}$$

$$\frac{\partial \boldsymbol{B}}{\partial t} + \operatorname{rot} \boldsymbol{E} = \boldsymbol{0}. \tag{1.86}$$

It is a *first-order system of partial differential equations*. Now let us consider its mathematical structure.

Introducing vector

$$U = (E_1, E_2, E_3, B_1, B_2, B_3)^{\mathrm{T}},$$

where the superscript "T" represents the transpose of the matrix, the system (1.85), (1.86) can be written as

$$A_0 \frac{\partial U}{\partial t} + A_1 \frac{\partial U}{\partial x_1} + A_2 \frac{\partial U}{\partial x_2} + A_3 \frac{\partial U}{\partial x_3} = F, \tag{1.87}$$

where

$$A_0 = \operatorname{diag}\{\varepsilon_0\mu_0, \varepsilon_0\mu_0, \varepsilon_0\mu_0, 1, 1, 1\}$$

is a diagonal matrix, and

$$A_1 = \begin{pmatrix} 0 & 0 & 0 & 0 & 0 & 0 \\ 0 & 0 & 0 & 0 & 0 & 1 \\ 0 & 0 & 0 & 0 & -1 & 0 \\ 0 & 0 & 0 & 0 & 0 & 0 \\ 0 & 0 & -1 & 0 & 0 & 0 \\ 0 & 1 & 0 & 0 & 0 & 0 \end{pmatrix},$$

$$A_2 = \begin{pmatrix} 0 & 0 & 0 & 0 & 0 & -1 \\ 0 & 0 & 0 & 0 & 0 & 0 \\ 0 & 0 & 0 & 1 & 0 & 0 \\ 0 & 0 & 1 & 0 & 0 & 0 \\ 0 & 0 & 0 & 0 & 0 & 0 \\ -1 & 0 & 0 & 0 & 0 & 0 \end{pmatrix},$$

$$A_3 = \begin{pmatrix} 0 & 0 & 0 & 0 & 1 & 0 \\ 0 & 0 & 0 & -1 & 0 & 0 \\ 0 & 0 & 0 & 0 & 0 & 0 \\ 0 & -1 & 0 & 0 & 0 & 0 \\ 1 & 0 & 0 & 0 & 0 & 0 \\ 0 & 0 & 0 & 0 & 0 & 0 \end{pmatrix},$$

$$F = -\mu_0 (j_1, j_2, j_3, 0, 0, 0)^{\mathrm{T}}.$$

1.5 Mathematical Structure of Maxwell's Equations

From the above we can find the following characters of the coefficient matrices of the first-order terms in the first-order system (1.87) of partial differential equations: the coefficient matrix A_0 before $\frac{\partial U}{\partial t}$ is a symmetric positively definite matrix, the coefficient matrices A_1, A_2, and A_3 before $\frac{\partial U}{\partial x_1}$, $\frac{\partial U}{\partial x_2}$, and $\frac{\partial U}{\partial x_3}$, respectively, are symmetric matrices. Such a system of first-order partial differential equations is called a *first-order symmetric hyperbolic system of partial differential equations*, which is a special type of hyperbolic system of partial differential equations, similar to general second-order hyperbolic equations. Systematic theories about it have been established by Friedrichs and Chaohao for the linear and quasi-linear cases, respectively (see [14] and [6]). Later we will see that many important mathematical physics partial differential equations can be written as linear or quasi-linear first-order symmetric hyperbolic systems. These are mathematical physics equations of significant importance, which will be introduced briefly in the following.

1.5.2 First-Order Symmetric Hyperbolic System of Partial Differential Equations

For simplicity, we consider only the linear case, which is the case for Maxwell's equations.

A general first-order linear system of partial differential equations can be written as

$$LU \stackrel{\mathrm{d}}{=} A_0 \frac{\partial U}{\partial t} + \sum_{k=1}^{n} A_k \frac{\partial U}{\partial x_k} + CU = F, \qquad (1.88)$$

where $U = (u_1, u_2, \ldots, u_N)^\mathrm{T}$, A_i ($i = 0, 1, \ldots, n$), and C are suitably smooth $N \times N$ matrix functions of t, and $x = (x_1, \ldots, x_n)$, F is an N-dimensional column vector with respect to t and x.

If A_0 is a positively definite symmetric matrix and A_k ($k = 1, \ldots, n$) are symmetric matrices, then system (1.88) is a *first-order symmetric hyperbolic system of partial differential equations*.

The fundamental method for studying such systems is the *method of energy integral*. For first-order symmetric hyperbolic system (1.88) of partial differential equations, we can pose two kinds of problems: the *Cauchy problem* (initial value problem) and the *initial-boundary value problem*.

The so-called Cauchy problem is to solve the system (1.88) in $(0, T) \times \mathbb{R}^n$ ($T > 0$) such that its solution $U = U(t, x)$ satisfies

$$U(0, x) = U_0(x), \qquad (1.89)$$

where $U_0(x)$ is a given vector function.

Next we will prove the uniqueness of solutions to Cauchy problem (1.88), (1.89) by the method of energy integral. For this, we first introduce an energy inequality. For any point $P(t^P, x^P)$ in the domain under consideration, similar to the characteristic cone of wave equations, we construct a surface S_P through point P such that it is smooth except at most at point P, and is a *weakly space-like surface* regarding operator L; that is, for any point Q other than point P on surface S_P, the matrix

$$n_0 A_0 + \sum_{k=1}^{n} n_k A_k$$

is always positively semidefinite, where $(n_0, n_1, \ldots, n_n)$ is a normal vector to the surface at this point. Obviously, the hyperplane $t = $ constant must be weakly space-like. Suppose that the initial plane $t = 0$ and S_P enclose a bounded domain. For any given $t < t^P$, we denote by Ω_t the portion of this bounded domain cut by the plane $t = t$, by Σ_t the portion of S_P between $t = 0$ and $t = t$, and by Q_t the domain enclosed by $t = 0$, $t = t$, and Σ_t. We consider the system (1.88) in domain Q_t. Taking the scalar product of U with system (1.88), it is easy to get

$$\frac{\partial}{\partial t}(U^T A_0 U) + \sum_{k=1}^{n} \frac{\partial}{\partial x_k}(U^T A_k U) + 2U^T K U = 2U^T F, \tag{1.90}$$

where

$$K = C - \frac{1}{2}\frac{\partial A_0}{\partial t} - \frac{1}{2}\sum_{k=1}^{n}\frac{\partial A_k}{\partial x_k}. \tag{1.91}$$

Integrating (1.90) over Q_t, and from Green's formula, we have

$$\int_{\Omega_t} U^T A_0 U \, dx - \int_{\Omega_0} U^T A_0 U \, dx$$
$$+ \int_{\Sigma_t} (U^T A_0 U \cos(\boldsymbol{n}, t) + \sum_{k=1}^{n} U^T A_k U \cos(\boldsymbol{n}, x_k)) dS$$
$$+ 2\int_{Q_t} U^T K U \, dx \, d\tau = 2\int_{Q_t} U^T F \, dx \, d\tau, \tag{1.92}$$

where $\boldsymbol{n}$ is the outward unit normal vector to Σ_t.

From the positivity and boundedness of A_0, there exist positive constants σ_1 and σ_2 such that

$$\int_{\Omega_t} U^T A_0 U \, dx \geq \sigma_1 \|U(t, \cdot)\|^2_{L^2(\Omega_t)}, \tag{1.93}$$

$$\left|\int_{\Omega_0} U^T A_0 U \, dx\right| \leq \sigma_2 \|U(0, \cdot)\|^2_{L^2(\Omega_0)}. \tag{1.94}$$

Since Σ_t is weakly space-like regarding L, we have

$$\int_{\Sigma_t} (U^T A_0 U \cos(\boldsymbol{n}, t) + \sum_{k=1}^{n} U^T A_k U \cos(\boldsymbol{n}, x_k)) dS \geq 0. \tag{1.95}$$

We assume temporarily that K is positively definite in Q_t; that is, it holds in Q_t that

$$U^T K U \geq \sigma_3 |U|^2, \tag{1.96}$$

where σ_3 is a positive constant. Under this assumption, we have

$$\int_{Q_t} U^T K U \, dx \, d\tau \geq \sigma_3 \|U\|^2_{L^2(Q_t)}. \tag{1.97}$$

1.5 Mathematical Structure of Maxwell's Equations

Moreover, it is obvious that

$$\left| \int_{Q_t} 2U^\mathrm{T} F \mathrm{d}x \mathrm{d}\tau \right| \leq 2\sigma_3 \|U\|^2_{L^2(Q_t)} + (2\sigma_3)^{-1} \|F\|^2_{L^2(Q_t)}. \tag{1.98}$$

From (1.92)–(1.95), (1.97), and (1.98), it is not difficult to obtain the following estimate:

$$\|U(t,\cdot)\|^2_{L^2(\Omega_t)} \leq \sigma \left(\|U_0\|^2_{L^2(\Omega)} + \int_0^t \|F(\tau,\cdot)\|^2_{L^2(\Omega_\tau)} \mathrm{d}\tau \right), \tag{1.99}$$

where σ is a positive constant.

Now let us explain the reasonableness of assumption (1.96). In fact, if (1.96) is not satisfied, we introduce, instead of U, a new unknown vector function

$$V = e^{-\lambda t} U,$$

where λ is a positive constant to be determined. It is easy to see that V satisfies the system

$$A_0 \frac{\partial V}{\partial t} + \sum_{k=1}^n A_k \frac{\partial V}{\partial x_k} + (C + \lambda A_0) V = e^{-\lambda t} F.$$

From the positive definiteness of A_0, it is easy to verify that condition (1.96) always holds for this system as long as λ is chosen to be large enough. Therefore, we can always obtain energy estimates like (1.99).

The uniqueness of solutions to Cauchy problem (1.88)–(1.89) follows immediately from the energy estimate (1.99). The existence of solutions can also be obtained by the method of energy estimate, and we skip it here. We refer the reader to [11], [4], and [10].

From energy estimate (1.99) we can also obtain the fact that the solution to Cauchy problem of system (1.88) has finite domain of dependence. In fact, it is easy to verify that for any given point $P(t^P, x^P)$, the conical surface through point P

$$|x - x^P| = a(t^P - t), \qquad 0 < t < t^P$$

is weakly space-like when a is sufficiently large. Thus the value of the solution at point P depends only on the initial value U_0 in the sphere $|x - x^P| \leq at^P$ in x space. From the fact that the solution has a finite domain of dependence, we conclude that the perturbation has a finite speed of propagation, which implies the hyperbolicity of system (1.88).

Now we discuss the initial-boundary value problem of system (1.88). Let Ω be a bounded domain in x space with smooth boundary Γ. Denote $Q = (0,T) \times \Omega$, where T is any given positive number. The so-called initial-boundary value problem is to find a solution of system (1.88) in Q such that it satisfies the initial condition at $t = 0$,

$$U(0,x) = U_0(x), \qquad x \in \Omega, \tag{1.100}$$

and the boundary condition for $t \in (0,T)$,

$$MU = 0 \quad \text{on } \Sigma = (0,T) \times \partial \Omega, \tag{1.101}$$

where $\partial \Omega$ is the boundary of Ω, and M is a $p \times N$ matrix with rank $p \leq N$.

Equation (1.101) is the general form of the linear homogeneous boundary condition, which implies that U satisfies p ($\leq N$) linearly independent linear equations at each point on lateral Σ. At each point on Σ, the solutions of linear homogeneous equations (1.101) constitute a $N-p$ dimensional subspace π of $\mathbb{R}^N$. Thus the requirement that U satisfy the boundary condition (1.101) on Σ is equivalent to the requirement that U belongs to the corresponding subspace π of $\mathbb{R}^N$ at each point of Σ. So the boundary condition (1.101) can also be written into the following form:

$$U \in \pi \quad \text{on } \Sigma. \tag{1.102}$$

Now we consider the requirement on the subspace π in order to obtain the uniqueness of solutions to the initial-boundary problem (1.88) and (1.100)–(1.101). Integrating (1.90) over the region $Q_t = (0,t) \times \Omega$ (where $t < T$), and using Green's formula, we can obtain, similar to (1.92), that

$$\int_\Omega U^T A_0 U(t,x)dx - \int_\Omega U^T A_0 U(0,x)dx$$
$$+ \int_{\Sigma_t} U^T A U dS + \int_{Q_t} 2U^T K U dx d\tau$$
$$= \int_{Q_t} 2U^T F dx d\tau, \tag{1.103}$$

where $\Sigma_t = (0,t) \times \partial\Omega$, and

$$A = \sum_{k=1}^n \cos(\boldsymbol{n},x_k)A_k, \tag{1.104}$$

where $\boldsymbol{n}$ is the outward unit normal vector to $\partial\Omega$. Similarly to the estimates for the Cauchy problem, it is not difficult to see that, for such U under condition (1.101), the quadratic form

$$U^T A U \tag{1.105}$$

is nonnegative on Σ; i.e., π is a nonnegative subspace of the above quadratic form for any point on Σ, and the following estimate follows immediately from (1.103):

$$\|U(t,\cdot)\|_{L^2(\Omega)}^2 \leq \sigma \left(\|U_0\|_{L^2(\Omega)}^2 + \int_0^t \|F(\tau,\cdot)\|_{L^2(\Omega)}^2 d\tau \right),$$
$$0 < t < T, \tag{1.106}$$

where σ is a positive constant possibly depending on T.

From the energy estimate (1.106) we obtain immediately the uniqueness of solutions of the initial-boundary value problem. However, can the condition that π is a nonnegative subspace of quadratic form (1.105) guarantee the existence of solutions of the initial-boundary value problem (1.88) and (1.100)–(1.101)? The answer is negative in general. In fact, if there exists a group of boundary conditions such that π is a nonnegative subspace of quadratic form (1.105) and the solution of the initial-boundary value problem is unique, then by adding some linear homogeneous conditions artificially to the boundary Σ, the space π will be made smaller, in which the nonnegativity of the quadratic form (1.105) will

1.5 Mathematical Structure of Maxwell's Equations

be kept accordingly. But generally speaking, we now cannot expect the existence of solutions of the corresponding problem. For this reason, to ensure the existence of solutions, we should ask the boundary space π to be the largest one in all of the subspaces which keep the nonnegativity of quadratic form $U^T A U$; that is, there does not exist a space which keeps the nonnegativity of quadratic form $U^T A U$ with π as its genuine subspace. Such a π is called the *largest nonnegative subspace* of $U^T A U$. If the corresponding boundary space π of boundary condition (1.101) at each point on Σ is the largest nonnegative subspace of $U^T A U$, then this boundary condition is called the *admissible boundary condition* with respect to operator L. It can be proved that, if the lateral Σ is noncharacteristic with respect to operator L, i.e., matrix A is nonsingular at each point on Σ, then the dimension of the largest nonnegative subspace of $U^T A U$ is equal to the number of positive eigenvalues of A. Under some additional assumptions we can also prove the existence of solutions to the initial-boundary value problem with an admissible boundary condition. For the above argument, interested readers can refer to [14] and [10].

Hence, from the previous discussion, we know that the Cauchy problem (1.85), (1.86) and the initial-boundary value problem with admissible boundary conditions to Maxwell's equations are well-posed. The energy integrals on the left-hand side of the corresponding energy estimates (1.99) and (1.106) are exactly the electromagnetic energy of the electromagnetic field in Ω_t and Ω (and may differ up to a constant factor), respectively. Besides, the perturbation in electromagnetic field has finite speed of propagation, that is, electromagnetic field has wave effect, about which we will discuss further in next subsection.

1.5.3 Wave Effect of Electromagnetic Field; Free Electromagnetic Wave

Now we further reveal the wave effect of the electromagnetic field starting from Maxwell's equations. This kind of electromagnetic field propagating in the form of a wave is called the *electromagnetic wave*. This discovery marked a new stage in the development of physics. On one hand, light and the electromagnetic field were unified on this basis, and this unity was extended to heat rays, X-rays and γ-rays, and this unity plays a significant role in revealing the microstructure of matter. On the other hand, an increasingly wide range of important applications in practice was achieved through electromagnetic waves, which also promoted more rapid development of electrodynamics.

Differentiating equation (1.85) with respect to t once, and eliminating $\boldsymbol{B}$ from equation (1.86), we can get

$$\frac{1}{c^2}\frac{\partial^2 \boldsymbol{E}}{\partial t^2} + \text{rot rot } \boldsymbol{E} = -\mu_0 \frac{\partial \boldsymbol{j}}{\partial t}, \tag{1.107}$$

where we used $c = \frac{1}{\sqrt{\varepsilon_0 \mu_0}}$. From the formula in vector analysis

$$\text{rot rot } \boldsymbol{E} = \text{grad div } \boldsymbol{E} - \Delta \boldsymbol{E} \tag{1.108}$$

and equation (1.54), (1.107) can be rewritten into

$$\frac{1}{c^2}\frac{\partial^2 \boldsymbol{E}}{\partial t^2} - \Delta \boldsymbol{E} = -\left(\frac{1}{\varepsilon_0}\text{grad } \rho + \mu_0 \frac{\partial \boldsymbol{j}}{\partial t}\right), \tag{1.109}$$

which says that $\boldsymbol{E}$ satisfies an inhomogeneous *wave equation* when ρ and $\boldsymbol{j}$ are given.

Similarly, eliminating E instead of B, we can obtain that B satisfies the corresponding inhomogeneous wave equation

$$\frac{1}{c^2}\frac{\partial^2 B}{\partial t^2} - \Delta B = \mu_0 \operatorname{rot} j. \tag{1.110}$$

In fact, differentiating (1.86) with respect to t once, using equation (1.85), formula (1.108) in vector analysis (replacing B with E), and (1.56), we can immediately obtain (1.110).

Thus, both E and B satisfy the properties described by wave equations. This implies that E and B move and change in the form of a wave, while charge and current as inhomogeneous terms are their sources, which can excite or absorb electromagnetic waves. According to the feature of wave motion, electromagnetic waves that have been excited and spread out still exist and propagate even though the sources which excite them vanish. Therefore, we have the following conclusions: **the electromagnetic field may exist alone without charge and current, and move and propagate in the form of a wave in general cases; it may interact with charge and current (the force upon charge and current due to the electromagnetic field is the Lorentz force), but its existence does not presuppose that of charge and current.** Furthermore, from the properties of wave equations, we know that **the propagation speed of electromagnetic waves in a vacuum is the light speed** c, which is one of the important foundations of the electromagnetic theory of light.

From the above we know that there may exist electromagnetic waves even if there is no charge or current. This kind of electromagnetic wave is called a *free electromagnetic wave*. At this time, the following homogeneous wave equations are satisfied:

$$\frac{1}{c^2}\frac{\partial^2 E}{\partial t^2} - \Delta E = \mathbf{0}, \tag{1.111}$$

$$\frac{1}{c^2}\frac{\partial^2 B}{\partial t^2} - \Delta B = \mathbf{0}. \tag{1.112}$$

To solve this equation, a variety of skills from mathematical physics equations can be copied here.

Here we particularly consider the *plane electromagnetic wave solution*, from the superposition of which solutions of wave equations in the general case can be obtained. Thus this kind of discussion is of general significance.

The so-called *plane wave* is the solution of the following form:

$$E = E_0 e^{i(k \cdot r - \omega t + \theta)}, \tag{1.113}$$

where $r = (x_1, x_2, x_3)$, and E_0 is a constant vector. For fixed t, it takes constant values on any hyperplane in (x_1, x_2, x_3) space

$$k \cdot r - \omega t + \theta = \text{constant}, \tag{1.114}$$

where k is the normal direction to this hyperplane; when t changes, the hyperplane with the same constant changes into another hyperplane parallel to it, on which the solution takes the same value. Therefore, the solution propagates in the form of a plane wave along the direction of k. We take ω in the above formula as the angular frequency of a wave: $\omega = 2\pi \nu$. When t passes one period $T = \frac{1}{\nu} = \frac{2\pi}{\omega}$, a wave passes the distance of one wave length λ, while the original hyperplane

$$k \cdot r - \omega t + \theta = a \text{ (constant)}$$

1.5 Mathematical Structure of Maxwell's Equations

changes into the hyperplane $\boldsymbol{k} \cdot \boldsymbol{r} - \omega(t+T) + \theta = a$, i.e.,

$$\boldsymbol{k} \cdot \boldsymbol{r} - \omega t + \theta = a + \omega T,$$

on which $\boldsymbol{E}$ takes the same value. The distance between these two hyperplanes should be $\lambda = \frac{\omega T}{|\boldsymbol{k}|} = \frac{2\pi}{k}$, where $k = |\boldsymbol{k}|$. Thus in the above-mentioned plane wave solutions, if ω is taken as an angular frequency, k should be $\frac{2\pi}{\lambda}$, where λ is wave length. $\boldsymbol{k}$ is called a *wave vector*, whose direction is the propagation direction of a wave.

Plugging plane wave solution (1.113) into wave equation (1.111), we obtain

$$\frac{1}{c^2}\omega^2 - k^2 = 0, \text{ i.e., } k = \frac{\omega}{c}, \tag{1.115}$$

from which, wave velocity = wave length × frequency = $\lambda \times \frac{\omega}{2\pi} = \frac{\omega}{k} = c$. This also proves that the propagation speed of the electromagnetic wave is the light speed c.

Now we consider the plane wave solutions

$$\boldsymbol{E} = \boldsymbol{E}_0 e^{i(\boldsymbol{k} \cdot \boldsymbol{r} - \omega t + \theta)}, \tag{1.116}$$

$$\boldsymbol{B} = \boldsymbol{B}_0 e^{i(\boldsymbol{k} \cdot \boldsymbol{r} - \omega t + \theta)}. \tag{1.117}$$

Because $\boldsymbol{B}$ and $\boldsymbol{E}$ do not change independently but are instead related by Maxwell's equations, the values of $\boldsymbol{k}$, ω, and θ in the above two formulas have to be taken the same, which can be seen by inserting them into Maxwell's equations. Furthermore, since $\rho \equiv 0$ for the free electromagnetic wave, we have from (1.54) that

$$\boldsymbol{k} \cdot \boldsymbol{E}_0 = 0. \tag{1.118}$$

Similarly, we have from (1.56) that

$$\boldsymbol{k} \cdot \boldsymbol{B}_0 = 0. \tag{1.119}$$

Inserting (1.116) and (1.117) into (1.55) and (1.57) (noticing $\boldsymbol{j} \equiv 0$ here), respectively, we can obtain

$$\boldsymbol{B}_0 = \frac{1}{\omega} \boldsymbol{k} \times \boldsymbol{E}_0, \tag{1.120}$$

$$\boldsymbol{E}_0 = -\frac{c^2}{\omega} \boldsymbol{k} \times \boldsymbol{B}_0. \tag{1.121}$$

In these three formulas (1.119), (1.120), and (1.121), (1.120) is fundamental because it obviously contains (1.119). It contains (1.121) as well, since

$$-\frac{c^2}{\omega} \boldsymbol{k} \times \boldsymbol{B}_0 = -\frac{c^2}{\omega^2} \boldsymbol{k} \times (\boldsymbol{k} \times \boldsymbol{E}_0) = \frac{c^2}{\omega^2} (\boldsymbol{k} \times \boldsymbol{E}_0) \times \boldsymbol{k}$$
$$= \frac{c^2}{\omega^2} [(\boldsymbol{k} \cdot \boldsymbol{k}) \boldsymbol{E}_0 - (\boldsymbol{k} \cdot \boldsymbol{E}_0) \boldsymbol{k}] = \boldsymbol{E}_0,$$

where we used (1.115) and (1.118). Then, for plane electromagnetic wave solution (1.116)–(1.117), we have

$$\boldsymbol{k} \cdot \boldsymbol{E}_0 = 0, \qquad \boldsymbol{B}_0 = \frac{1}{\omega} \boldsymbol{k} \times \boldsymbol{E}_0, \tag{1.122}$$

from which we know that $\boldsymbol{k}$, $\boldsymbol{E}_0$, and $\boldsymbol{B}_0$ (then $\boldsymbol{k}$, $\boldsymbol{E}$, and $\boldsymbol{B}$) are always perpendicular to each other (see Figure 1.1). Therefore we conclude that, for the free electromagnetic wave, the directions of the electric field intensity and magnetic induction intensity are perpendicular everywhere, and also perpendicular to the propagation direction of the electromagnetic wave there. So the vibration directions of the electric field and the magnetic field are both perpendicular to the propagation direction of wave, and then the electromagnetic wave is a *transverse wave*.

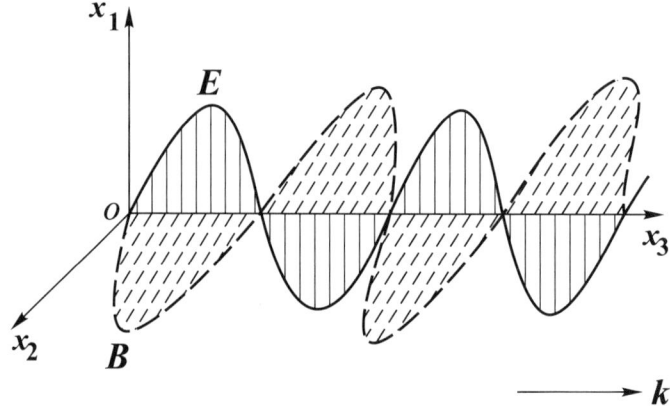

Figure 1.1.

Here we point out that the feature of the free electromagnetic wave being a transverse wave is the consequence of

$$\operatorname{div} \boldsymbol{E} = 0 \tag{1.123}$$

and

$$\operatorname{div} \boldsymbol{B} = 0. \tag{1.124}$$

Thus the above conditions are called *transverse wave conditions*; that is, the field considered is a source-free field. A field satisfying transverse wave conditions is a source-free field, also called a *transverse field*.

We mention here that if a plane wave

$$\boldsymbol{F} = \boldsymbol{F}_0 e^{i(\boldsymbol{k}\cdot\boldsymbol{r}-\omega t+\theta)}$$

satisfies the condition

$$\operatorname{rot} \boldsymbol{F} = \boldsymbol{0}, \tag{1.125}$$

then

$$\boldsymbol{k} \times \boldsymbol{F}_0 = \boldsymbol{0}, \tag{1.126}$$

which shows that the vibration direction of a wave is parallel to the propagation direction of a wave, and thus it is a *longitudinal wave*. Equation (1.125) is called a *longitudinal*

1.6 Scalar Potential and Vector Potential of an Electromagnetic Field

wave condition; that is, the field considered is an irrotational field. A field satisfying the longitudinal wave condition is an irrotational field, also called a *longitudinal field*.

1.6 Scalar Potential and Vector Potential of an Electromagnetic Field

In electrostatic fields, we can use electrostatic potential ϕ to describe an electrostatic field, and the electric field intensity $\boldsymbol{E} = -\operatorname{grad} \phi$. Since ϕ is a scalar function, while $\boldsymbol{E}$ is a vector function, it brings great convenience to our discussion to describe the electrostatic field with the electrostatic potential. Certainly, electrostatic potential ϕ is not determined uniquely and may differ up to an arbitrary additive constant, but this does not affect the determination of the electric field.

In this section, we are going to demonstrate that, for general electromagnetic fields in a vacuum, potential functions can also be introduced to describe the motion of the field and its interaction with charge and current; however, we should use both *scalar potential* and *vector potential*.

1.6.1 Preliminaries

Lemma 1.1. *If a vector field $\boldsymbol{B}$ is a transverse field, that is, it satisfies*

$$\operatorname{div} \boldsymbol{B} = 0, \tag{1.127}$$

then it can be represented as the curl of another vector field $\boldsymbol{A}$,

$$\boldsymbol{B} = \operatorname{rot} \boldsymbol{A}, \tag{1.128}$$

and vice versa.

Proof. Without loss of generality we take $A_3 = 0$, i.e., to find $\boldsymbol{A} = (A_1(x_1, x_2, x_3), A_2(x_1, x_2, x_3), 0)$ such that

$$-\frac{\partial A_2}{\partial x_3} = B_1, \quad \frac{\partial A_1}{\partial x_3} = B_2, \quad \frac{\partial A_2}{\partial x_1} - \frac{\partial A_1}{\partial x_2} = B_3. \tag{1.129}$$

Integrating the first two equations, we get

$$A_2(x_1, x_2, x_3) = -\int_{x_3^0}^{x_3} B_1(x_1, x_2, x_3) \mathrm{d}x_3 + f(x_1, x_2), \tag{1.130}$$

$$A_1(x_1, x_2, x_3) = \int_{x_3^0}^{x_3} B_2(x_1, x_2, x_3) \mathrm{d}x_3 + g(x_1, x_2), \tag{1.131}$$

where (x_1^0, x_2^0, x_3^0) is any fixed point in the field, and f and g are any functions of (x_1, x_2). Inserting (1.130), (1.131) into the third equation of (1.129), and using condition (1.127), it is clear that

$$\frac{\partial f(x_1, x_2)}{\partial x_1} - \frac{\partial g(x_1, x_2)}{\partial x_2} = B_3(x_1, x_2, x_3^0). \tag{1.132}$$

Thus, so long as f and g are chosen to satisfy the above formula, $\boldsymbol{A}$ can be given by (1.130) and (1.131). We can especially take

$$f(x_1,x_2) = \int_{x_1^0}^{x_1} B_3(x_1,x_2,x_3^0)dx_1, \quad g \equiv 0.$$

Thus taking

$$A_1(x_1,x_2,x_3) = \int_{x_3^0}^{x_3} B_2(x_1,x_2,x_3)dx_3,$$

$$A_2(x_1,x_2,x_3) = -\int_{x_3^0}^{x_3} B_1(x_1,x_2,x_3)dx_3 + \int_{x_1^0}^{x_1} B_3(x_1,x_2,x_3^0)dx_1,$$

$$A_3(x_1,x_2,x_3) = 0,$$

we get (1.128). The inverse is obvious. We complete the proof. □

Remark 1.1. The determined $\boldsymbol{A}$ is certainly not unique. Obviously, if $\boldsymbol{A}$ satisfies the requirement in the above lemma, then for an arbitrary scalar function ψ, $\boldsymbol{A}' = \boldsymbol{A} + \operatorname{grad}\psi$ still satisfies the same requirement. Lemma 1.2 following shows that the choice of $\boldsymbol{A}$ has only one such degree of freedom.

Lemma 1.2. *If a vector field $\boldsymbol{A}$ is a longitudinal field, that is, it satisfies*

$$\operatorname{rot}\boldsymbol{A} = \boldsymbol{0}, \tag{1.133}$$

then $\boldsymbol{A}$ must be the gradient of a certain scalar field,

$$\boldsymbol{A} = \operatorname{grad}\psi,$$

and vice versa.

Proof. Condition (1.133) shows that

$$A_1 dx_1 + A_2 dx_2 + A_3 dx_3$$

is a total differential, so we have

$$\psi(x_1,x_2,x_3) = \int_{(x_1^0,x_2^0,x_3^0)}^{(x_1,x_2,x_3)} A_1 dx_1 + A_2 dx_2 + A_3 dx_3 + \psi_0,$$

where ψ_0 is an arbitrary constant. The inverse is obvious. We complete the proof. □

Lemma 1.3. *Any vector field can be decomposed into the superposition of a longitudinal field and a transverse field, that is, the superposition of an irrotational field and a solenoidal field.*

Proof. For any vector field $\boldsymbol{A}$, we want to prove that there exist a scalar field ψ and a vector field $\boldsymbol{C}$ such that

$$\boldsymbol{A} = \operatorname{grad}\psi + \operatorname{rot}\boldsymbol{C}. \tag{1.134}$$

1.6 Scalar Potential and Vector Potential of an Electromagnetic Field

Noticing that div rot $C = 0$, acting on both sides of the above formula with divergence operator div, we have

$$\Delta \psi = \text{div}\, A.$$

Take ψ to be a special solution of the above Poisson equation; since

$$\text{div}(A - \text{grad}\,\psi) = 0,$$

from Lemma 1.1, there must exist C such that

$$A - \text{grad}\,\psi = \text{rot}\, C,$$

which proves (1.134). The lemma is proved. □

From Lemma 1.3 and Remark 1.1, we can conclude the following: if a vector field B is a transverse field, i.e., it satisfies div $B = 0$, then it can be represented as the curl of another vector field A: $B = \text{rot}\, A$. Such an A is not uniquely determined, which may have a degree of freedom up to an additive gradient function grad ψ, where ψ is an arbitrary scalar function. Since grad ψ is a longitudinal field, $A + \text{grad}\,\psi$ has the same properties as A, which indicates that the nature of A does not change when added by an arbitrary longitudinal field. Thus, **the longitudinal field component of vector field A in Lemma 1.1 can be taken as zero; that is, A can be assumed to be a transverse field, div A = 0.** Later we will deal with magnetostatic field like this (see section 1.8.3).

1.6.2 Scalar Potential and Vector Potential of Electromagnetic Field

In an electromagnetic field, the divergence of magnetic induction intensity B is always zero, that is, div $B = 0$. From Lemma 1.1, magnetic induction intensity B can be represented as the curl of a certain vector field A,

$$B = \text{rot}\, A. \tag{1.135}$$

Here, the transverse field part of A has definite meaning, while the longitudinal field part can be chosen arbitrarily with a degree of freedom as in Remark 1.1.

For electric field intensity E, we can not introduce electrostatic potential as in the electrostatic field, since rot $E = -\frac{\partial B}{\partial t}$ is not zero in general. But from equation (1.55) in the Maxwell's equations and (1.135), we have

$$\text{rot}\left(E + \frac{\partial A}{\partial t}\right) = \mathbf{0}, \tag{1.136}$$

and then from Lemma 1.2, there must exist a scalar field ϕ such that

$$E + \frac{\partial A}{\partial t} = -\text{grad}\,\phi. \tag{1.137}$$

Thus, to sum up, we have

$$B = \text{rot}\, A, \tag{1.138}$$

$$E = -\text{grad}\,\phi - \frac{\partial A}{\partial t}, \tag{1.139}$$

where ϕ and A are called the *scalar potential* and *vector potential* of the electromagnetic field, respectively. The magnetic induction intensity depends only on vector potential A, while, different from the case of the electrostatic field, in general the electric field intensity E depends not only on scalar potential ϕ but also on vector potential A.

From the aforementioned, if A satisfies (1.138), then for arbitrarily given function $\psi = \psi(t, x_1, x_2, x_3)$, $A' = A + \text{grad}\,\psi$ satisfies the same condition. From (1.139), we know that A' and $\phi' = \phi - \frac{\partial \psi}{\partial t}$ satisfy conditions (1.138) and (1.139). That is to say, if A and ϕ are the vector potential and scalar potential of the electromagnetic field, then

$$A' = A + \text{grad}\,\psi, \tag{1.140}$$

$$\phi' = \phi - \frac{\partial \psi}{\partial t} \tag{1.141}$$

are also the corresponding vector potential and scalar potential, where ψ is an arbitrarily given function. In other words, the vector potential and scalar potential are not uniquely determined, with some degree of freedom implied in the above formulas. The transform defined by (1.140) and (1.141) is called the *gauge transform*. ψ determines the given gauge. Different ψ determines different gauge; special ψ suitably chosen can simplify the problem. When we use the gauge transform on A and ϕ, both E and B remain unchanged even if both vector and scalar potentials change at this time. This kind of invariance is called the *gauge invariance*. Thanks to this invariance, we can select a suitable gauge to simplify the treatment when describing the electromagnetic field with scalar potential ϕ and vector potential A. The potential may vary under gauge transform while the field is invariant under gauge transform. Actually, the electromagnetic field is the simplest gauge field, which may be generalized nontrivially to the gauge field in modern physics.

Now we consider the equations for the electromagnetic field represented by scalar potential ϕ and vector potential A. Inserting (1.139) into Maxwell's equation (1.54), it is easy to see that

$$\frac{1}{c^2} \frac{\partial^2 \phi}{\partial t^2} - \Delta \phi - \frac{\partial}{\partial t}\left(\text{div}\,A + \frac{1}{c^2}\frac{\partial \phi}{\partial t}\right) = \frac{\rho}{\varepsilon_0}. \tag{1.142}$$

Plugging (1.138) and (1.139) into (1.57), we have

$$\text{rot rot}\,A = -\frac{1}{c^2}\frac{\partial}{\partial t}(\text{grad}\,\phi) - \frac{1}{c^2}\frac{\partial^2 A}{\partial t^2} + \mu_0 j,$$

which, from formula (1.108) in vector analysis, can be written as

$$\frac{1}{c^2}\frac{\partial^2 A}{\partial t^2} - \Delta A + \text{grad}\left(\text{div}\,A + \frac{1}{c^2}\frac{\partial \phi}{\partial t}\right) = \mu_0 j. \tag{1.143}$$

As for the other two Maxwell's equations (1.55) and (1.56), they are automatically satisfied since they are used to define A and ϕ. Thus we obtain equations (1.142) and (1.143) satisfied by ϕ and A, which are still equations coupled by ϕ and A.

If we choose ϕ and A such that

$$\text{div}\,A + \frac{1}{c^2}\frac{\partial \phi}{\partial t} = 0, \tag{1.144}$$

1.6 Scalar Potential and Vector Potential of an Electromagnetic Field

then (1.142), (1.143) can be reduced to the following two independent equations:

$$\frac{1}{c^2}\frac{\partial^2 \phi}{\partial t^2} - \Delta \phi = \frac{\rho}{\varepsilon_0}, \tag{1.145}$$

$$\frac{1}{c^2}\frac{\partial^2 \mathbf{A}}{\partial t^2} - \Delta \mathbf{A} = \mu_0 \mathbf{j}; \tag{1.146}$$

that is, ϕ and $\mathbf{A}$ satisfy the wave equations with source ρ and $\mathbf{j}$, respectively.

Equation (1.144) is called the *Lorentz condition*. In general, the Lorentz condition can always be satisfied through an appropriate gauge transform; that is, we can find a ψ such that, under gauge transform (1.140)–(1.141), it holds that

$$\operatorname{div} \mathbf{A}' + \frac{1}{c^2}\frac{\partial \phi'}{\partial t} = 0. \tag{1.147}$$

In fact, inserting (1.140) and (1.141) into the above formula, we have

$$\frac{1}{c^2}\frac{\partial^2 \psi}{\partial t^2} - \Delta \psi = \operatorname{div} \mathbf{A} + \frac{1}{c^2}\frac{\partial \phi}{\partial t}, \tag{1.148}$$

which is the wave equation with the right-hand side term $\operatorname{div} \mathbf{A} + \frac{1}{c^2}\frac{\partial \phi}{\partial t}$. As long as we take ψ to be a solution of this equation, $\mathbf{A}'$ and ϕ' will satisfy (1.147) accordingly. Such a ψ always exists.

$\mathbf{A}$ and ϕ satisfying the Lorentz condition are called $\mathbf{A}$ and ϕ under *Lorentz gauge*.

It is worth noticing that, Lorentz condition (1.144) cannot determine $\mathbf{A}$ and ϕ completely. It is not difficult to see from (1.148) that $\mathbf{A}'$ and ϕ' obtained from ψ satisfying (1.148) plus an arbitrary function satisfying the homogeneous wave equation still satisfy Lorentz condition (1.147). Therefore, in the Lorentz gauge there still exist the problems of gauge variability and gauge invariance.

In modern physics, people more often take $\mathbf{A}$ and ϕ instead of $\mathbf{E}$ and $\mathbf{B}$ to describe the electromagnetic field, and usually take the Lorentz gauge.

Especially in the case of the free electromagnetic field, $\rho = 0$, $\mathbf{j} = \mathbf{0}$, under the Lorentz gauge, both ϕ and $\mathbf{A}$ satisfy the homogeneous wave equation. It is not difficult to verify that now we can choose a suitable ψ such that, under the gauge transform keeping the Lorentz condition, the new scalar potential $\phi' = \phi - \frac{\partial \psi}{\partial t} \equiv 0$. To this end we need only take ψ such that it satisfies $\frac{\partial \psi}{\partial t} = \phi$ and the homogeneous wave equation. At this time, equations satisfied by the scalar potential and the vector potential are reduced to

$$\frac{1}{c^2}\frac{\partial^2 \mathbf{A}}{\partial t^2} - \Delta \mathbf{A} = \mathbf{0} \tag{1.149}$$

only, while the Lorentz condition is reduced to

$$\operatorname{div} \mathbf{A} = 0;$$

that is, $\mathbf{A}$ has only a transverse field part. Correspondingly,

$$\mathbf{E} = -\frac{\partial \mathbf{A}}{\partial t} \tag{1.150}$$

and

$$B = \operatorname{rot} A \tag{1.151}$$

are both transverse fields, which is the simplest form.

In general, the gauge that satisfies

$$\operatorname{div} A = 0 \tag{1.152}$$

is called the *Coulomb gauge*. Under the Coulomb gauge, we skip the detailed discussion about equations satisfied by the scalar potential ϕ and vector potential A of the electromagnetic field.

1.6.3 Example: Electric Dipole Radiation

In this subsection we will illustrate, through a simple example, how electromagnetic waves are generated. From Maxwell's equations we know that changing the electric field may generate the magnetic field, and thus generate the electromagnetic wave as well. Next we will consider the electromagnetic field caused by the dipole whose electric dipole moment varies with time.

The so-called *dipole* is the system consisting of a pair of point charges $-q$ and $+q$ with distance l. What we are interested in and is of significance is the case when the distance l is very small. Suppose l is the vector from the point charge $-q$ to $+q$; then

$$m = ql \tag{1.153}$$

is called the *electric dipole moment* of this dipole.

The dipole in steady state cannot stimulate the electromagnetic wave. Now we consider the time-dependent dipole caused by the point charge reciprocating periodically along the direction of l. Suppose that the point charge $-q$ is fixed at the origin O and that the direction of l is identical to the direction of the z axis. Then we can assume

$$l(t) = l_0 e^{-i\omega t} e_3, \tag{1.154}$$

where e_3 is the unit vector in the x_3 direction. The complex form in (1.154) is mainly for convenience; for practical use, we usually take its real or imaginary part. The dipole moment corresponding to (1.154) can generate the electromagnetic field. Next we will determine the vector potential and the scalar potential of this electromagnetic field.

Under Lorentz condition (1.144), vector potential A satisfies the inhomogeneous wave equation (1.146). We can find a special solution of the inhomogeneous wave equation through retarded potential (for example, see [4]), and the general solution is the sum of this special solution and the solution of the homogeneous wave equation. Through a detailed discussion (see [11]) we can prove that the special solution given by the retarded potential on the right-hand side of (1.155) following is indeed the vector potential A of the generated electromagnetic field satisfying the Lorentz condition. So

$$A(t, P) = \frac{\mu_0}{4\pi} \int_{r_{P'P} \leq ct} \frac{j\left(t - \frac{r_{P'P}}{c}, P'\right)}{r_{P'P}} dV_{P'}. \tag{1.155}$$

1.6 Scalar Potential and Vector Potential of an Electromagnetic Field

Since the current is produced only by point charge $+q$ reciprocating periodically, the current density j on the right-hand side of the above formula is actually a δ-function. When $r \gg l$, (1.155) can also be represented as

$$A(t, P) = \frac{\mu_0}{4\pi r} \int_{r_{P'P} \leq ct} j\left(t - \frac{r_{P'P}}{c}, P'\right) dV_{P'}, \quad (1.156)$$

where $r = |\overline{OP}|$ is the distance between the origin O and point P (see Figure 1.2).

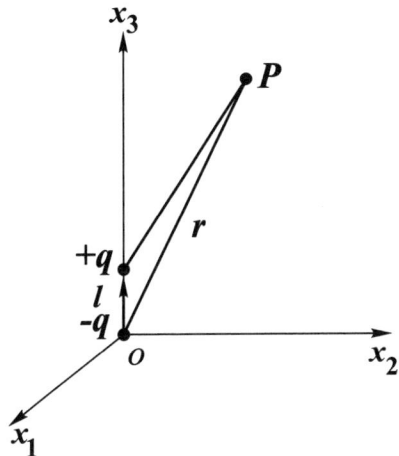

Figure 1.2.

Suppose the velocity of the point charge $+q$ is v; then

$$j = \rho v = \rho \frac{dl}{dt} = -i\omega l_0 \rho e^{-i\omega t} e_3,$$

where ρ is the charge density at point $+q$, which is a δ-function. Inserting it into (1.156), we obtain

$$A(t, P) = -\frac{i\mu_0 \omega l_0}{4\pi r} e^{-i\omega t + i\frac{\omega r}{c}} \int_{r_{P'P} \leq ct} \rho \, dV_{P'} e_3,$$

which, if t is sufficiently large (satisfying $r < ct$), will be

$$A(t, P) = -i\frac{\mu_0 \omega}{4\pi r} e^{ikr} m(t), \quad (1.157)$$

where $m(t) = ql(t)$, and $k = \omega/c$ is the wave number. This is exactly the vector potential of this electromagnetic field.

In order to get the scalar potential ϕ of the electromagnetic field, we substitute the vector potential given by (1.157) into Lorentz condition (1.144) to obtain

$$\frac{\partial}{\partial x_3}\left(-i\frac{\mu_0 \omega}{4\pi r} e^{ikr} m(t)\right) + \frac{1}{c^2} \frac{\partial \phi}{\partial t} = 0,$$

which, by $\frac{dm}{dt} = -i\omega m(t)$ and $\varepsilon_0\mu_0 = c^{-2}$, can be rewritten as

$$\frac{\partial}{\partial t}\left(\frac{\partial}{\partial x_3}\left(\frac{1}{4\pi\varepsilon_0 r}e^{ikr}\right)m(t) + \phi\right) = 0.$$

Then

$$\phi = -\frac{\partial}{\partial x_3}\left(\frac{1}{4\pi\varepsilon_0 r}e^{ikr}\right)m(t) + \phi_0(x_1,x_2,x_3),$$

where $\phi_0(x_1,x_2,x_3)$ is a function to be determined. Noting that ϕ should have the form $\phi = \tilde{\phi}(x_1,x_2,x_3)e^{-i\omega t}$, we take $\phi_0(x_1,x_2,x_3) \equiv 0$. Then we have

$$\phi(t,x_1,x_2,x_3) = -\frac{1}{4\pi\varepsilon_0}\frac{\partial}{\partial x_3}\left(\frac{1}{r}e^{ikr}\right)m(t), \qquad (1.158)$$

which is exactly the scalar potential of this electromagnetic field.

It is easy to verify directly that, for an arbitrary differentiable function $f(r)$, it holds that

$$\mathrm{rot}(f(r)\boldsymbol{m}) = \mathrm{grad}\, f(r) \times \boldsymbol{m} = \frac{f'(r)}{r}\boldsymbol{r} \times \boldsymbol{m}.$$

From this relation, the definition (1.138) of vector potential, and formula (1.157), we have

$$\boldsymbol{B} = \frac{\mu_0 c k^2}{4\pi r}e^{ikr}\left(1 - \frac{1}{ikr}\right)\boldsymbol{n} \times \boldsymbol{m}, \qquad (1.159)$$

where $\boldsymbol{n} = \frac{\boldsymbol{r}}{r}$. Then noting that

$$x_3 m(t) = \boldsymbol{r} \cdot \boldsymbol{m}, \qquad m(t)\,\mathrm{grad}\, x_3 = \boldsymbol{m},$$

and

$$(\boldsymbol{n} \times \boldsymbol{m}) \times \boldsymbol{n} = \boldsymbol{m} - (\boldsymbol{m} \cdot \boldsymbol{n})\boldsymbol{n},$$

and from the definition (1.139) of scalar potential and formulas (1.157), (1.158), we can calculate directly to obtain that

$$\boldsymbol{E} = \frac{k^2}{4\pi\varepsilon_0 r}e^{ikr}(\boldsymbol{n} \times \boldsymbol{m}) \times \boldsymbol{n}$$
$$+ \frac{1}{4\pi\varepsilon_0}\left(\frac{1}{r^3} - \frac{ik}{r^2}\right)e^{ikr}(3(\boldsymbol{n} \cdot \boldsymbol{m})\boldsymbol{n} - \boldsymbol{m}). \qquad (1.160)$$

1.6 Scalar Potential and Vector Potential of an Electromagnetic Field

In a neighboring region of a dipole oscillator, that is, when $r \ll \lambda$ (wave length) $= \frac{2\pi c}{\omega} = \frac{2\pi}{k}$, we see from (1.159) and (1.160) that magnetic induction intensity and electric field intensity can be taken approximately as

$$\boldsymbol{B} = \frac{\mu_0 c k}{4\pi r^2}(\boldsymbol{n} \times \boldsymbol{m}), \qquad (1.161)$$

$$\boldsymbol{E} = \frac{1}{4\pi \varepsilon_0 r^3}(3(\boldsymbol{n} \cdot \boldsymbol{m})\boldsymbol{n} - \boldsymbol{m}). \qquad (1.162)$$

Now the electric field is just the electric field generated by the corresponding steady-state dipole except that it changes with time (see exercise 5). In such a region, $kr \ll 1$, it follows from (1.161) and (1.162) that the magnetic induction intensity is much smaller than the electric field intensity (up to a quantity with factor kr). Thus in a neighboring region, the feature of the field is dominated by the electric field, and this region is called the *static field region*, which is not very important for studying the phenomenon that electric dipole oscillator radiates electromagnetic waves

In the region far away from the dipole oscillator, that is, when $r \gg \lambda$, we see from (1.159) and (1.160) that magnetic induction intensity and electric field intensity can be taken approximately as

$$\boldsymbol{B} = \frac{\mu_0 c k^2}{4\pi r} e^{ikr} \boldsymbol{n} \times \boldsymbol{m}, \qquad (1.163)$$

$$\boldsymbol{E} = c\boldsymbol{B} \times \boldsymbol{n}, \qquad (1.164)$$

from which, and by noticing the definition of $\boldsymbol{m}$ and $\boldsymbol{n}$, it can be seen in this region that the three vectors $\boldsymbol{E}$, $\boldsymbol{B}$, and $\boldsymbol{r}$ are orthogonal to each other, and the direction of $\boldsymbol{E}$ is along that of the meridian, and the direction of $\boldsymbol{B}$ is along that of the latitude (see Figure 1.3), whose value is

$$E = cB = \frac{\mu_0 \omega^2 m_0}{4\pi r}\sin\theta \left|\cos\left(\omega\left(t - \frac{r}{c}\right)\right)\right|, \qquad (1.165)$$

where $m_0 = ql_0$, and θ is the angle between $\boldsymbol{r}$ and the x_3-axis, and taking the real part of $e^{-i\omega(t-\frac{r}{c})}$. This shows that in such a region, field intensities E and B have the same

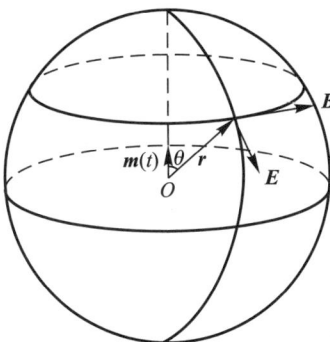

Figure 1.3.

phase $\omega(t - \frac{r}{c})$ and propagate along the direction of radius vector r with the speed of light. Thus overall the electromagnetic wave is the spherical wave with the center at the dipole oscillator, while the field intensity may be different on those points on the wave surface with the same phase. Actually, field intensity is related to the polar angle θ. Field intensity achieves its maximum at the equator. The region satisfying the condition $r \geq \lambda$ is called the *wave field region* or *radiation region*.

Now consider the energy radiating from the electric dipole. From the expression of electromagnetic field energy density in section 1.4, for E and B given by (1.165), we have

$$\frac{1}{2}\left(\varepsilon_0 E^2 + \frac{1}{\mu_0} B^2\right) = \frac{\mu_0 k^2 \omega^2 m_0^2}{16\pi r^2} \sin^2\theta \cos^2\left(\omega\left(t - \frac{r}{c}\right)\right).$$

From the expression (1.83) of energy flux density vector S, we know that the direction of S is the same as r, and its magnitude is

$$S = \frac{c\mu_0 k^2 \omega^2 m_0^2}{16\pi^2 r^2} \sin^2\theta \cos^2\left(\omega\left(t - \frac{r}{c}\right)\right). \tag{1.166}$$

Then, the total energy across the sphere with center at the oscillator and radius r during unit time is

$$\begin{aligned} P(t) &= \int_0^\pi S \cdot 2\pi r^2 \sin\theta \, d\theta \\ &= \frac{1}{6\pi} c\mu_0 k^2 \omega^2 m_0^2 \cos^2\left(\omega\left(t - \frac{r}{c}\right)\right). \end{aligned} \tag{1.167}$$

So, the average energy radiating during unit time is

$$\begin{aligned} \overline{P} &= \frac{1}{T} \int_0^T P(t) \, dt \\ &= \frac{1}{12\pi} c\mu_0 k^2 \omega^2 m_0^2 = \frac{1}{12\pi c} \mu_0 \omega^4 m_0^2, \end{aligned} \tag{1.168}$$

where T is the period: $T = \frac{2\pi}{\omega}$. Thus it is clear that the oscillator keeps radiating energy into the surrounding space, and the average radiation energy per unit time is proportional to the square of the amplitude m_0 of the dipole moment, while it is also proportional to the biquadratic of angular frequency ω, i.e., inversely proportional to the biquadratic of the wave length. This is why it is necessary to use electromagnetic waves with higher frequency and shorter wave length in radio communications and broadcasting.

1.7 Maxwell's Equations in a Medium

1.7.1 Maxwell's Equations in a Medium

In the previous sections we discussed the electromagnetic field in a vacuum and obtained its general law. But for the electromagnetic problems, we often encounter cases when medium exists; then accordingly, the forms of the corresponding Maxwell's equations and Lorentz force formula have to be suitably modified.

1.7 Maxwell's Equations in a Medium

A medium can be either dielectric or conductible. The following cases of the medium in the electromagnetic field may occur:

(1) *Polarization of medium*. All media consist of atoms, while an atom has a positively charged nucleus and negatively charged electrons. In usual cases, an atom remains neutral since positive and negative charges are equal. If the medium is insulated, then all the positive and negative charges are bound in their original positions. If the medium is conductive, parts of the charges are still bound to their original positions, which are called *bound charges*. If there exists an electromagnetic field in the medium, then under the electric field force, bound charges will move slightly and generate a relatively small displacement whose magnitude and direction are determined by that of the electric field intensity, as well as by the electric property of bound charges. Thus, the displacement between positive and negative charges generates an electric dipole moment. This phenomenon is called the *polarization of medium*, which can generate an additional electric field.

Suppose that the intensity of the electric polarization occurring in the medium is P (i.e., the electric dipole moment per unit volume) and that the volume density of bound charges is ρ'. According to Gauss's theorem, we have

$$\text{div}\, \boldsymbol{E} = \frac{1}{\varepsilon_0}(\rho_f + \rho'), \qquad (1.169)$$

where ρ_f is the volume density of free charges. It can be proved that

$$\rho' = -\text{div}\, \boldsymbol{P}. \qquad (1.170)$$

Inserting this into (1.169) gives

$$\text{div}\, \boldsymbol{D} = \rho_f, \qquad (1.171)$$

where

$$\boldsymbol{D} = \varepsilon_0 \boldsymbol{E} + \boldsymbol{P} \qquad (1.172)$$

is called the *electric flux density* or *electric displacement vector*. When the field intensity is not very large, experiments show that there is a linear relationship between $\boldsymbol{P}$ and $\boldsymbol{E}$. In particular, when the medium is isotropic,

$$\boldsymbol{P} = \chi_e \varepsilon_0 \boldsymbol{E}, \qquad (1.173)$$

where χ_e is called the *electric polarizability*. Then from (1.172) and (1.173) we have

$$\boldsymbol{D} = \varepsilon \boldsymbol{E}, \qquad (1.174)$$

where $\varepsilon = \varepsilon_r \varepsilon_0$ is called the *dielectric constant*, and $\varepsilon_r = 1 + \chi_e$ is called the *relative dielectric constant*. For nonuniform media, ε may be a function of (x_1, x_2, x_3).

(2) *Magnetization of medium*. The magnetization phenomenon will occur if there exists an electromagnetic field in the medium. The continuous movement of electrons around the atomic nucleus as well as electron spin generate currents, called molecular currents (current limited within the molecules). Each can be regarded as a small magnetic needle. When there is no external magnetic field in the medium (except for ferromagnetic media), from the macroscopic point of view, these molecular currents and the magnetic fields

generated by them will be cancelled out by each other, and thus there are no macroscopic currents and corresponding macroscopic magnetic fields visible in the medium. However, when the medium is placed in an external magnetic field, under the effect of the magnetic field force there will appear a regular arrangement of these small magnetic needles to some extent, and thus an additional macroscopic magnetic field is generated. This phenomenon is called *magnetization of medium*.

Assume that the magnetization current density in the medium is $\boldsymbol{j}'$; according to Ampère theorem in the magnetic field generated by steady current, we have

$$\operatorname{rot}\boldsymbol{B} = \mu_0(\boldsymbol{j}_f + \boldsymbol{j}'), \tag{1.175}$$

where $\boldsymbol{j}_f$ is the conduction current density. It can be proved that

$$\operatorname{rot}\boldsymbol{M} = \boldsymbol{j}', \tag{1.176}$$

where $\boldsymbol{M}$ is the intensity of magnetization in the medium (i.e., the magnetic dipole moment per unit volume). Inserting the above formula into (1.175), we obtain

$$\operatorname{rot}\boldsymbol{H} = \boldsymbol{j}_f, \tag{1.177}$$

where

$$\boldsymbol{H} = \frac{1}{\mu_0}\boldsymbol{B} - \boldsymbol{M} \tag{1.178}$$

is called the *magnetic field intensity*. Experiments show that when the field intensity is not very large, for an isotropic medium we have

$$\boldsymbol{M} = \chi_m \boldsymbol{H}, \tag{1.179}$$

where χ_m is called the *magnetic susceptibility of medium*. Then from (1.178) we obtain

$$\boldsymbol{B} = \mu \boldsymbol{H}, \tag{1.180}$$

where $\mu = \mu_r \mu_0$ is called the *magnetic conductivity*, and $\mu_r = 1 + \chi_m$ is called the *relative magnetic permeability*. For nonuniform media, μ may be a function of (x_1, x_2, x_3).

Similarly to the electromagnetic field in a vacuum, for unsteady cases, (1.177) should be modified as

$$\operatorname{rot}\boldsymbol{H} = \frac{\partial \boldsymbol{D}}{\partial t} + \boldsymbol{j}_f. \tag{1.181}$$

It is easy to verify directly that (1.181) is compatible with the conservation law of charges

$$\frac{\partial \rho_f}{\partial t} + \operatorname{div}\boldsymbol{j}_f = 0. \tag{1.182}$$

(3) When the medium is conductible, *conduction current* may also appear. It is the result when free charges (electrons or ions) move under the effect of the electromagnetic field.

Overall, charges in the medium under the effect of the electromagnetic field may have three motion patterns, polarization, magnetization, and conduction, which may change the original electromagnetic field and the forms of the corresponding equations. From the above arguments we know that Maxwell's equations in an isotropic medium should take the following form:

1.7 Maxwell's Equations in a Medium

$$\text{div } \boldsymbol{D} = \rho_f, \tag{1.183}$$

$$\text{rot } \boldsymbol{E} = -\frac{\partial \boldsymbol{B}}{\partial t}, \tag{1.184}$$

$$\text{div } \boldsymbol{B} = 0, \tag{1.185}$$

$$\text{rot } \boldsymbol{H} = \frac{\partial \boldsymbol{D}}{\partial t} + \boldsymbol{j}_f, \tag{1.186}$$

where ρ_f and $\boldsymbol{j}_f$ should satisfy the charge conservation (1.182).

The distinguishing feature of this system of equations is that there appears only free charge density and conduction current density instead of charge and current corresponding to polarization and magnetization of the medium. $\boldsymbol{D}$ and $\boldsymbol{H}$ are introduced such that the effects of polarization and magnetization are taken into consideration, while only free charges and conduction currents remain in the equations. Introducing these will bring great convenience. Because the polarization and magnetization situation can be determined directly by the local electric field and magnetic field, so can the bound charges and induced currents (generated by polarization and magnetization), while free charges and conduction currents may come from the outside. Furthermore, conduction currents can be measured in the usual way by a galvanometer, while induced currents cannot. It is convenient to deal with them separately.

The first term on the right-hand side of (1.186),

$$\boldsymbol{j}_d \stackrel{\text{d}}{=} \frac{\partial \boldsymbol{D}}{\partial t}, \tag{1.187}$$

is called the *displacement current*, which is not a genuine current.

The integral forms corresponding to (1.183)–(1.186) and (1.182) are

$$\int_{\partial \Omega} \boldsymbol{D} \cdot \boldsymbol{n} \, \mathrm{d}S = \int_{\Omega} \rho_f \, \mathrm{d}V, \tag{1.188}$$

$$\oint_l \boldsymbol{E} \cdot \mathrm{d}\boldsymbol{l} = -\int_S \frac{\partial \boldsymbol{B}}{\partial t} \cdot \boldsymbol{n} \, \mathrm{d}S, \tag{1.189}$$

$$\int_{\partial \Omega} \boldsymbol{B} \cdot \boldsymbol{n} \, \mathrm{d}S = 0, \tag{1.190}$$

$$\oint_l \boldsymbol{H} \cdot \mathrm{d}\boldsymbol{l} = \int_S \frac{\partial \boldsymbol{D}}{\partial t} \cdot \boldsymbol{n} \, \mathrm{d}S + \int_S \boldsymbol{j}_f \cdot \boldsymbol{n} \, \mathrm{d}S, \tag{1.191}$$

and

$$\frac{\mathrm{d}}{\mathrm{d}t} \int_{\Omega} \rho_f \, \mathrm{d}V = -\int_{\partial \Omega} \boldsymbol{j}_f \cdot \boldsymbol{n} \, \mathrm{d}S, \tag{1.192}$$

where Ω is an arbitrarily given domain in the electromagnetic field, with boundary $\partial \Omega$; l is any given closed curve in the electromagnetic field, and S is any surface enclosed by l whose direction of the outward unit normal $\boldsymbol{n}$ constitutes the right-handed coordinate system with the rotary direction of l.

1.7.2 Conditions on the Interface of Media

In practical problems we often encounter interfaces between different media (including the surface of a medium, which can be viewed as the interface between the medium and

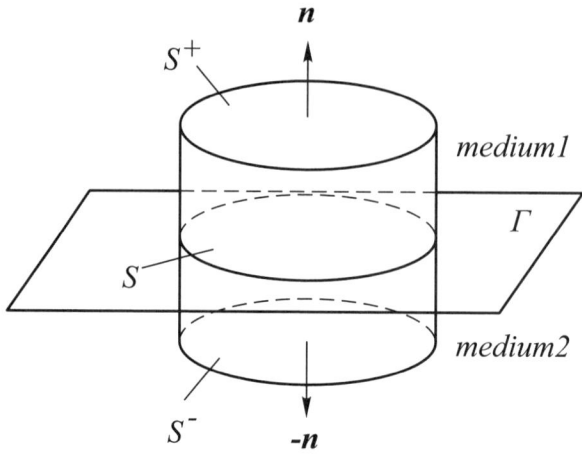

Figure 1.4.

a vacuum). On such interfaces, since the nature of the media experiences a sudden change, so does the electromagnetic field, and thus some electromagnetic field variables may be discontinuous on the interface of media. Therefore, the differential form of Maxwell's equations will lose its meaning on the interface, while their integral form is still valid near the interface. Now we are going to use these integral forms to derive the relations that the electromagnetic field variables on both sides of the interface should satisfy, that is, the *interface condition* which should be satisfied on the interface of media. This is a kind of inner condition, which is the representation of Maxwell's equations on the interface. On both sides, away from the interface, quantities of the electromagnetic field are normal, and the original differential forms of Maxwell's equations are still valid. We will obtain a complete description if we combine Maxwell's equations in the interior of media with the interface condition on the interface of media.

First, let us look at equation (1.188). Let S_- and S_+ be two surfaces parallel to the interface Γ on both sides, respectively, and let Ω be the cylinder formed accordingly (see Figure 1.4). From (1.188) on Ω, we immediately obtain

$$\int_{\partial \Omega} \boldsymbol{D} \cdot \boldsymbol{n} \mathrm{d}S = Q_f, \tag{1.193}$$

where Q_f is the total amount of charges contained in this cylinder. First, let the height of the cylinder tend to zero; then the surfaces S^+ and S^- tend to the surface block S on the interface from both sides, respectively, while the limit values of $\boldsymbol{D}$ on both sides are different, which are assumed to be $\boldsymbol{D}_1$ and $\boldsymbol{D}_2$, respectively. It is stipulated that the unit normal vector $\boldsymbol{n}$ takes the same orientation for both sides. To be specific, we assume that $\boldsymbol{n}$ points to the interior of medium 1, and then the limit of the left-hand side in (1.193) is

$$\int_{S} (\boldsymbol{D}_1 - \boldsymbol{D}_2) \cdot \boldsymbol{n} \mathrm{d}S,$$

1.7 Maxwell's Equations in a Medium

while the total charge amount on the right-hand side is contributed only by surface charges. Let ω_f be the surface density of free charges on Γ; then the limit of the right-hand side in (1.193) is

$$\int_S \omega_f \, dS.$$

From this we have

$$\int_S (\boldsymbol{D}_1 - \boldsymbol{D}_2) \cdot \boldsymbol{n} \, dS = \int_S \omega_f \, dS. \tag{1.194}$$

Now by asking surface block S to shrink to one point, or noticing that the above formula holds for any surface block S on the interface Γ, we obtain

$$(\boldsymbol{D}_1 - \boldsymbol{D}_2) \cdot \boldsymbol{n} = \omega_f, \tag{1.195}$$

which is the representation of (1.188) on the interface. Denote the jump magnitude of $\boldsymbol{D}$ on the interface by

$$[\boldsymbol{D}] = \boldsymbol{D}_1 - \boldsymbol{D}_2;$$

then (1.195) can be written as

$$[\boldsymbol{D}] \cdot \boldsymbol{n} = \omega_f. \tag{1.196}$$

This implies that **when there is no free surface charge on the interface ($\omega_f = 0$), the normal component of electric flux density D is continuous on the interface; otherwise, it has a jump ω_f.**

Similarly, the representation of equation (1.190) on the interface can be given by

$$[\boldsymbol{B}] \cdot \boldsymbol{n} = 0, \tag{1.197}$$

where $[\boldsymbol{B}] = \boldsymbol{B}_1 - \boldsymbol{B}_2$. This implies that **the normal component of magnetic induction intensity B is continuous on the interface**.

Now consider equation (1.189). Compose a loop on both sides of the interface, whose two bottom sides l^+ and l^- are parallel to the interface and two lateral sides are perpendicular to it (see Figure 1.5). Using (1.189) for this loop and the surface enclosed by it, we have

$$\oint_l \boldsymbol{E} \cdot d\boldsymbol{l} = -\int_S \frac{\partial \boldsymbol{B}}{\partial t} \cdot \boldsymbol{n} \, dS. \tag{1.198}$$

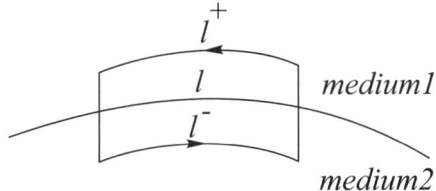

Figure 1.5.

First, let the lateral side (height) of this loop tend to zero, and then l^+ and l^- tend to the curve segment l on the interface. The limit of the left-hand side in (1.198) is

$$\int_l (E_1 - E_2) \cdot dl,$$

where the direction of dl is the same as l^+, and E_1 and E_2 are the limit values of E on both sides of the interface. Since the limit of the right-hand side in (1.198) is zero, we obtain

$$\int_l (E_1 - E_2) \cdot dl = 0. \tag{1.199}$$

Then by asking this curve segment to shrink to one point, or noticing the arbitrariness of this curve segment, from the above formula we know that

$$[E] \cdot m = 0 \tag{1.200}$$

holds on the interface, where m is any direction tangential to the interface, $[E] = E_1 - E_2$. This implies that **the tangential component of electric field intensity E is continuous on the interface**. Equation (1.200) can also be written as

$$[E] \times n = \mathbf{0}, \tag{1.201}$$

which is the representation of (1.189) on the interface.

Similarly, the representation of equation (1.191) on the interface is given by

$$[H] \times n = \mathbf{0}; \tag{1.202}$$

that is, **the tangential component of magnetic field intensity H is continuous on the interface**. Here it is assumed that there is only volume conduction current density but no surface conduction current density. However, if there is a surface conduction current density f_s, (1.202) should be replaced by

$$[H] \times n = f_s \tag{1.202}'$$

instead.

Therefore the inner boundary conditions on the interface are

$$[D] \cdot n = \omega_f, \tag{1.203}$$
$$[B] \cdot n = 0, \tag{1.204}$$
$$[E] \times n = \mathbf{0}, \tag{1.205}$$
$$[H] \times n = \mathbf{0}, \tag{1.206}$$

where the unit normal vector n takes the same direction on both sides of the interface. Thus, **on the interface of media, the normal component of B is continuous, and the tangential components of E and H are continuous, while the normal component of D has a jump** ω_f.

Similarly, from (1.192) we obtain the representation of the conservation law of charges on the interface as

$$[j_f] \cdot n = -\frac{\partial \omega_f}{\partial t}, \tag{1.207}$$

where $[j_f]$ is the jump of j_f across the interface, and ω_f is the area density of free charges on the interface, while n takes the same direction on both sides of the interface.

1.7.3 Representations of Electromagnetic Field Variables in a Medium

Now we derive the representations of electromagnetic field variables in a medium. From Maxwell's equations (1.184) and (1.186), we have

$$0 = \operatorname{rot} \boldsymbol{E} + \frac{\partial \boldsymbol{B}}{\partial t}, \tag{1.208}$$

$$-\boldsymbol{j}_f = \frac{\partial \boldsymbol{D}}{\partial t} - \operatorname{rot} \boldsymbol{H}. \tag{1.209}$$

Taking the scalar products of (1.208) with $\boldsymbol{H}$ and of (1.209) with $\boldsymbol{E}$, adding the corresponding results, and then noticing that $\boldsymbol{D} = \varepsilon \boldsymbol{E}$, $\boldsymbol{B} = \mu \boldsymbol{H}$, while ε and μ are functions of only (x_1, x_2, x_3) and independent of t, we obtain

$$-\boldsymbol{j}_f \cdot \boldsymbol{E} = \frac{1}{2} \frac{\partial}{\partial t} (\boldsymbol{E} \cdot \boldsymbol{D} + \boldsymbol{B} \cdot \boldsymbol{H}) + \operatorname{rot} \boldsymbol{E} \cdot \boldsymbol{H} - \operatorname{rot} \boldsymbol{H} \cdot \boldsymbol{E}. \tag{1.210}$$

From the formula in vector analysis,

$$\operatorname{div}(\boldsymbol{E} \times \boldsymbol{H}) = \boldsymbol{H} \cdot \operatorname{rot} \boldsymbol{E} - \boldsymbol{E} \cdot \operatorname{rot} \boldsymbol{H}, \tag{1.211}$$

(1.210) can be written as

$$-\boldsymbol{j}_f \cdot \boldsymbol{E} = \frac{1}{2} \frac{\partial}{\partial t} (\boldsymbol{E} \cdot \boldsymbol{D} + \boldsymbol{B} \cdot \boldsymbol{H}) + \operatorname{div}(\boldsymbol{E} \times \boldsymbol{H}). \tag{1.212}$$

Hence, we can similarly define the *electromagnetic energy density* by

$$\frac{1}{2}(\boldsymbol{E} \cdot \boldsymbol{D} + \boldsymbol{B} \cdot \boldsymbol{H}), \tag{1.213}$$

while the *electromagnetic energy flux density vector* is defined by

$$\boldsymbol{S} = \boldsymbol{E} \times \boldsymbol{H}, \tag{1.214}$$

which is still called the *Poynting vector*. Equation (1.212) is then the differential form of the conservation and transformation law of energy in a medium.

The *electromagnetic momentum density vector* and the *electromagnetic momentum flux density tensor* of the electromagnetic field in medium are, respectively,

$$\frac{\boldsymbol{S}}{c^2} \tag{1.215}$$

and

$$\frac{1}{2}(\varepsilon E^2 + \mu H^2)\boldsymbol{I} - \varepsilon \boldsymbol{E} \otimes \boldsymbol{E} - \mu \boldsymbol{H} \otimes \boldsymbol{H}, \tag{1.216}$$

where we omit the process of derivations.

1.8 Electrostatic Fields and Magnetostatic Fields

1.8.1 Electrostatic Fields

The steady distribution of electric fields generated by static charges is called the *electrostatic field*. Since there is no charge flowing, and the electric field intensity $\boldsymbol{E}$ does not change with time, i.e., $\boldsymbol{E} = \boldsymbol{E}(x_1, x_2, x_3)$, the magnetic induction intensity is zero, i.e., $\boldsymbol{B} = \boldsymbol{0}$.

According to Maxwell's equations (1.183) and (1.184), we have

$$\operatorname{div} \boldsymbol{D} = \rho_f, \tag{1.217}$$

$$\operatorname{rot} \boldsymbol{E} = \boldsymbol{0}. \tag{1.218}$$

From (1.218), there exists an electrostatic potential ϕ (determined up to an arbitrary additive constant) such that

$$\boldsymbol{E} = -\operatorname{grad} \phi. \tag{1.219}$$

Thus, equation (1.217) is reduced to

$$-\frac{\partial}{\partial x_1}\left(\varepsilon \frac{\partial \phi}{\partial x_1}\right) - \frac{\partial}{\partial x_2}\left(\varepsilon \frac{\partial \phi}{\partial x_2}\right) - \frac{\partial}{\partial x_3}\left(\varepsilon \frac{\partial \phi}{\partial x_3}\right) = \rho_f. \tag{1.220}$$

That is, ϕ satisfies a second-order elliptic equation in the interior of the medium. Equation (1.220) is an inhomogeneous *quasi-harmonic equation*.

Now we discuss the condition that should be satisfied by ϕ on the interface of media. From (1.203) and (1.205), we have on the interface of media

$$[\boldsymbol{D}] \cdot \boldsymbol{n} = \omega_f, \tag{1.221}$$

$$[\boldsymbol{E}] \times \boldsymbol{n} = \boldsymbol{0}. \tag{1.222}$$

Noticing that $\boldsymbol{D} = -\varepsilon \operatorname{grad} \phi$, (1.221) can be written as

$$\varepsilon_1 \left(\frac{\partial \phi}{\partial n}\right)_1 - \varepsilon_2 \left(\frac{\partial \phi}{\partial n}\right)_2 = -\omega_f, \tag{1.223}$$

where subscripts 1 and 2 represent the corresponding values on the interface at the sides of medium 1 and 2, respectively. Equation (1.223) can be written as

$$\left[\varepsilon \frac{\partial \phi}{\partial n}\right] = -\omega_f. \tag{1.224}$$

From (1.222), for any direction $\boldsymbol{m}$ tangential to the interface, we have

$$(\boldsymbol{E}_1 - \boldsymbol{E}_2) \cdot \boldsymbol{m} = 0,$$

i.e.,

$$\frac{\partial}{\partial m}(\phi_1 - \phi_2) = 0,$$

which implies that $\phi_1 - \phi_2$ is a constant on the interface. Since ϕ is determined up to an arbitrary additive constant, we can choose a suitable value of this constant such that we have on the interface

$$\phi_1 = \phi_2 \quad \text{or} \quad [\phi] = 0. \tag{1.225}$$

1.8 Electrostatic Fields and Magnetostatic Fields

Then the conditions that the electrostatic potential should satisfy on the interface of media are

$$\phi_1 = \phi_2, \tag{1.226}$$

$$\varepsilon_1 \left(\frac{\partial \phi}{\partial n}\right)_1 - \varepsilon_2 \left(\frac{\partial \phi}{\partial n}\right)_2 = -\omega_f; \tag{1.227}$$

that is, it is continuous itself, but its first-order normal derivative is discontinuous, while in the case without surface charge, we have

$$\phi_1 = \phi_2, \tag{1.228}$$

$$\varepsilon_1 \left(\frac{\partial \phi}{\partial n}\right)_1 = \varepsilon_2 \left(\frac{\partial \phi}{\partial n}\right)_2. \tag{1.229}$$

Now we consider the boundary conditions on other boundaries. If we want to determine the electrostatic field in the outer space of a charged conductor, we have to know the boundary conditions on the surface of the conductor. We know that when we endow the conductor with free charges, the charges will redistribute automatically to reach the equilibrium so that the electric field intensity becomes zero everywhere in the interior of the conductor, and then the electrostatic potential on the whole conductor becomes a constant. This is because when the electric field intensity is not zero in the interior of the conductor, charges must flow gradually under the effect of the electric field, while only when the electric field intensity is zero is there a static situation. Therefore, the electrostatic distribution on the conductor satisfies the following conditions:

(1) The total amount of charge distribution on each conductor is equal to the total amount of endowed charges.

(2) The charges of the conductor distribute on the surface in the form of surface charges, and there is no charge in the interior of the conductor (this phenomenon is called *skin effect*). This is because $\boldsymbol{D} = \boldsymbol{E} = \boldsymbol{0}$ in the interior of the conductor, and then the charge density is zero according to (1.217).

(3) Each conductor is an equipotential body on which the electrostatic potential is a constant.

(4) The total electric flux across the boundary of the conductor issued by free charges is always equal to the total amount of free charges on the conductor:

$$\int_{\partial \Omega} \boldsymbol{D} \cdot \boldsymbol{n} \, \mathrm{d}S = Q_f, \tag{1.230}$$

where $\boldsymbol{n}$ is the outward unit normal to the boundary of the conductor Ω.

So the boundary conditions on the surface of the charged conductor should be

$$\phi = \text{constant (to be determined)}, \tag{1.231}$$

$$\int_{\partial \Omega} \varepsilon \frac{\partial \phi}{\partial n} \, \mathrm{d}S = Q_f, \tag{1.232}$$

where $\boldsymbol{n}$ points to the interior of the conductor. Here, the boundary conditions consist of two formulas. The first condition (1.231) shows that ϕ is a constant to be determined, which is the mathematical expression that the surface $\partial \Omega$ of the conductor is an equipotential surface. Because this constant is to be determined, it is not a Dirichlet-type boundary condition, and another condition needs to be supplemented so as to restrict its degree of freedom. The second condition (1.232) is the integral of $\varepsilon \frac{\partial \phi}{\partial n}$ over $\partial \Omega$, implying that the total electric flux is given. This is a nonlocal boundary condition. Conditions (1.231), (1.232) is a new type of boundary condition called the *equivalued surface boundary condition* or *total flux boundary condition*. There is a series of studies concerning this type of boundary condition (see [5], [8], [12], [17], [21], and the references cited in [21]).

Furthermore, if the electrostatic field is in an unbounded domain, we should also consider the boundary condition at infinity. Usually it is assumed that the electric potential at infinity is zero,

$$\phi = 0, \qquad (1.233)$$

which is a *Dirichlet boundary condition*. When actually solving a problem, we can take a considerably large approximate bounded domain, instead of the unbounded domain, and set the above condition on its "infinity boundary."

There is further the condition on the symmetric surface of the electrostatic field,

$$\frac{\partial \phi}{\partial n} = 0, \qquad (1.234)$$

which is a *Neumann boundary condition*.

Thus the determination of the electrostatic field can always be reduced to solving the well-posed boundary value problem of a second-order elliptic equation and can easily achieve its numerical solutions by the finite element method (see [7]).

All electrostatic field variables can be described by electrostatic potential ϕ. For example, the energy density is

$$\frac{1}{2} \boldsymbol{E} \cdot \boldsymbol{D} = \frac{\varepsilon}{2} E^2 = \frac{\varepsilon}{2} |\operatorname{grad} \phi|^2,$$

while the electromagnetic energy is

$$U_{e,m} = \frac{1}{2} \int_{\Omega} \varepsilon |\operatorname{grad} \phi|^2 \mathrm{d}V. \qquad (1.235)$$

Equation (1.235) is the well-known *Dirichlet integral*, which plays a fundamental role in the study of the variational method.

Noticing (1.217), we have

$$\begin{aligned} \varepsilon |\operatorname{grad} \phi|^2 &= -\boldsymbol{D} \cdot \operatorname{grad} \phi \\ &= -\operatorname{div}(\phi \boldsymbol{D}) + \phi \operatorname{div} \boldsymbol{D} \\ &= -\operatorname{div}(\phi \boldsymbol{D}) + \phi \rho_f; \end{aligned}$$

then (1.235) can be rewritten as

$$U_{e,m} = \frac{1}{2} \int_{\Omega} \phi \rho_f \mathrm{d}V, \qquad (1.236)$$

where Ω is the whole domain occupied by the electrostatic field.

1.8 Electrostatic Fields and Magnetostatic Fields

1.8.2 Electric Fields Caused by Steady Currents

An electric field caused by steady currents has the same features as the electrostatic field. Now ρ_f and $\boldsymbol{j}_f$ are functions with respect to only (x_1, x_2, x_3), and independent of t, as are $\boldsymbol{E}$ and $\boldsymbol{B}$. From Maxwell's equations (1.183)–(1.186) we obtain

$$\operatorname{div} \boldsymbol{D} = \rho_f, \tag{1.237}$$

$$\operatorname{rot} \boldsymbol{E} = \boldsymbol{0}, \tag{1.238}$$

$$\operatorname{div} \boldsymbol{B} = 0, \tag{1.239}$$

$$\operatorname{rot} \boldsymbol{H} = \boldsymbol{j}_f. \tag{1.240}$$

Furthermore, equation (1.182) of conservation law of charges leads to

$$\operatorname{div} \boldsymbol{j}_f = 0. \tag{1.241}$$

As the electrostatic field, the first two equations (1.237) and (1.238) are enough to determine the electric field. That is to say, we can introduce electric potential ϕ such that (1.219) and (1.220) hold. On the interface of media, we have the same conditions, (1.226) and (1.227), as before, and thus we can do the same as for the electrostatic field.

However, in practical applications, particularly in resistivity well-logging (see [7], [21]), for electric fields caused by steady currents we usually establish the corresponding differential equation models by the following method, which may be different to some extent.

Potential function ϕ is still introduced by (1.238):

$$\boldsymbol{E} = -\operatorname{grad} \phi. \tag{1.242}$$

From the *differential form of Ohm's law*, the electric field intensity $\boldsymbol{E}$ and the current density $\boldsymbol{j}_f$ have the following relation:

$$\boldsymbol{j}_f = \sigma \boldsymbol{E} = \frac{1}{\gamma} \boldsymbol{E}, \tag{1.243}$$

that is,

$$\boldsymbol{E} = \gamma \boldsymbol{j}_f, \tag{1.244}$$

where σ and $\gamma = \frac{1}{\sigma}$ are called *electric conductivity* and *resistivity*, respectively. This is because the direction of the field intensity determines that of the current flow in the medium, while the velocity of the current flow is proportional to $\boldsymbol{E}$ and inversely proportional to resistivity.

Thus, from (1.242) and (1.244) we have

$$\boldsymbol{j}_f = -\frac{1}{\gamma} \operatorname{grad} \phi. \tag{1.245}$$

Inserting this into the continuity equation (1.241) of current, we obtain the following second-order elliptic equation satisfied by ϕ:

$$-\operatorname{div}\left(\frac{1}{\gamma} \operatorname{grad} \phi\right) = 0, \tag{1.246}$$

that is

$$-\frac{\partial}{\partial x_1}\left(\frac{1}{\gamma}\frac{\partial \phi}{\partial x_1}\right) - \frac{\partial}{\partial x_2}\left(\frac{1}{\gamma}\frac{\partial \phi}{\partial x_2}\right) - \frac{\partial}{\partial x_3}\left(\frac{1}{\gamma}\frac{\partial \phi}{\partial x_3}\right) = 0, \quad (1.247)$$

which is a *quasi-harmonic equation*.

On the interface of media, we can obtain corresponding conditions on the interface from the corresponding integral forms of (1.238) and (1.241) as in section 1.7.2,

$$[\boldsymbol{E}] \times \boldsymbol{n} = \boldsymbol{0}, \quad (1.248)$$
$$[\boldsymbol{j}_f] \cdot \boldsymbol{n} = 0. \quad (1.249)$$

Then it follows from (1.242) and (1.245) that

$$\phi_1 = \phi_2, \quad (1.250)$$
$$\frac{1}{\gamma_1}\left(\frac{\partial \phi}{\partial n}\right)_1 = \frac{1}{\gamma_2}\left(\frac{\partial \phi}{\partial n}\right)_2. \quad (1.251)$$

Here, (1.250) is also obtained by adjusting an additional constant. The above two formulas imply that **on the interface of media, both the potential and the normal component of the current are continuous**.

At this time, the boundary conditions we may encounter are as follows:

(1) Dirichlet boundary condition:

$$\phi = \text{a given function.} \quad (1.252)$$

For example, we can assume $\phi = 0$ on the grounded boundary, and the electric potential can also be taken as zero on the "infinity boundary."

(2) Neumann boundary condition:

$$\frac{\partial \phi}{\partial n} = \text{a given function.} \quad (1.253)$$

For example, $\frac{\partial \phi}{\partial n} = 0$ on the symmetric surface; on the boundary consisting of insulation materials, $\boldsymbol{j}_f \cdot \boldsymbol{n} = -\frac{1}{\gamma}\frac{\partial \phi}{\partial n} = 0$ since no current flows in the boundary, and hence the above boundary condition still holds.

(3) Equivalued surface boundary condition: ϕ should be a constant (to be determined) on the electrode surface Γ_0, since the electrode surface is a conductor. Furthermore, a constant current I_0 is launched into the medium through the electrode surface, which generates the steady current. This is the case for electric well-logging in petroleum exploitation. Let $\boldsymbol{n}$ be the outward unit normal to Γ_0. Since the total current launched into the medium is I_0, we have

$$-\int_{\Gamma_0} \boldsymbol{j}_f \cdot \boldsymbol{n}\, dS = I_0. \quad (1.254)$$

Then from (1.245), we have

$$\int_{\Gamma_0} \frac{1}{\gamma}\frac{\partial \phi}{\partial n}\, dS = I_0. \quad (1.255)$$

1.8 Electrostatic Fields and Magnetostatic Fields

Thus on the electrode surface Γ_0 we obtain once more the following equivalued surface boundary condition or total flux boundary condition:

$$\phi|_{\Gamma_0} = \text{constant to be determined}, \quad (1.256)$$

$$\int_{\Gamma_0} \frac{1}{\gamma} \frac{\partial \phi}{\partial n} dS = I_0 \text{ (a given constant)}. \quad (1.257)$$

So the whole problem is reduced to solving the corresponding *boundary value problem with equivalued surface* or *total flux boundary value problem* (see [7], [17], [21]).

1.8.3 Magnetostatic Fields

As stated before, the magnetic field generated by steady current is called the *magnetostatic field*. Now ρ_f and j_f are functions with respect to only (x_1, x_2, x_3) and independent of t. Maxwell's equations and the equation for conservation law of charges have the forms (1.237)–(1.241).

The electric field generated by steady current has been treated in the previous subsection. We can also deal separately with the corresponding magnetic field which satisfies

$$\text{div } \boldsymbol{B} = 0, \quad (1.258)$$

$$\text{rot } \boldsymbol{H} = \boldsymbol{j}_f, \quad (1.259)$$

and

$$\text{div } \boldsymbol{j}_f = 0. \quad (1.260)$$

Next we will present two possible methods dealing with magnetostatic fields.

First, we discuss the *vector potential method*. From (1.258), we can introduce vector potential $\boldsymbol{A}$ such that

$$\boldsymbol{B} = \text{rot } \boldsymbol{A}. \quad (1.261)$$

Because $\boldsymbol{A}$ may differ up to the gradient of an arbitrary function ψ (that is, a longitudinal field), we can assume that $\boldsymbol{A}$ is a transverse field, i.e.,

$$\text{div } \boldsymbol{A} = 0 \quad (1.262)$$

(see section 1.6.1). The $\boldsymbol{A}$ introduced above is called the *vector potential of magnetostatic field*.

Substituting $\boldsymbol{H} = \frac{1}{\mu} \boldsymbol{B} = \frac{1}{\mu} \text{rot } \boldsymbol{A}$ into (1.259), we obtain

$$\text{rot} \left(\frac{1}{\mu} \text{rot } \boldsymbol{A} \right) = \boldsymbol{j}_f, \quad (1.263)$$

which is the equation that vector potential $\boldsymbol{A}$ should satisfy. From vector analysis formula, the above formula can be written as

$$\text{grad} \left(\frac{1}{\mu} \right) \times \text{rot } \boldsymbol{A} + \frac{1}{\mu} \text{rot rot } \boldsymbol{A} = \boldsymbol{j}_f. \quad (1.264)$$

Together with vector analysis formula (1.35), and noticing (1.262), the above formula can be rewritten as

$$-\frac{1}{\mu}\Delta A + \mathrm{grad}\left(\frac{1}{\mu}\right) \times \mathrm{rot}\, A = j_f. \qquad (1.265)$$

This is a second-order elliptic equation with decoupled principal part that the vector potential A should satisfy.

Considering the special case when μ is a constant (this is, the case usually occurred in applications), now we have

$$-\Delta A = \mu j_f. \qquad (1.266)$$

That is, the vector potential A satisfies the Poisson equation when it is a transverse field. Using volume potential, a special solution of this equation can be taken as

$$A(x_1, x_2, x_3) = \frac{\mu}{4\pi}\int_\Omega \frac{j_f(P')\mathrm{d}V'}{r_{P'P}}. \qquad (1.267)$$

Moreover, by (1.260), it is not hard to prove that A given by the above formula does satisfy condition (1.262) for the transverse field.

Next let us look at an expression for the magnetic field energy (neglecting the electric field). From (1.213) we know that

$$U_{e,m} = \frac{1}{2}\int_\Omega B \cdot H \, \mathrm{d}V. \qquad (1.268)$$

Noticing that

$$\begin{aligned} B \cdot H &= \mathrm{rot}\, A \cdot H \\ &= \mathrm{div}(A \times H) + A \cdot \mathrm{rot}\, H \\ &= \mathrm{div}(A \times H) + A \cdot j_f, \end{aligned}$$

and that the integral of $\mathrm{div}(A \times H)$ over the whole domain Ω is zero after being transformed into an integral on the boundary by Green's formula, we obtain

$$U_{e,m} = \frac{1}{2}\int_\Omega j_f \cdot A \, \mathrm{d}V. \qquad (1.269)$$

Thus we can see that there are many similarities between the electrostatic field E and the magnetostatic field B; see Table 1.1.

We now introduce the *scalar potential method* for magnetostatic fields. There will appear three functions A_1, A_2, and A_3 when we deal with the magnetostatic field with vector potential A; moreover, when μ is not a constant, the system has a complicated structure and is not easy to handle. We will see that the magnetostatic field can also be treated with the scalar potential, which is different from general electromagnetic fields.

From (1.260), together with Lemma 1.1, we know that there exists a vector F such that

$$j_f = \mathrm{rot}\, F. \qquad (1.270)$$

1.8 Electrostatic Fields and Magnetostatic Fields

Table 1.1.

	Irrotational field (longitudinal field)
Electro-	There exists a scalar potential ϕ such that $\boldsymbol{E} = -\operatorname{grad}\phi$
	(ϕ may differ up to an arbitrary additive constant),
static	ϕ satisfies
	$-\varepsilon\Delta\phi - \operatorname{grad}\varepsilon \cdot \operatorname{grad}\phi = \rho_f;$
field	$-\Delta\phi = \frac{1}{\varepsilon}\rho_f$ (if ε is a constant)
	Energy $U = \frac{1}{2}\int_\Omega \rho_f \phi \, dV$
	Solenoidal field (transverse field)
Magneto-	There exists a vector potential $\boldsymbol{A}$ such that $\boldsymbol{B} = \operatorname{rot}\boldsymbol{A}$
	($\boldsymbol{A}$ may differ up to a gradient of an arbitrary scalar field),
static	$\boldsymbol{A}$ (when $\operatorname{div}\boldsymbol{A} = 0$) satisfies
	$-\frac{1}{\mu}\Delta\boldsymbol{A} + \operatorname{grad}\left(\frac{1}{\mu}\right) \times \operatorname{rot}\boldsymbol{A} = \boldsymbol{j}_f;$
field	$-\Delta\boldsymbol{A} = \mu\boldsymbol{j}_f$ (if μ is constant)
	Energy $U = \frac{1}{2}\int_\Omega \boldsymbol{j}_f \cdot \boldsymbol{A} \, dV$

Moreover, from Remark 1.1, we can always choose $\boldsymbol{F}$ such that

$$\operatorname{div}\boldsymbol{F} = 0. \tag{1.271}$$

Plugging (1.270) into (1.259), we get

$$\operatorname{rot}(\boldsymbol{H} - \boldsymbol{F}) = \boldsymbol{0}.$$

Then from Lemma 1.2, there must exist a scalar function ϕ such that

$$\boldsymbol{H} - \boldsymbol{F} = -\operatorname{grad}\phi. \tag{1.272}$$

We call ϕ on the right-hand side of the above the *scalar potential of magnetostatic field*. Thus it follows that

$$\boldsymbol{B} = -\mu\operatorname{grad}\phi + \mu\boldsymbol{F}. \tag{1.273}$$

Plugging this into (1.258), we obtain the following equation, which the scalar potential ϕ satisfies:

$$-\frac{\partial}{\partial x_1}\left(\mu\frac{\partial\phi}{\partial x_1}\right) - \frac{\partial}{\partial x_2}\left(\mu\frac{\partial\phi}{\partial x_2}\right) - \frac{\partial}{\partial x_3}\left(\mu\frac{\partial\phi}{\partial x_3}\right)$$
$$= -\operatorname{div}(\mu\boldsymbol{F}). \tag{1.274}$$

But

$$\operatorname{div}(\mu\boldsymbol{F}) = \mu\operatorname{div}\boldsymbol{F} + \boldsymbol{F} \cdot \operatorname{grad}\mu = \boldsymbol{F} \cdot \operatorname{grad}\mu,$$

where we used (1.271). Plugging the above formula into the right-hand side of (1.274), we obtain the inhomogeneous quasi-harmonic equation that the scalar potential ϕ satisfies,

$$-\frac{\partial}{\partial x_1}\left(\mu\frac{\partial\phi}{\partial x_1}\right) - \frac{\partial}{\partial x_2}\left(\mu\frac{\partial\phi}{\partial x_2}\right) - \frac{\partial}{\partial x_3}\left(\mu\frac{\partial\phi}{\partial x_3}\right) = -\boldsymbol{F}\cdot\operatorname{grad}\mu, \qquad (1.275)$$

which, especially when μ is a constant, becomes *harmonic equation*

$$-\Delta\phi = 0. \qquad (1.276)$$

Next let us discuss the conditions on the interface of media. Since the tangential component of magnetic field intensity $\boldsymbol{H}$ is continuous across the interface (see (1.206)), from (1.272) we have

$$[-\operatorname{grad}\phi + \boldsymbol{F}]\cdot\boldsymbol{m} = 0, \quad \forall \text{ tangent vector } \boldsymbol{m}. \qquad (1.277)$$

Noticing that, in Lemma 1.1, if $\boldsymbol{B}$ and its first-order partial derivatives have discontinuity of the first type on the surface S, while the normal component of $\boldsymbol{B}$ is continuous on S, then we can prove that $\boldsymbol{A}$ given in the lemma is continuous in the domain containing S (see exercise 15 for the case that S is the plane $x_3 = 0$). Together with (1.249), we know that the normal component of $\boldsymbol{j}_f$ is continuous across the interface. So, $\boldsymbol{F}$ on the right-hand side of (1.270) is continuous on the interface, and then (1.277) gives

$$\frac{\partial}{\partial m}[\phi] = 0, \quad \forall \text{ tangent vector } \boldsymbol{m}. \qquad (1.278)$$

Noticing that ϕ can be determined up to an arbitrary additive constant, it follows from above that

$$[\phi] = 0, \quad \text{i.e.,} \quad \phi_1 = \phi_2. \qquad (1.279)$$

Furthermore, since the normal component of magnetic induction intensity $\boldsymbol{B}$ is continuous on the interface (see (1.204)), from (1.273) we have

$$[-\mu\operatorname{grad}\phi + \mu\boldsymbol{F}]\cdot\boldsymbol{n} = 0,$$

which implies, from the continuity of $\boldsymbol{F}$, that

$$\left[\mu\frac{\partial\phi}{\partial n}\right] = [\mu]\boldsymbol{F}\cdot\boldsymbol{n}, \qquad (1.280)$$

that is,

$$\mu_1\left(\frac{\partial\phi}{\partial n}\right)_1 - \mu_2\left(\frac{\partial\phi}{\partial n}\right)_2 = (\mu_1 - \mu_2)\boldsymbol{F}\cdot\boldsymbol{n}, \qquad (1.281)$$

whose right-hand side term is a known function, since $\boldsymbol{F}$ can be determined by the current density $\boldsymbol{j}_f$. Thus, the conditions on the interface of media are

$$\phi_1 = \phi_2, \qquad (1.282)$$

$$\mu_1\left(\frac{\partial\phi}{\partial n}\right)_1 - \mu_2\left(\frac{\partial\phi}{\partial n}\right)_2 = (\mu_1 - \mu_2)\boldsymbol{F}\cdot\boldsymbol{n}, \qquad (1.283)$$

which can be easily treated by the variational method similarly.

Now we will discuss the expression of the magnetic field energy. From (1.272) and (1.273), we have

$$\boldsymbol{B}\cdot\boldsymbol{H} = \mu(-\operatorname{grad}\phi + \boldsymbol{F})\cdot(-\operatorname{grad}\phi + \boldsymbol{F})$$
$$= \mu|\operatorname{grad}\phi|^2 - 2\mu\boldsymbol{F}\cdot\operatorname{grad}\phi + \mu|\boldsymbol{F}|^2. \quad (1.284)$$

But

$$\mu\boldsymbol{F}\cdot\operatorname{grad}\phi = \operatorname{div}(\mu\phi\boldsymbol{F}) - \phi\operatorname{div}(\mu\boldsymbol{F})$$
$$= \operatorname{div}(\mu\phi\boldsymbol{F}) - \phi(\mu\operatorname{div}\boldsymbol{F} + \boldsymbol{F}\cdot\operatorname{grad}\mu)$$
$$= \operatorname{div}(\mu\phi\boldsymbol{F}) - \phi\boldsymbol{F}\cdot\operatorname{grad}\mu, \quad (1.285)$$

where we used (1.271). Plugging (1.285) into (1.284), integrating over the whole magnetic field Ω, and noticing that the integral of the divergence term $\operatorname{div}(\mu\phi\boldsymbol{F})$ over Ω will vanish after being transformed into the integral on the boundary by Green's formula, we get

$$U_{e,m} = \frac{1}{2}\int_\Omega \boldsymbol{B}\cdot\boldsymbol{H}\,\mathrm{d}V$$
$$= \frac{1}{2}\int_\Omega (\mu|\operatorname{grad}\phi|^2 + 2\phi\boldsymbol{F}\cdot\operatorname{grad}\mu + \mu|\boldsymbol{F}|^2)\mathrm{d}V, \quad (1.286)$$

which, since μ and $\boldsymbol{F}$ are known, can be regarded essentially as

$$U_{e,m} = \frac{1}{2}\int_\Omega (\mu|\operatorname{grad}\phi|^2 + 2\phi\boldsymbol{F}\cdot\operatorname{grad}\mu)\mathrm{d}V. \quad (1.287)$$

This is the Dirichlet integral corresponding to the inhomogeneous second-order elliptic equation (1.275) under consideration.

1.9 Darwin Model

1.9.1 Quasi-electrostatic Model and Its Amended Form

In the previous section we discussed the problem concerning the electrostatic field and the magnetostatic field. This kind of problem is relatively simple, while the general well-posed problem of Maxwell's equations is much more complicated in both theory and in numerical computation. However, in practical problems we usually face the situation in which the electromagnetic field changes at a relatively low frequency (for example, the electromagnetic field generated by the usual alternating current) and the geometric dimension of the space under consideration is not very large. At this moment Maxwell's equations can be simplified. Suppose that the angular frequency of the electromagnetic field is ω; then $\frac{1}{c}\frac{\partial \boldsymbol{B}}{\partial t} \sim \frac{\omega}{c}$ holds since $\frac{\partial \boldsymbol{B}}{\partial t}$ is the derivative of magnetic induction intensity with respect to t. Noting that $\operatorname{rot}\boldsymbol{E}$ contains only the partial derivative with respect to spatial variables, we have $\operatorname{rot}\boldsymbol{E} \sim \frac{1}{l}$; here l is the characteristic length of the region under discussion. If $\frac{\omega}{c} \ll \frac{1}{l}$, that is, $\omega \ll \frac{c}{l}$, then we can neglect $\frac{1}{c}\frac{\partial \boldsymbol{B}}{\partial t}$ in (1.184) of Maxwell's equations so that it is reduced to $\operatorname{rot}\boldsymbol{E} = \boldsymbol{0}$. This is the equation satisfied by the electric field intensity in the electrostatic field. This situation is called the *quasi-electrostatic state*, while Maxwell's equations corresponding to this state are called the *quasi-electrostatic model* or *steady-like field equations*. From Maxwell's equations (1.183)–(1.186) together with (1.174) and (1.180), the quasi-electrostatic model should have the following form:

$$\varepsilon \frac{\partial \boldsymbol{E}}{\partial t} - \frac{1}{\mu} \operatorname{rot} \boldsymbol{B} = -\boldsymbol{j}, \qquad (1.288)$$

$$\operatorname{rot} \boldsymbol{E} = \boldsymbol{0}, \qquad (1.289)$$

$$\operatorname{div} \boldsymbol{E} = \frac{\rho}{\varepsilon}, \qquad (1.290)$$

$$\operatorname{div} \boldsymbol{B} = 0. \qquad (1.291)$$

Here, we assume that the dielectric constant ε and the magnetic permeability μ are both constants. For the sake of convenience, we denote j_f and ρ_f by $\boldsymbol{j}$ and ρ.

From Lemma 1.3, any vector field can be decomposed into the superposition of a transverse field and a longitudinal field, and then we have

$$\boldsymbol{E} = \boldsymbol{E}_T + \boldsymbol{E}_L, \qquad (1.292)$$

where $\boldsymbol{E}_T$ and $\boldsymbol{E}_L$ are the transverse and longitudinal field part of $\boldsymbol{E}$, respectively, that is, they satisfy

$$\operatorname{div} \boldsymbol{E}_T = 0, \qquad \operatorname{rot} \boldsymbol{E}_L = \boldsymbol{0}.$$

Thus in the quasi-electrostatic model, the transverse field part $\boldsymbol{E}_T$ generated by varying the magnetic field are actually neglected. As a natural amendment, we can neglect the t-derivative $\varepsilon \frac{\partial \boldsymbol{E}_T}{\partial t}$ of the transverse field part of $\boldsymbol{E}$ in only (1.186) instead of in the whole system of Maxwell's equations. Now the model obtained is called the *Darwin model* (see [15], [16], and [18]). The equations describing the Darwin model are given in section 1.9.3.

1.9.2 A Well-Posed Problem of Maxwell's Equations

Assume that Ω is a bounded spatial domain, whose external is an ideal conductor, that is, $\sigma = +\infty$. The Maxwell's equations satisfied by the electromagnetic field in Ω are

$$\varepsilon \frac{\partial \boldsymbol{E}}{\partial t} - \frac{1}{\mu} \operatorname{rot} \boldsymbol{B} = -\boldsymbol{j}, \qquad (1.293)$$

$$\frac{\partial \boldsymbol{B}}{\partial t} + \operatorname{rot} \boldsymbol{E} = \boldsymbol{0}, \qquad (1.294)$$

$$\operatorname{div} \boldsymbol{E} = \frac{\rho}{\varepsilon}, \qquad (1.295)$$

$$\operatorname{div} \boldsymbol{B} = 0, \qquad (1.296)$$

where $\boldsymbol{j}$ and ρ should further satisfy the continuity equation

$$\frac{\partial \rho}{\partial t} + \operatorname{div} \boldsymbol{j} = 0. \qquad (1.297)$$

Now we investigate the boundary conditions satisfied by $\boldsymbol{E}$ and $\boldsymbol{B}$ on the boundary $\partial \Omega$. For ideal conductors, $\boldsymbol{E} = \boldsymbol{0}$ holds from Ohm's law $\boldsymbol{j} = \sigma \boldsymbol{E}$. Thus $\boldsymbol{E} = \frac{\partial \boldsymbol{B}}{\partial t} = \boldsymbol{0}$ holds

1.9 Darwin Model

on the outer side of $\partial\Omega$, where $\partial\Omega$ is regarded as the interface of different media, and then by using conditions (1.205) and (1.204) that the electromagnetic field should satisfy on the interface of the media, we can obtain immediately

$$\boldsymbol{E} \times \boldsymbol{n} = \boldsymbol{0} \quad \text{on} \quad \partial\Omega, \tag{1.298}$$

$$\frac{\partial}{\partial t}(\boldsymbol{B} \cdot \boldsymbol{n}) = 0 \quad \text{on} \quad \partial\Omega. \tag{1.299}$$

Additionally we assume that $\boldsymbol{E}$ and $\boldsymbol{B}$ also satisfy the initial conditions

$$t = 0 : \boldsymbol{E} = \boldsymbol{E}_0, \boldsymbol{B} = \boldsymbol{B}_0, \tag{1.300}$$

where $\boldsymbol{E}_0$ and $\boldsymbol{B}_0$ are the vector functions satisfying the following compatibility conditions:

$$\text{div}\,\boldsymbol{E}_0 = \frac{\rho_0}{\varepsilon}, \quad \rho_0 = \rho(0, x_1, x_2, x_3), \tag{1.301}$$

$$\text{div}\,\boldsymbol{B}_0 = 0, \tag{1.302}$$

$$\boldsymbol{E}_0 \times \boldsymbol{n} = \boldsymbol{0} \quad \text{on} \quad \partial\Omega. \tag{1.303}$$

Next, taking problem (1.293)–(1.296), (1.298)–(1.300) as an example, we will explore how to use the Darwin model to solve the problem.

1.9.3 Darwin Model

From the proof of Lemma 1.3 we can see that the decomposition (1.292) of $\boldsymbol{E}$ is not unique. In fact, once we take $\boldsymbol{E}_L = -\text{grad}\,\phi$ we can get such a decomposition, where ϕ is any solution to the Poisson equation

$$-\Delta\phi = \text{div}\,\boldsymbol{E}.$$

However, we can guarantee the uniqueness of the decomposition (1.292) only by imposing some conditions on $\boldsymbol{E}_T$. Lemma 1.3 can be improved as follows.

Lemma 1.4. *Any vector field $\boldsymbol{A}$ in Ω can be decomposed into the superposition of a transverse field $\boldsymbol{A}_T$ and a longitudinal field $\boldsymbol{A}_L$:*

$$\boldsymbol{A} = \boldsymbol{A}_T + \boldsymbol{A}_L. \tag{1.304}$$

If $\boldsymbol{A}_L$ is required to satisfy the condition

$$\boldsymbol{A}_L \times \boldsymbol{n} = \boldsymbol{0} \quad on \quad \partial\Omega, \tag{1.305}$$

then the above decomposition is unique, and $\boldsymbol{A}_L$ has the following form:

$$\boldsymbol{A}_L = -\text{grad}\,\psi,$$

where ψ is the solution to the following problem:

$$-\Delta\psi = \text{div}\,\boldsymbol{A} \quad in \quad \Omega, \tag{1.306}$$

$$\psi = C \quad on \quad \partial\Omega, \tag{1.307}$$

where C is an arbitrary constant.

Proof. According to the proof of Lemma 1.3, the decomposition (1.304) of A is equivalent to $A_L = -\text{grad}\,\psi$, where ψ is the solution of equation (1.306). Note that a sufficient and necessary condition of $\text{grad}\,\psi \times n = 0$ on $\partial\Omega$ is that the tangential component of $\text{grad}\,\psi$ is zero, i.e., (1.307) holds. So the decomposition (1.304) of A with boundary condition (1.305) is equivalent to $A_L = -\text{grad}\,\psi$, where ψ is the solution of boundary value problem (1.306)–(1.307). But for a different constant C, the solutions of boundary value problem (1.306)–(1.307) differ only up to an additive constant. Then the A_L determined by its solutions is unique. The proof is finished. □

Now we assume that the decomposition (1.292) of E satisfies the requirement of the above lemma; that is, $E_L \times n = 0$ holds on the boundary $\partial\Omega$. Neglecting $\varepsilon\frac{\partial E_T}{\partial t}$ in equation (1.293), Maxwell's equations (1.293)–(1.296) can be reduced to the following Darwin model:

$$\varepsilon\frac{\partial E_L}{\partial t} - \frac{1}{\mu}\text{rot}\,B = -j, \tag{1.308}$$

$$\frac{\partial B}{\partial t} + \text{rot}\,E_T = 0, \tag{1.309}$$

$$\text{div}\,E_L = \frac{\rho}{\varepsilon}, \tag{1.310}$$

$$\text{div}\,E_T = 0, \tag{1.311}$$

$$\text{rot}\,E_L = 0, \tag{1.312}$$

$$\text{div}\,B = 0. \tag{1.313}$$

The following discussions are based on boundary value problem (1.308)–(1.313).

Noticing that E_L satisfies the condition on the boundary required in Lemma 1.4, together with the second condition in (1.300), boundary conditions (1.298), (1.299) can be written as

$$E_L \times n = 0 \quad \text{on} \quad \partial\Omega, \tag{1.314}$$

$$E_T \times n = 0 \quad \text{on} \quad \partial\Omega, \tag{1.315}$$

$$B \cdot n = B_0 \cdot n \quad \text{on} \quad \partial\Omega. \tag{1.316}$$

As for the initial condition, concerning E, now we need only give the value of E_L at $t = 0$, and then

$$t = 0 : E_L = E_{0L}, \tag{1.317}$$

where E_{0L} is the longitudinal field part of E_0 according to the decomposition in Lemma 1.4, which satisfies the compatibility condition

$$\text{div}\,E_{0L} = \frac{\rho_0}{\varepsilon}, \quad \rho_0 = \rho(0, x_1, x_2, x_3),$$

as well as, from (1.305),

$$E_{0L} \times n = 0 \quad \text{on} \quad \partial\Omega.$$

From the discussion in section 1.5 we know that, the initial-boundary value problem of Maxwell's equations is a mixed problem of hyperbolic equations. However, although

1.9 Darwin Model

Darwin model (1.308)–(1.313) is formally a system containing the partial derivatives with respect to time t, we will see later that its problem can be regarded essentially as the boundary value problem of an elliptic equation. In fact, we have the following result (see [13]).

Theorem 1.1. *Solving the initial-boundary value problem* (1.308)–(1.317) *of the Darwin model is equivalent to solving the following three elliptic boundary value problems for any given t:*

(1°) $E_L = -\operatorname{grad}\phi$, *where ϕ is the solution of the following Dirichlet problem of the Poisson equation:*

$$-\Delta\phi = \frac{\rho}{\varepsilon} \quad \text{in} \quad \Omega, \tag{1.318}$$

$$\phi = C(t) \quad \text{on} \quad \partial\Omega, \tag{1.319}$$

while $C(t)$ is an arbitrary function depending only on t.

(2°) ***B** is the solution of the following problem:*

$$-\Delta \boldsymbol{B} = \mu \operatorname{rot} \boldsymbol{j} \quad \text{in} \quad \Omega, \tag{1.320}$$

$$\operatorname{div} \boldsymbol{B} = 0 \quad \text{in} \quad \Omega, \tag{1.321}$$

$$\boldsymbol{B} \cdot \boldsymbol{n} = \boldsymbol{B}_0 \cdot \boldsymbol{n} \quad \text{on} \quad \partial\Omega, \tag{1.322}$$

$$(\operatorname{rot} \boldsymbol{B}) \times \boldsymbol{n} = \mu \boldsymbol{j} \times \boldsymbol{n} \quad \text{on} \quad \partial\Omega. \tag{1.323}$$

(3°) E_T *is the solution of the following problem:*

$$\Delta \boldsymbol{E}_T = \frac{\partial}{\partial t} \operatorname{rot} \boldsymbol{B} \quad \text{in} \quad \Omega, \tag{1.324}$$

$$\operatorname{div} \boldsymbol{E}_T = 0 \quad \text{in} \quad \Omega, \tag{1.325}$$

$$\boldsymbol{E}_T \times \boldsymbol{n} = \boldsymbol{0} \quad \text{on} \quad \partial\Omega. \tag{1.326}$$

Proof. Let E and B be the solution of Darwin model (1.308)–(1.317). From equations (1.310) and (1.312), together with boundary condition (1.314) satisfied by E_L, according to Lemma 1.4 we know that there exists a ϕ such that

$$E_L = -\operatorname{grad}\phi,$$

where ϕ satisfies (1.318) and (1.319); i.e., (1°) holds.

Acting the curl operator rot on both sides of (1.308), from formula

$$\operatorname{rot}\operatorname{rot} A = \operatorname{grad}\operatorname{div} A - \Delta A, \tag{1.327}$$

together with (1.312) and (1.313), we have

$$-\Delta \boldsymbol{B} = \operatorname{rot} \boldsymbol{j};$$

that is, (1.320) holds. From (1.313) and (1.316), in order to prove (2°), it remains to prove boundary condition (1.323). For this, we need only take the vector product of both sides of (1.308) with n on $\partial\Omega$, and (1.323) follows immediately from the given boundary condition (1.314). So, (2°) holds.

As for (3°), by virtue of (1.311) and (1.315), we need only prove (1.324). Acting the curl operator rot on both sides of equation (1.309), from (1.327) and noting (1.311) we obtain (1.324) immediately.

Therefore, if $\boldsymbol{E}$ and $\boldsymbol{B}$ are the solution to problem (1.308)–(1.317) of the Darwin model, then $\boldsymbol{E}_L$, $\boldsymbol{B}$, and $\boldsymbol{E}_T$ satisfy problem (1°), (2°), and (3°), respectively.

On the other hand, let $\boldsymbol{E}_L$, $\boldsymbol{B}$, and $\boldsymbol{E}_T$ be the solutions of problems (1°), (2°), and (3°), respectively. Now we prove that $\boldsymbol{E} = \boldsymbol{E}_T + \boldsymbol{E}_L$ and $\boldsymbol{B}$ must be the solution of problem (1.308)–(1.317) of the Darwin model.

According to (1°), it is easy to have (1.310), (1.312), and (1.314), while (1.311), (1.313), (1.315), and (1.316) are (1.325), (1.321), (1.326), and (1.322), respectively. Next it remains to prove (1.308), (1.309) and (1.317).

First, we show that $\boldsymbol{E}_L$ given by (1°) satisfies the initial condition (1.317). In fact, from (1.318), at $t = 0$ we have

$$-\operatorname{div}\operatorname{grad}\phi_0 = \frac{\rho_0}{\varepsilon};$$

here a quantity with subscript zero means the value of this quantity at $t = 0$. From the compatibility condition (1.301), together with (1.311), we have

$$\operatorname{div} \boldsymbol{E}_{0L} = \frac{\rho_0}{\varepsilon}.$$

From the above two formulas we get

$$\operatorname{div}(\operatorname{grad}\phi_0 + \boldsymbol{E}_{0L}) = 0,$$

which implies that $\operatorname{grad}\phi_0 + \boldsymbol{E}_{0L}$ is a transverse field. But on the other hand, both $-\operatorname{grad}\phi_0 = \boldsymbol{E}_L|_{t=0}$ and $\boldsymbol{E}_{0L}$ are the longitudinal field parts satisfying the requirements in Lemma 1.4, and thus $\operatorname{grad}\phi_0 + \boldsymbol{E}_{0L}$ is still a longitudinal field, and

$$(\operatorname{grad}\phi_0 + \boldsymbol{E}_{0L}) \times \boldsymbol{n} = \boldsymbol{0} \quad \text{on } \partial\Omega.$$

Then from Lemma 1.4 we obtain

$$-\operatorname{grad}\phi_0 = \boldsymbol{E}_{0L},$$

which implies that the initial condition (1.317) holds.

Now we prove that $\boldsymbol{E}_L$ and $\boldsymbol{B}$ satisfy equation (1.308). Using formula (1.327), together with (1.321), equation (1.320) can be rewritten as

$$\operatorname{rot}(\operatorname{rot}\boldsymbol{B} - \mu\boldsymbol{j}) = \boldsymbol{0} \quad \text{in} \quad \Omega,$$

so $\operatorname{rot}\boldsymbol{B} - \mu\boldsymbol{j}$ is a longitudinal field. Besides, boundary condition (1.323) gives

$$(\operatorname{rot}\boldsymbol{B} - \mu\boldsymbol{j}) \times \boldsymbol{n} = \boldsymbol{0} \quad \text{on} \quad \partial\Omega.$$

Thus from Lemma 1.4 we know that

$$\operatorname{rot}\boldsymbol{B} - \mu\boldsymbol{j} = -\operatorname{grad}\psi, \tag{1.328}$$

1.9 Darwin Model

where ψ satisfies

$$-\Delta \psi = \text{div}(\text{rot } \boldsymbol{B} - \mu \boldsymbol{j}) \quad \text{in} \quad \Omega, \tag{1.329}$$

$$\psi = C_1(t) \quad \text{on} \quad \partial \Omega, \tag{1.330}$$

in which $C_1(t)$ is an arbitrary function with respect to only t. But noticing div rot $\boldsymbol{B} = 0$, and using the continuity equation (1.297), (1.329) can be rewritten as

$$-\Delta \psi = \mu \frac{\partial \rho}{\partial t}.$$

Differentiating both sides of equation (1.318) with respect to t, comparing with the above formula, and noticing the boundary conditions that they satisfy, we can obtain immediately

$$\psi = \varepsilon \mu \frac{\partial \phi}{\partial t} + C_2(t),$$

where $C_2(t)$ is an arbitrary function depending only on t. Since $\boldsymbol{E}_L = -\text{grad}\,\phi$, the above formula implies

$$-\text{grad}\,\psi = \varepsilon \mu \frac{\partial \boldsymbol{E}_L}{\partial t}.$$

Substituting this into (1.328), we obtain (1.308) immediately.

Now we prove that $\boldsymbol{E}_T$ and $\boldsymbol{B}$ satisfy (1.309). Using the formulas (1.327) and (1.325), equation (1.324) can be rewritten as

$$\text{rot}\left(\text{rot } \boldsymbol{E}_T + \frac{\partial \boldsymbol{B}}{\partial t}\right) = \boldsymbol{0}. \tag{1.331}$$

From Lemma 1.2, there exists ψ such that

$$\text{rot } \boldsymbol{E}_T + \frac{\partial \boldsymbol{B}}{\partial t} = -\text{grad}\,\psi. \tag{1.332}$$

Noticing (1.321), it is obvious that $\text{rot } \boldsymbol{E}_T + \frac{\partial \boldsymbol{B}}{\partial t}$ is also a transverse field, and then ψ on the right-hand side of the above formula must be a harmonic function:

$$-\Delta \psi = 0 \quad \text{in} \quad \Omega. \tag{1.333}$$

In addition, we point out that, by virtue of the boundary condition (1.326), we have

$$\text{rot } \boldsymbol{E}_T \cdot \boldsymbol{n} = 0 \quad \text{on} \quad \partial \Omega. \tag{1.334}$$

In fact, for any surface block S on $\partial \Omega$ enclosed by l, from the Stokes formula we have

$$\int_S \text{rot } \boldsymbol{E}_T \cdot \boldsymbol{n}\,dS = \int_l \boldsymbol{E}_T \cdot d\boldsymbol{l}. \tag{1.335}$$

Equation (1.326) means that $\boldsymbol{E}_T$ is parallel to the normal vector $\boldsymbol{n}$ on $\partial \Omega$, and thus is orthogonal to all the tangent vectors, which leads to $\boldsymbol{E}_T \cdot d\boldsymbol{l} = 0$. Then from (1.335), we obtain that

$$\int_S \text{rot } \boldsymbol{E}_T \cdot \boldsymbol{n}\,dS = 0$$

holds for an arbitrary surface block S on $\partial\Omega$. Equation (1.334) follows immediately from the arbitrariness of S.

From (1.334) and (1.322), we have

$$\left(\text{rot } E_T + \frac{\partial B}{\partial t}\right) \cdot n = 0 \quad \text{on} \quad \partial\Omega.$$

Then from (1.332), we obtain

$$\frac{\partial \psi}{\partial n} = 0 \quad \text{on} \quad \partial\Omega. \tag{1.336}$$

From our knowledge of mathematical physics equations, we know that the homogeneous Neumann problem (1.333), (1.336) of the harmonic equation has only a constant solution (which may depend on t). Then (1.332) implies (1.309). Theorem 1.1 is proved. □

Remark 1.2. In order to solve the system (1.308)–(1.313), in addition to the boundary conditions (1.314)–(1.316) and the initial condition (1.317), we should also give the initial condition concerning B,

$$t = 0: \quad B = B_0. \tag{1.337}$$

Therefore, in order to prove the equivalence of the problem of the Darwin model and the corresponding problem given in Theorem 1.1, we need to illustrate that B given by problem (1.320)–(1.323) satisfies the initial condition (1.337). Certainly B_0 is assumed to satisfy some compatibility conditions. It is obvious from (1.320), (1.321), and (1.323) that these compatibility conditions are

$$\text{div } B_0 = 0 \quad \text{in} \quad \Omega, \tag{1.338}$$

$$-\Delta B_0 = \mu \text{ rot } j_0 \quad \text{in} \quad \Omega, \tag{1.339}$$

and

$$(\text{rot } B_0) \times n = \mu j_0 \times n \quad \text{on} \quad \partial\Omega, \tag{1.340}$$

where $j_0 = j(0, x_1, x_2, x_3)$. It can be proved that, the solution B of problem (1.320)–(1.323) must satisfy the initial condition (1.337) provided that it satisfies the above compatibility conditions.

In order to prove this, from compatibility conditions (1.338)–(1.340), we need only prove the uniqueness of solutions to problem (1.320)–(1.323) at $t = 0$, namely, prove that this problem has only a trivial solution $B = 0$ when $j = B_0 = 0$. In fact, using (1.327) and (1.321), (1.320) can be written as

$$\text{rot rot } B = 0 \quad \text{in} \quad \Omega, \tag{1.341}$$

that is, rot B is a longitudinal field. But rot B is also a transverse field, and from (1.323) it satisfies the boundary condition

$$(\text{rot } B) \times n = 0 \quad \text{on} \quad \partial\Omega. \tag{1.342}$$

Then from Lemma 1.4 we obtain

$$\text{rot } B = 0 \quad \text{in} \quad \Omega; \tag{1.343}$$

that is, $\boldsymbol{B}$ is a longitudinal field. But from (1.321) and (1.322) we know that $\boldsymbol{B}$ is also a transverse field and satisfies the boundary condition

$$\boldsymbol{B} \cdot \boldsymbol{n} = 0 \quad \text{on} \quad \partial \Omega. \tag{1.344}$$

Thus we get $\boldsymbol{B} = \boldsymbol{0}$ (see Lemma 2.3 in Chapter 2).

At last we point out that, under certain conditions, the Darwin model is indeed a good approximation for Maxwell's equations. Let $\boldsymbol{E}$, $\boldsymbol{B}$ and $\boldsymbol{E}^D$, $\boldsymbol{B}^D$ represent the solutions to the problems (1.293)–(1.300) for Maxwell's equations and (1.308)–(1.317) and (1.337) for the Darwin model, respectively. Then through further discussion we can prove that, as $\eta = \omega l/c \to 0$, $\boldsymbol{E}^D$ and $\boldsymbol{B}^D$ tend to $\boldsymbol{E}$ and $\boldsymbol{B}$ in a certain sense (see [13]). And actually, for the longitudinal field part of the electric field intensity, the solution of the Darwin model gives the exact solution of the original Maxwell's equations, namely, $\boldsymbol{E}_L^D = \boldsymbol{E}_L$. Now we prove the latter conclusion. From equations (1.293) and (1.308), we obtain

$$\varepsilon \frac{\partial \boldsymbol{E}_T}{\partial t} + \varepsilon \frac{\partial \boldsymbol{E}_L}{\partial t} - \frac{1}{\mu} \operatorname{rot} \boldsymbol{B}$$
$$= \varepsilon \frac{\partial \boldsymbol{E}_L^D}{\partial t} - \frac{1}{\mu} \operatorname{rot} \boldsymbol{B}^D.$$

It is easy to see that both $\varepsilon \frac{\partial \boldsymbol{E}_T}{\partial t} - \frac{1}{\mu} \operatorname{rot} \boldsymbol{B}$ and $-\frac{1}{\mu} \operatorname{rot} \boldsymbol{B}^D$ are transverse fields; while both $\varepsilon \frac{\partial \boldsymbol{E}_L}{\partial t}$ and $\varepsilon \frac{\partial \boldsymbol{E}_L^D}{\partial t}$ are longitudinal fields and satisfy

$$\varepsilon \frac{\partial \boldsymbol{E}_L}{\partial t} \times \boldsymbol{n} = \varepsilon \frac{\partial \boldsymbol{E}_L^D}{\partial t} \times \boldsymbol{n} = \boldsymbol{0} \quad \text{on} \quad \partial \Omega.$$

Then from Lemma 1.4, we have

$$\frac{\partial \boldsymbol{E}_L}{\partial t} = \frac{\partial \boldsymbol{E}_L^D}{\partial t}.$$

Regarding the initial conditions (1.300) and (1.317), we obtain immediately

$$\boldsymbol{E}_L = \boldsymbol{E}_L^D.$$

Exercises

1. Suppose that there is an infinitely long line uniformly distributed by charges with line charge density σ (namely, the charge amount per unit length). Find the electric field intensity and electric potential of the electric field generated by this line.
2. Suppose that there is a sphere of radius R distributed uniformly by charges with surface charge density σ (namely, the charge amount per unit area). Find the electric field intensity and electric potential of the electric field generated by this sphere.
3. Put a conductive ball with radius R in an electric field with intensity $\boldsymbol{E}_0$ ($\boldsymbol{E}_0$ is a constant vector). Derive the equations and the corresponding boundary conditions satisfied by the electric potential in the exterior of the ball.

4. For the electric dipole generated by the point charges $-q$ and $+q$ located at P_0 and P_1, respectively, its dipole moment is $\boldsymbol{m} = q\boldsymbol{l}$, where $\boldsymbol{l} = \overrightarrow{P_0 P_1}$. Prove that the potential of the electric field generated by this dipole is

$$\phi(P) = -\frac{1}{4\pi\varepsilon_0} \boldsymbol{m} \cdot \operatorname{grad}_P \left(\frac{1}{r_{P_0 P}}\right),$$

provided that $l \to 0$, $q \to +\infty$, while $m = ql$ remains unchanged, where grad_P represents the gradient at point P.

5. Calculate the electric field intensity generated by the electric dipole given in exercise 4.

6. Calculate the magnetic induction intensity of the magnetic field generated by the infinitely long wire with current intensity I.

7. Suppose that there is a circular circuit of radius R with current intensity I. On the straight line through the center of the circle and perpendicular to the plane where the circuit lies, calculate the magnetic induction intensity of the magnetic field generated by this circular circuit.

8. Suppose that there is a cylindrical magnetic field in a vacuum,

$$B(P) = \begin{cases} \dfrac{2I}{Cr} & \text{when} \quad r \geq R, \\ \dfrac{2I}{CR^2} r & \text{when} \quad r < R, \end{cases}$$

where r is the distance from point P under consideration to the symmetric axis, I and R are constant, and the direction of the magnetic field at any point is in accordance with that of the rotary direction around the symmetric axis through this point. Find the current distribution which generates this magnetic field.

9. Suppose that there is a ring current with current intensity I on the plane with normal $\boldsymbol{n}$, whose direction and $\boldsymbol{n}$ constitute the right-hand coordinate system. Suppose furthermore that the area surrounded by the ring current is S_0; then

$$\boldsymbol{m} = I S_0 \boldsymbol{n}$$

is called the *magnetic dipole moment* of this ring current. Prove that the vector potential of the magnetic field generated by this magnetic dipole moment is

$$\boldsymbol{A}(P) = -\frac{\mu_0 \boldsymbol{m}}{4\pi} \operatorname{rot}_P \frac{1}{r_{OP}},$$

provided that $S_0 \to 0$ (the ring shrinks to one point), $I \to +\infty$, while $\boldsymbol{n}$ and $m = I S_0$ remain unchanged, where rot_P is the curl at point P.

10. In the case of a free electromagnetic field, prove that the scalar potential can be identically equal to zero under the gauge transform which keeps the Lorentz condition.

11. For the isotropic conductor, Ohm's law has the following form:

$$\boldsymbol{j} = \sigma \boldsymbol{E},$$

where σ is called the *electric conductivity*. Prove that the continuity equation of the conductor in a vacuum is given by

$$\frac{\partial \rho}{\partial t} + \frac{\sigma}{\varepsilon_0}\rho = 0,$$

from which we prove that any charge distribution in the conductor decays exponentially to zero with the increase of time.

12. Establish the equations satisfied by the scalar potential ϕ and the vector potential A of the electromagnetic field under the Coulomb gauge.

13. Discuss how to determine the conditions on the interface of media for the vector potential of the static magnetic field.

14. Suppose that the vector function $B(x_1, x_2, x_3) = (B_1, B_2, B_3)$ has continuous first-order partial derivatives at $x_3 \neq 0$ and discontinuity of the first type at $x_3 = 0$, and that
$$\text{div}\, B = 0, \quad x_3 \neq 0.$$

If B_3 is continuous at $x_3 = 0$, prove that there exists a continuous vector function $A(x_1, x_2, x_3)$ such that
$$B = \text{rot}\, A.$$

15. For the electromagnetic field in a medium, derive the expressions of the electromagnetic momentum density vector and electromagnetic momentum flux density tensor (7.47), (7.48).

Bibliography

[1] Cao C. *Electrodynamics* (in Chinese). Beijing: People's Education Press, 1962.

[2] Jackson J. D. *Classical Electrodynamics*, Third Edition. New York: John Wiley & Sons, 1998.

[3] Yan J. *Electromagnetics* (in Chinese). Beijing: Higher Education Press, 1988.

[4] Gu C., Li T. et al. *Mathematical Physics Equations* (in Chinese). Shanghai: Shanghai Science and Technology Press, 1987.

[5] Gu C., Li T., Shen W. *Applied Partial Differential Equations* (in Chinese). Beijing: Higher Education Press, 1993.

[6] Gu C. *Some developments and applications to the theory of positive symmetric system* (in Chinese). In Collected Mathematical Papers, Institute of Mathematics of Fudan University, 1964, 42–58.

[7] Li T. et al. *The Application of Finite Element Method in Electric Well-Logging* (in Chinese). Beijing: Petroleum Industry Press, 1980.

[8] Li T. et al. *The boundary value problem with equivalued surface for the self-conjugate elliptic equations* (I), (II) (in Chinese). J. Fudan University. 1 (1976), 61–71; 3/4 (1976), 136–145.

[9] Courant R., Hilbert D. *Methods of Mathematical Physics*, Volume II. New York: John Wiley & Sons, 1989.

[10] Chen S. *Introduction to Partial Differential Equations* (in Chinese). Beijing: People's Education Press, 1981.

[11] Bers L., John F., Schechter M. *Partial Differential Equations*. New York: Wiley-Interscience, 1964.

[12] Damlamian A., Li T. *Boundary homogenization for elliptic problems*. J. Math. Pures Appl., 66 (1987), 351–361.

[13] Degond P., Raviart P. A. *An analysis of the Darwin model of approximation to Maxwell's equations*. Forum Math., 4 (1992), 13–44.

[14] Friedrichs K. O. *Symmetric hyperbolic linear differential equations*. Comm. Pure Appl. Math., 7 (1954), 345–392.

[15] Hewett D. W., Nielson C. *A multidimensional quasineutral plasma simulation model*. J. Comput. Phys., 29 (1978), 219–236.

[16] Hewett D. W., Boyd J. K. *Streamlined Darwin simulation of nonneutral plasmas*. J. Comput. Phys., 70 (1987). 166–181.

[17] Li T. *A class of non-local boundary value problems for partial differential equations and its applications in numerical analysis*. J. Comput. Appl. Math., 28 (1989), 49–62.

[18] Nielson C. W., Lewis H. R. *Particle code models in the nonradiative limit*. In Methods of Computational Physics, Vol. 16, 367–388. New York: Academic Press, 1976.

[19] Feynman R. P., Leiden R. B., Sands M. *The Feynman Lectures on Physics*, The New Millennium Edition. New York: Basic Books, 2011.

[20] Maxwell J. C. *A Treatise on Electricity and Magnetism*. Oxford: Clarendon Press, 1873.

[21] Li T., Zheng S., Tan Y., Shen W. *Boundary Value Problems with Equivalued Surface and Resistivity Well-Logging*. Pitman Research Notes in Mathematics Series 382, Harlow, UK: Longman Scientific and Technical, 1998.

Chapter 2
Fluid Dynamics

2.1 System of Ideal Fluid Dynamics

2.1.1 Preliminaries

In this chapter, we will first establish the system of ideal fluid dynamics. The *ideal fluid* is the fluid that exists when viscosity and heat conduction are neglected, while the actual fluid always has viscosity and heat conduction, which we will discuss further in the next section. It has to be pointed out that the ideal fluid is a reasonable approximation in many cases. For example, in the study of the distribution of a flow field around an aircraft, any other part of the flow field can be regarded as an ideal fluid except a thin layer near the surface of the aircraft in which viscosity and heat conduction have to be taken into consideration; even if we assume that the whole flow field is an ideal fluid, we can also obtain quite reasonable results. Therefore, the investigation of the ideal fluid has not only theoretical significance, but also great value in practice. In addition, we discuss compressible fluid, namely, gas or liquid under high pressure. Discussion about incompressible fluid (the fluid under normal pressure) will be carried out in the next section.

We describe the state of the fluid (state of motion and thermodynamical state) by $\boldsymbol{u} = (u_1, u_2, u_3)$, ρ, p, T, and so on. In the case of unsteady motion, these are all functions of time t and position $\boldsymbol{x} = (x_1, x_2, x_3)$ in Cartesian coordinates. Now we explain their meanings, respectively, as follows.

$\boldsymbol{u}$ is the *velocity vector*. This refers to the macroscopic velocity of the infinitesimal element of the fluid and not to the velocity of the irregular motion of individual fluid molecules. Now let dS be an infinitesimal area element passing through point $\boldsymbol{x} = (x_1, x_2, x_3)$ with unit normal vector $\boldsymbol{n}$. Then the volume of the fluid flowing across dS along the $\boldsymbol{n}$ direction in the time interval $[t, t+dt]$ is

$$\boldsymbol{u}(t,\boldsymbol{x}) \cdot \boldsymbol{n} dS dt. \tag{2.1}$$

ρ is the *mass density*, namely, the mass of the fluid per unit volume. So, the mass of the fluid flowing across dS along the $\boldsymbol{n}$ direction in the time interval $[t, t+dt]$ is

$$\rho(t,\boldsymbol{x})\boldsymbol{u}(t,\boldsymbol{x}) \cdot \boldsymbol{n} dS dt. \tag{2.2}$$

We call $\rho\boldsymbol{u}$ the *mass flux vector*. From the above, during unit time, the mass of the fluid flowing along any direction $\boldsymbol{n}$ across a unit area vertical to $\boldsymbol{n}$ is $\rho\boldsymbol{u}\cdot\boldsymbol{n}$.

$\rho\boldsymbol{u}$ is also called the *momentum density vector*, namely, the momentum of the fluid per unit volume. The momentum of the fluid flowing across $\mathrm{d}S$ along the $\boldsymbol{n}$ in the time interval $[t,t+\mathrm{d}t]$ is

$$\rho\boldsymbol{u}(\boldsymbol{u}\cdot\boldsymbol{n})\mathrm{d}S\mathrm{d}t = \rho(\boldsymbol{u}\otimes\boldsymbol{u})\boldsymbol{n}\mathrm{d}S\mathrm{d}t, \tag{2.3}$$

where $\boldsymbol{u}\otimes\boldsymbol{u}$ is the tensor product of velocity vector, i.e.,

$$\boldsymbol{u}\otimes\boldsymbol{u} = \begin{pmatrix} u_1^2 & u_1u_2 & u_1u_3 \\ u_2u_1 & u_2^2 & u_2u_3 \\ u_3u_1 & u_3u_2 & u_3^2 \end{pmatrix},$$

while $(\boldsymbol{u}\otimes\boldsymbol{u})\boldsymbol{n}$ represents the product of matrix $\boldsymbol{u}\otimes\boldsymbol{u}$ and vector $\boldsymbol{n}$ in the usual sense, hereinafter the same. We call $\rho\boldsymbol{u}\otimes\boldsymbol{u}$ the *momentum flux tensor*.

p is the *pressure*, namely, the pressure of the fluid per unit area, whose direction is vertical to this area by ignoring the internal friction (viscosity) of the fluid. Under the assumption of isotropy, the pressure is equal in any unit area through the same point with a different normal direction. Here, the so-called isotropy refers to the fact that the physical properties of fluid have nothing to do with any pregiven spatial direction. Fluids usually meet this property. Then, the pressure on the area element $\mathrm{d}S$, received from the side of the fluid in a positive direction of unit normal vector $\boldsymbol{n}$, is

$$-p\boldsymbol{n}\mathrm{d}S, \tag{2.4}$$

in which the minus sign means pressing.

T is the *absolute temperature*. According to thermodynamics, only two variables are independent from among all the variables describing the thermodynamical states. Therefore, there is a relation among the three thermodynamical variables ρ, p, and T determined by the function

$$p = f(\rho, T), \tag{2.5}$$

which has different forms for different fluids. Equation (2.5) is called the *equation of state* of the fluid. In particular, if the equation of state is

$$p = R\rho T, \tag{2.6}$$

then we call the fluid an *ideal gas*, where R is a positive constant.

Now we introduce another thermodynamical variable. We denote by e the *internal energy* of the fluid per unit mass, i.e., the sum of the kinetic energy possessed by the irregular thermal motion of fluid molecules and the potential energy determined by the intermolecular relative position. For the ideal gas, it can be proved that there is no intermolecular interaction, and thus no molecular potential energy, while its internal energy depends only on temperature; i.e., e is only a function of T. In particular, if

$$e = c_V T, \tag{2.7}$$

then we call the gas a *polytropic gas*, where c_V is a positive constant called the *specific heat at constant volume*.

2.1 System of Ideal Fluid Dynamics

The energy of the fluid consists of the internal energy and the macroscopic kinetic energy, so $\rho e + \frac{1}{2}\rho u^2$ is the *energy density*, i.e., the energy of the fluid per unit volume, where $u^2 = u_1^2 + u_2^2 + u_3^2$. Therefore, the energy of the fluid flowing across dS along the direction of $\boldsymbol{n}$ in the time interval $[t, t+dt]$ is

$$\left(\rho e + \frac{1}{2}\rho u^2\right) \boldsymbol{u} \cdot \boldsymbol{n} \, dS \, dt. \tag{2.8}$$

We call $(\rho e + \frac{1}{2}\rho u^2)\boldsymbol{u}$ the *energy flux vector*.

2.1.2 System of Ideal Fluid Dynamics

For an ideal fluid, according to the conservation laws of mass, momentum, and energy in the motion process, we can derive the fundamental system of ideal fluid dynamics.

(1) Conservation law of mass.

In the domain under consideration, take any closed smooth surface Γ surrounding a domain denoted by Ω. According to the conservation law of mass, the increment of the fluid mass in the time interval $[t_1, t_2]$,

$$\int_\Omega \rho(t_2, \boldsymbol{x}) dx - \int_\Omega \rho(t_1, \boldsymbol{x}) dx$$

(where $dx = dx_1 dx_2 dx_3$), should be equal to the mass of the fluid flowing into Ω through the boundary Γ during this period of time, while the latter, according to (2.2), should be

$$-\int_{t_1}^{t_2} \int_\Gamma \rho \boldsymbol{u} \cdot \boldsymbol{n} \, dS \, dt.$$

Then the conservation law of mass can be written into the following integral form:

$$\int_\Omega (\rho(t_2, \boldsymbol{x}) - \rho(t_1, \boldsymbol{x})) dx = -\int_{t_1}^{t_2} \int_\Gamma \rho \boldsymbol{u} \cdot \boldsymbol{n} \, dS \, dt,$$
$$\forall \Omega, \forall [t_1, t_2]. \tag{2.9}$$

If the functions under consideration are assumed to be continuously differentiable, then according to Green's formula, the above formula can be rewritten as

$$\int_{t_1}^{t_2} \int_\Omega \left(\frac{\partial \rho}{\partial t} + \mathrm{div}(\rho \boldsymbol{u})\right) dx \, dt = 0, \quad \forall \Omega, \forall [t_1, t_2].$$

So, from the arbitrariness of Ω and $[t_1, t_2]$, together with the continuity of the integrand, we obtain the following differential form of the conservation law of mass:

$$\frac{\partial \rho}{\partial t} + \mathrm{div}(\rho \boldsymbol{u}) = 0, \tag{2.10}$$

which is usually called the *continuity equation*.

Hereinafter it should be noted that the differential forms of conservation laws make sense only when the relevant state functions are continuously differentiable. Otherwise, in the cases when these functions are not continuously differentiable (corresponding to shocks in gas dynamics, which are the most important cases in either theory or applications), the relevant integral forms of the conservation laws still make sense, which will be regarded as a starting point for any further discussion (see section 2.4).

(2) Conservation law of momentum.

According to the conservation law of momentum, the increment of the fluid momentum in domain Ω in the time interval $[t_1,t_2]$,

$$\int_\Omega \rho \boldsymbol{u}(t_2,\boldsymbol{x})\mathrm{d}x - \int_\Omega \rho \boldsymbol{u}(t_1,\boldsymbol{x})\mathrm{d}x,$$

should be equal to the momentum of the fluid flowing into Ω through the boundary Γ during this period of time, plus the impulse of the force acting upon Ω during this period of time, where the former, according to (2.3), should be

$$-\int_{t_1}^{t_2}\int_\Gamma \rho(\boldsymbol{u}\otimes\boldsymbol{u})\boldsymbol{n}\,\mathrm{d}S\mathrm{d}t,$$

while the latter consists of the impulse of the volume force acting upon Ω and the impulse of the surface force acting upon the boundary of Ω. Let $\boldsymbol{F}(t,\boldsymbol{x})$ be the *volume force density*, namely, the external force upon the unit volume fluid, then the first part of the impulse is

$$\int_{t_1}^{t_2}\int_\Omega \rho \boldsymbol{F}\,\mathrm{d}x\mathrm{d}t,$$

while the only surface force upon Γ is the force received from the fluid outside of Ω; then from (2.4) the second part of the impulse is

$$-\int_{t_1}^{t_2}\int_\Gamma p\boldsymbol{n}\,\mathrm{d}S\mathrm{d}t.$$

Therefore, the conservation law of momentum can be written into the following integral form:

$$\int_\Omega (\rho\boldsymbol{u}(t_2,\boldsymbol{x}) - \rho\boldsymbol{u}(t_1,\boldsymbol{x}))\mathrm{d}x$$
$$= -\int_{t_1}^{t_2}\int_\Gamma \rho(\boldsymbol{u}\otimes\boldsymbol{u})\boldsymbol{n}\,\mathrm{d}S\mathrm{d}t$$
$$-\int_{t_1}^{t_2}\int_\Gamma p\boldsymbol{n}\,\mathrm{d}S\mathrm{d}t + \int_{t_1}^{t_2}\int_\Omega \rho\boldsymbol{F}\,\mathrm{d}x\mathrm{d}t,$$
$$\forall\Omega, \forall[t_1,t_2]. \tag{2.11}$$

If the functions under consideration are assumed to be continuously differentiable, then according to Green's formula, the above formula can be rewritten as

$$\int_{t_1}^{t_2}\int_\Omega \frac{\partial}{\partial t}(\rho\boldsymbol{u})\mathrm{d}x\mathrm{d}t = -\int_{t_1}^{t_2}\int_\Omega \mathrm{div}(\rho\boldsymbol{u}\otimes\boldsymbol{u} + p\boldsymbol{I})\mathrm{d}x\mathrm{d}t$$
$$+\int_{t_1}^{t_2}\int_\Omega \rho\boldsymbol{F}\,\mathrm{d}x\mathrm{d}t, \quad \forall\Omega, \forall[t_1,t_2], \tag{2.12}$$

2.1 System of Ideal Fluid Dynamics

where $\boldsymbol{I}$ is the second-order unit tensor. From above, and using the continuity of the integrand and the arbitrariness of Ω and $[t_1, t_2]$, we obtain the following differential form of the conservation law of momentum:

$$\frac{\partial}{\partial t}(\rho \boldsymbol{u}) + \mathrm{div}(\rho \boldsymbol{u} \otimes \boldsymbol{u} + p \boldsymbol{I}) = \rho \boldsymbol{F}, \tag{2.13}$$

or the corresponding representation in components,

$$\frac{\partial}{\partial t}(\rho u_i) + \sum_{k=1}^{3} \frac{\partial}{\partial x_k}(\rho u_k u_i + p \delta_{ki}) = \rho F_i \quad (i = 1, 2, 3), \tag{2.14}$$

where δ_{ki} is the Kronecker symbol.

Here we point out that, by taking advantage of the continuity equation (2.10), the above equation can be simplified to

$$\frac{\partial u_i}{\partial t} + \sum_{k=1}^{3} u_k \frac{\partial u_i}{\partial x_k} + \frac{1}{\rho} \frac{\partial p}{\partial x_i} = F_i \quad (i = 1, 2, 3), \tag{2.15}$$

which can also be written as

$$\frac{\mathrm{d} \boldsymbol{u}}{\mathrm{d} t} + \frac{1}{\rho} \mathrm{grad}\, p = \boldsymbol{F}, \tag{2.16}$$

where

$$\frac{\mathrm{d}}{\mathrm{d} t} = \frac{\partial}{\partial t} + \sum_{k=1}^{3} u_k \frac{\partial}{\partial x_k} \tag{2.17}$$

refers to the derivative with respect to t when fixing the fluid particle (other than the spatial point!). In fact, since $\dot{x}_k(t) = u_k$ ($k = 1, 2, 3$) for the motion law $x_k = x_k(t)$ ($k = 1, 2, 3$) of any fixed fluid particle, the derivative of a function $f(t, \boldsymbol{x})$ with respect to t, when fixing the fluid particle, is

$$\frac{\mathrm{d}}{\mathrm{d} t} f(t, x_1(t), x_2(t), x_3(t)) = \frac{\partial f}{\partial t} + \sum_{k=1}^{3} \frac{\partial f}{\partial x_k} \dot{x}_k(t)$$

$$= \frac{\partial f}{\partial t} + \sum_{k=1}^{3} u_k \frac{\partial f}{\partial x_k}.$$

Equation (2.15) or (2.16) is usually called the *Euler equation*, which, different from (2.14), does not take the conserved form of divergence type.

(3) Conservation law of energy.

According to the conservation law of energy, the increment of the fluid energy in domain Ω in the time interval $[t_1, t_2]$,

$$\int_\Omega \left(\rho e + \frac{1}{2} \rho u^2\right)(t_2, \boldsymbol{x}) \mathrm{d} x - \int_\Omega \left(\rho e + \frac{1}{2} \rho u^2\right)(t_1, \boldsymbol{x}) \mathrm{d} x,$$

should be equal to the energy of the fluid flowing into Ω through the boundary Γ during this period of time, plus the work done by the force acting upon Ω during this period of time, where the former, according to (2.8), should be

$$-\int_{t_1}^{t_2}\int_{\Gamma}\left(\rho e+\frac{1}{2}\rho u^2\right)\boldsymbol{u}\cdot\boldsymbol{n}\mathrm{d}S\mathrm{d}t,$$

while the latter consists of two parts: the first part is the work done by the volume force acting on Ω,

$$\int_{t_1}^{t_2}\int_{\Omega}\rho\boldsymbol{F}\cdot\boldsymbol{u}\mathrm{d}x\mathrm{d}t,$$

and the second part is the work done by the surface force (now the pressure received from the fluid outside of Ω) acting on Γ, which, from (2.4), should be

$$-\int_{t_1}^{t_2}\int_{\Gamma}p\boldsymbol{u}\cdot\boldsymbol{n}\mathrm{d}S\mathrm{d}t.$$

Then the conservation law of energy can be written into the following integral form:

$$\int_{\Omega}\left(\rho e+\frac{1}{2}\rho u^2\right)(t_2,\boldsymbol{x})\mathrm{d}x - \int_{\Omega}\left(\rho e+\frac{1}{2}\rho u^2\right)(t_1,\boldsymbol{x})\mathrm{d}x$$
$$= -\int_{t_1}^{t_2}\int_{\Gamma}\left(\rho e+\frac{1}{2}\rho u^2\right)\boldsymbol{u}\cdot\boldsymbol{n}\mathrm{d}S\mathrm{d}t - \int_{t_1}^{t_2}\int_{\Gamma}p\boldsymbol{u}\cdot\boldsymbol{n}\mathrm{d}S\mathrm{d}t$$
$$+\int_{t_1}^{t_2}\int_{\Omega}\rho\boldsymbol{F}\cdot\boldsymbol{u}\mathrm{d}x\mathrm{d}t, \quad \forall\Omega,\forall[t_1,t_2]. \tag{2.18}$$

If the relevant functions are assumed to be continuously differentiable, then according to Green's formula, the above formula can be rewritten as

$$\int_{t_1}^{t_2}\int_{\Omega}\frac{\partial}{\partial t}\left(\rho e+\frac{1}{2}\rho u^2\right)\mathrm{d}x\mathrm{d}t$$
$$= -\int_{t_1}^{t_2}\int_{\Omega}\mathrm{div}\left(\left(\rho e+\frac{1}{2}\rho u^2+p\right)\boldsymbol{u}\right)\mathrm{d}x\mathrm{d}t$$
$$+\int_{t_1}^{t_2}\int_{\Omega}\rho\boldsymbol{F}\cdot\boldsymbol{u}\mathrm{d}x\mathrm{d}t, \quad \forall\Omega,\forall[t_1,t_2].$$

So, by using the arbitrariness of Ω and $[t_1,t_2]$, together with the continuity of the integrand, from the above formula, the differential form of the conservation law of energy is given by the following equation:

$$\frac{\partial}{\partial t}\left(\rho e+\frac{1}{2}\rho u^2\right)+\mathrm{div}\left(\left(\rho e+\frac{1}{2}\rho u^2+p\right)\boldsymbol{u}\right)=\rho\boldsymbol{F}\cdot\boldsymbol{u}. \tag{2.19}$$

From the continuity equation (2.10), (2.19) can be simplified to

$$\rho\frac{\partial}{\partial t}\left(e+\frac{u^2}{2}\right)+\rho\boldsymbol{u}\cdot\mathrm{grad}\left(e+\frac{u^2}{2}\right)+\mathrm{div}(p\boldsymbol{u})=\rho\boldsymbol{F}\cdot\boldsymbol{u} \tag{2.20}$$

2.1 System of Ideal Fluid Dynamics

or

$$\rho \frac{d}{dt}\left(e + \frac{u^2}{2}\right) + \operatorname{div}(pu) = \rho \boldsymbol{F} \cdot \boldsymbol{u}, \qquad (2.21)$$

where $\frac{d}{dt}$ is defined by (2.17). Noting that

$$\operatorname{div}(pu) = p \operatorname{div} \boldsymbol{u} + \boldsymbol{u} \cdot \operatorname{grad} p$$

and $\frac{d}{dt}(\frac{u^2}{2}) = \boldsymbol{u} \cdot \frac{d\boldsymbol{u}}{dt}$, and using Euler equation (2.16), (2.21) can be rewritten as

$$\rho \frac{de}{dt} + p \operatorname{div} \boldsymbol{u} = 0. \qquad (2.22)$$

Noticing further that the continuity equation (2.10) can be written as

$$\frac{d\rho}{dt} + \rho \operatorname{div} \boldsymbol{u} = 0,$$

(2.22) can be rewritten further as

$$\frac{de}{dt} - \frac{p}{\rho^2}\frac{d\rho}{dt} = 0. \qquad (2.23)$$

Now we have obtained five equations of conservation laws of mass, momentum, and energy. We can take, say, $\boldsymbol{u} = (u_1, u_2, u_3)$, ρ, and T as unknown functions, including the equation of state $p = f(\rho, T)$ and corresponding $e = e(\rho, T)$, to constitute a closed system of partial differential equations, which is valid in a continuously differentiable flow field. This is a *first-order quasi-linear system of partial differential equations* with four independent variables (t, x_1, x_2, x_3) and five unknown functions. The study of this system, particularly that on discontinuous solutions (with shock as original model), has always been a very important research subject up to now. In 1986 the American National Research Council proposed six key projects for the development of mathematics, one of which is the "nonlinear hyperbolic system of conservation laws" (see [11]).

Now we show that, in a continuously differentiable flow field, the conservation equation of energy (2.23) can be rewritten in the following simple form:

$$\frac{dS}{dt} = 0, \qquad (2.24)$$

that is,

$$\frac{\partial S}{\partial t} + \boldsymbol{u} \cdot \operatorname{grad} S = 0, \qquad (2.25)$$

where S is the *entropy* of the fluid per unit mass, determined by

$$dS = \frac{1}{T}(de + pd\tau), \qquad (2.26)$$

where τ is the *specific volume*, namely, the volume per unit mass, and obviously $\tau = \frac{1}{\rho}$. Equation (2.24) implies that the entropy S remains a constant on any fixed particle point in a continuously differentiable flow field, which is consistent with the hypothesis that the fluid is ideal. But the fact is valid only for the continuously differentiable flow field, while the entropy of the particle will increase across the jump (shock) in the flow field with discontinuity (see section 2.4).

***Proof of* (2.24).** Plugging $\rho = \frac{1}{\tau}$ into (2.23), we have

$$\frac{de}{dt} + p\frac{d\tau}{dt} = 0,$$

which, by the definition (2.26) of entropy, can be turned into

$$T\frac{dS}{dt} = 0.$$

Then (2.24) is proved. □

Equation (2.24) or (2.25) can replace the conservation equation of energy in a continuously differentiable flow field (for this case only!). At this time, it is better to take $u = (u_1, u_2, u_3)$, take ρ and S as unknown functions, and make use of the equations of state such as $p = f(\rho, S)$ and $e = e(\rho, S)$.

Particularly for polytropic gases, the equation of state takes the following form:

$$p = A(S)\rho^\gamma, \qquad (2.27)$$

where

$$A(S) = (\gamma - 1)\exp\left(\frac{S - S_0}{c_V}\right); \qquad (2.28)$$

here $\gamma > 1$ is a constant, called the *adiabatic exponent*, and S_0 is also a constant. For common gases, the value of γ lies between 1 and $\frac{5}{3}$; for air, $\gamma = 1.4$. For polytropic gases, the entropy equation (2.25) can be written as

$$\frac{\partial}{\partial t}\left(\frac{p}{\rho^\gamma}\right) + u \cdot \mathrm{grad}\left(\frac{p}{\rho^\gamma}\right) = 0. \qquad (2.29)$$

At this moment, we can take $u = (u_1, u_2, u_3)$, take ρ and p as unknown functions, and make use of the equation of state with the form $e = e(\rho, p)$.

2.1.3 Mathematical Structure of the System of Ideal Fluid Dynamics

Now we illustrate that the system of ideal fluid dynamics can be written in the form of a *first-order quasi-linear symmetric hyperbolic system*.

For this purpose, we first rewrite the Euler equations (the conservation equations of momentum) (2.15) into

$$\rho\frac{\partial u_i}{\partial t} + \sum_{k=1}^{3}\rho u_k \frac{\partial u_i}{\partial x_k} + \frac{\partial p}{\partial x_i} = \rho F_i \quad (i = 1, 2, 3) \qquad (2.30)$$

2.1 System of Ideal Fluid Dynamics

and rewrite the conservation equation (2.25) of entropy into

$$\frac{\partial S}{\partial t} + \sum_{k=1}^{3} u_k \frac{\partial S}{\partial x_k} = 0. \tag{2.31}$$

Using the equation of state $p = p(\rho, S)$ and denoting

$$c^2 = \frac{\partial p}{\partial \rho} > 0,$$

where c is the *local speed of sound*, the conservation equation of mass (see (2.10))

$$\frac{\partial \rho}{\partial t} + \sum_{k=1}^{3} u_k \frac{\partial \rho}{\partial x_k} + \rho \sum_{k=1}^{3} \frac{\partial u_k}{\partial x_k} = 0$$

can be rewritten, from (2.31), as

$$\frac{\partial p}{\partial \rho} \left(\frac{\partial \rho}{\partial t} + \sum_{k=1}^{3} u_k \frac{\partial \rho}{\partial x_k} + \rho \sum_{k=1}^{3} \frac{\partial u_k}{\partial x_k} \right)$$
$$+ \frac{\partial p}{\partial S} \left(\frac{\partial S}{\partial t} + \sum_{k=1}^{3} u_k \frac{\partial S}{\partial x_k} \right) = 0,$$

i.e.,

$$\frac{\partial p}{\partial t} + \sum_{k=1}^{3} u_k \frac{\partial p}{\partial x_k} + \rho c^2 \sum_{k=1}^{3} \frac{\partial u_k}{\partial x_k} = 0$$

or

$$\frac{1}{\rho c^2} \frac{\partial p}{\partial t} + \sum_{k=1}^{3} \frac{\partial u_k}{\partial x_k} + \sum_{k=1}^{3} \frac{u_k}{\rho c^2} \frac{\partial p}{\partial x_k} = 0. \tag{2.32}$$

So, if we take (u_1, u_2, u_3, p, S) as unknown functions, arrange the above equations in the order of conservation equations of momentum (2.30), conservation equation of mass (2.32), and conservation equation of entropy (2.31), then the system of equations will be written in the following matrix form:

$$A_0 \frac{\partial U}{\partial t} + A_1 \frac{\partial U}{\partial x_1} + A_2 \frac{\partial U}{\partial x_2} + A_3 \frac{\partial U}{\partial x_3} = C, \tag{2.33}$$

where $U = (u_1, u_2, u_3, p, S)^T$,

$$A_0 = \begin{pmatrix} \rho & 0 & 0 & 0 & 0 \\ 0 & \rho & 0 & 0 & 0 \\ 0 & 0 & \rho & 0 & 0 \\ 0 & 0 & 0 & \rho^{-1}c^{-2} & 0 \\ 0 & 0 & 0 & 0 & 1 \end{pmatrix},$$

$$A_1 = \begin{pmatrix} \rho u_1 & 0 & 0 & 1 & 0 \\ 0 & \rho u_1 & 0 & 0 & 0 \\ 0 & 0 & \rho u_1 & 0 & 0 \\ 1 & 0 & 0 & u_1 \rho^{-1} c^{-2} & 0 \\ 0 & 0 & 0 & 0 & u_1 \end{pmatrix},$$

$$A_2 = \begin{pmatrix} \rho u_2 & 0 & 0 & 0 & 0 \\ 0 & \rho u_2 & 0 & 1 & 0 \\ 0 & 0 & \rho u_2 & 0 & 0 \\ 0 & 1 & 0 & u_2 \rho^{-1} c^{-2} & 0 \\ 0 & 0 & 0 & 0 & u_2 \end{pmatrix},$$

$$A_3 = \begin{pmatrix} \rho u_3 & 0 & 0 & 0 & 0 \\ 0 & \rho u_3 & 0 & 0 & 0 \\ 0 & 0 & \rho u_3 & 1 & 0 \\ 0 & 0 & 1 & u_3 \rho^{-1} c^{-2} & 0 \\ 0 & 0 & 0 & 0 & u_3 \end{pmatrix},$$

while

$$C = (\rho F_1, \rho F_2, \rho F_3, 0, 0)^{\mathrm{T}}.$$

In the region where $\rho > 0$ (i.e., no vacuum appears), A_0 is a positively definite symmetric matrix, and A_1, A_2, and A_3 are all symmetric matrices. So, system (2.33) is a *first-order quasi-linear symmetric hyperbolic system* of partial differential equations.

The Cauchy problem can be raised to the first-order quasi-linear symmetric hyperbolic system (2.33): find a solution $U = U(t, x_1, x_2, x_3)$ of this system in $t > 0$ such that it satisfies the initial condition

$$t = 0 : U = U^0(x_1, x_2, x_3), \tag{2.34}$$

where $U^0(x_1, x_2, x_3)$ is a given vector function. Under the assumption that $U^0(x_1, x_2, x_3)$ is smooth enough, the existence and uniqueness of solutions (in local range of time $0 \leq t \leq \delta$ ($\delta > 0$)) to this problem can be proved. We can also propose corresponding initial-boundary value problems to system (2.33) as stated in section 1.5 of Chapter 1.

System (2.33) is derived starting from the differential forms of conservation laws, which make sense only in the continuously differentiable flow field. But since the discontinuous flow field can usually be patched together from some continuously differentiable flow fields, it still makes sense to construct indirectly discontinuous solutions from this kind of system. As for the direct construction of discontinuous solutions, we then fall back on the integral form of conservation laws or the corresponding conservation equations of divergence form. From the point of view of numerical computations and theoretical research, if we enable the system of conservation laws with divergence form to be a symmetric hyperbolic system; that is, the *system of conservation laws* with divergence form

$$\frac{\partial G_i(V)}{\partial t} + \sum_{k=1}^{3} \frac{\partial G_i^k(V)}{\partial x_k} = C_i(V) \quad (i = 1, \ldots, 5), \tag{2.35}$$

where $V = (v_1, v_2, v_3, v_4, v_5)^{\mathrm{T}}$ is a unknown function, with expanded form

$$\sum_{l=1}^{5} \frac{\partial G_i(V)}{\partial v_l} \frac{\partial v_l}{\partial t} + \sum_{k=1}^{3} \sum_{l=1}^{5} \frac{\partial G_i^k(V)}{\partial v_l} \frac{\partial v_l}{\partial x_k} = C_i(V) \quad (i = 1, \ldots, 5) \tag{2.36}$$

2.1 System of Ideal Fluid Dynamics

is meanwhile a symmetric hyperbolic system, i.e.,

$\left(\dfrac{\partial G_i(V)}{\partial v_l}\right)$ is a positively definite symmetric matrix,

$\left(\dfrac{\partial G_i^k(V)}{\partial v_l}\right)$ $(k = 1, 2, 3)$ are symmetric matrices,

then things will become more convenient to some extent.

For the system of ideal fluid dynamics, an interesting research topic is whether it is possible to realize this requirement by choosing $V = (v_1, v_2, v_3, v_4, v_5)^T$ as a suitable combination or function of u_1, u_2, u_3, p, and S. This problem was solved by Soviet mathematician K. C. Godunov in the 1960s (see [10] and the references cited therein). American mathematicians K. O. Friderichs and P. D. Lax also did research on this problem (see [8] and [9]). We now proceed with further discussions on this issue.

We start from the system (2.10), (2.14), (2.19) of conservation laws, i.e.,

$$\frac{\partial \rho}{\partial t} + \sum_{k=1}^{3} \frac{\partial}{\partial x_k}(\rho u_k) = 0, \tag{2.37}$$

$$\frac{\partial}{\partial t}(\rho u_i) + \sum_{k=1}^{3} \frac{\partial}{\partial x_k}(\rho u_k u_i + p\delta_{ki}) = 0 \quad (i = 1, 2, 3), \tag{2.38}$$

$$\frac{\partial}{\partial t}\left(\rho e + \frac{1}{2}\rho u^2\right) + \sum_{k=1}^{3} \frac{\partial}{\partial x_k}\left(\left(\rho e + \frac{1}{2}\rho u^2 + p\right)u_k\right) = 0. \tag{2.39}$$

Here we assume $\boldsymbol{F} = 0$ for simplicity. From the continuity equation (2.37), the conservation equation (2.31) of entropy can be written in the following form of conservation law:

$$\frac{\partial}{\partial t}(\rho S) + \sum_{k=1}^{3} \frac{\partial}{\partial x_k}(\rho S u_k) = 0, \tag{2.40}$$

where ρS is the entropy per unit volume, while $\rho S \boldsymbol{u}$ is called the *entropy flux vector*.

Now we try to introduce a new unknown variable $V = (v_0, v_1, v_2, v_3, v_4)^T$ (here, for simplicity the components are counted starting from 0), and we reduce the system (2.37)–(2.39) into the following form of conservation laws:

$$\frac{\partial}{\partial t} L_{v_i}^0 + \sum_{k=1}^{3} \frac{\partial}{\partial x_k} L_{v_i}^k = 0, \quad i = 0, \ldots, 4, \tag{2.41}$$

where $L_{v_i}^j$ stands for the partial derivative of L^j with respect to v_i $(j = 0, 1, 2, 3)$. By rewriting system (2.41) in the form of (2.36), it is easy to see that

$$\left(\frac{\partial G_i}{\partial v_l}\right) = \left(L_{v_i v_l}^0\right),$$

$$\left(\frac{\partial G_i^k}{\partial v_l}\right) = \left(L_{v_i v_l}^k\right), \quad k = 1, 2, 3.$$

Obviously, since the above matrices are all symmetric, if we can find a system of conservation laws of the form (2.41), then (2.41) is a first-order symmetric hyperbolic system of conservation laws, provided that $(L^0_{v_i v_l})$ is positively definite; i.e., function $L^0(V)$ is strictly convex.

From the above arguments, the conservation equation (2.40) of entropy can be derived from the system (2.37)–(2.39) of conservation laws. Carefully examining the above procedure of derivation, we can see that there exist factors v_0, v_1, v_2, v_3, v_4 such that

$$v_0 \left(\frac{\partial \rho}{\partial t} + \sum_{k=1}^{3} \frac{\partial}{\partial x_k}(\rho u_k) \right)$$

$$+ \sum_{i=1}^{3} v_i \left(\frac{\partial}{\partial t}(\rho u_i) + \sum_{k=1}^{3} \frac{\partial}{\partial x_k}(\rho u_k u_i + p\delta_{ki}) \right)$$

$$+ v_4 \left(\frac{\partial}{\partial t}\left(\rho e + \frac{1}{2}\rho u^2\right) + \sum_{k=1}^{3} \frac{\partial}{\partial x_k}\left(\left(\rho e + \frac{1}{2}\rho u^2 + p\right) u_k\right) \right)$$

$$\equiv \frac{\partial}{\partial t}(\rho S) + \sum_{k=1}^{3} \frac{\partial}{\partial x_k}(\rho S u_k), \qquad (2.42)$$

where

$$v_0 = -\frac{e + \rho e_\rho - Se_S - \frac{1}{2}u^2}{e_S},$$

$$v_i = -\frac{u_i}{e_S} \ (i = 1, 2, 3), \quad v_4 = \frac{1}{e_S}. \qquad (2.43)$$

Here $e = e(\rho, S)$ is regarded as a function of ρ and S. In fact, it is easy to directly verify the correctness of (2.42) from the expression (2.43) of V. Let

$$L^0 = v_0 \rho + \sum_{i=1}^{3} v_i \rho u_i + v_4 \left(\rho e + \frac{1}{2}\rho u^2\right) - \rho S, \qquad (2.44)$$

$$L^k = v_0 \rho u_k + \sum_{i=1}^{3} v_i (\rho u_k u_i + p\delta_{ki})$$

$$+ v_4 \left(\rho e + \frac{1}{2}\rho u^2 + p\right) u_k - \rho S u_k, \quad k = 1, 2, 3. \qquad (2.45)$$

Using the expression (2.43) of V, it is easy to check that

$$v_0 d\rho + \sum_{i=1}^{3} v_i d(\rho u_i) + v_4 d\left(\rho e + \frac{1}{2}\rho u^2\right) = d(\rho S), \qquad (2.46)$$

and then from (2.44) we have

$$dL^0 = \rho dv_0 + \sum_{i=1}^{3} \rho u_i dv_i + \left(\rho e + \frac{1}{2}\rho u^2\right) dv_4. \qquad (2.47)$$

2.1 System of Ideal Fluid Dynamics

Similarly, we have

$$dL^k = \rho u_k dv_0 + \sum_{i=1}^{3}(\rho u_k u_i + p\delta_{ki})dv_i$$
$$+ \left(\rho e + \frac{1}{2}\rho u^2 + p\right)u_k dv_4, \quad k=1,2,3. \tag{2.48}$$

The above two formulas imply that

$$L^0_{v_0} = \rho, \quad L^0_{v_i} = \rho u_i \ (i=1,2,3), \quad L^0_{v_4} = \rho e + \frac{1}{2}\rho u^2 \tag{2.49}$$

and

$$L^k_{v_0} = \rho u_k, \quad L^k_{v_i} = \rho u_k u_i + p\delta_{ki} \ (i=1,2,3),$$
$$L^k_{v_4} = \left(\rho e + \frac{1}{2}\rho u^2 + p\right)u_k \quad (k=1,2,3). \tag{2.50}$$

Hence, system (2.37)–(2.39) can be written in the following form:

$$\frac{\partial}{\partial t}L^0_{v_0} + \sum_{k=1}^{3}\frac{\partial}{\partial x_k}L^k_{v_0} = 0, \tag{2.51}$$

$$\frac{\partial}{\partial t}L^0_{v_i} + \sum_{k=1}^{3}\frac{\partial}{\partial x_k}L^k_{v_i} = 0 \quad (i=1,2,3), \tag{2.52}$$

$$\frac{\partial}{\partial t}L^0_{v_4} + \sum_{k=1}^{3}\frac{\partial}{\partial x_k}L^k_{v_4} = 0. \tag{2.53}$$

This is exactly the system of conservation laws of the form (2.41).

v_0, v_1, v_2, v_3, v_4 and L^0, L^1, L^2, L^3 introduced above can be rewritten in a simpler forms by using the thermodynamical relations. In fact, from (2.26) we have

$$de = TdS + \frac{p}{\rho^2}d\rho, \tag{2.54}$$

so

$$e_S = T, \quad e_\rho = \frac{p}{\rho^2}. \tag{2.55}$$

Thus, V given by (2.43) can be written as

$$v_0 = -\frac{e + \rho e_\rho - TS - \frac{1}{2}u^2}{T}, \tag{2.56}$$

$$v_i = -\frac{u_i}{T} \ (i=1,2,3), \quad v_4 = \frac{1}{T}; \tag{2.57}$$

then, together with (2.44) and (2.45), we have

$$L^0 = -\frac{p}{T}, \quad L^k = -\frac{p}{T}u_k \ (k=1,2,3). \tag{2.58}$$

Now we show the strict convexity of $L^0(V)$ (or $-L^0$) with respect to

$$V = (v_0, v_1, v_2, v_3, v_4)^T.$$

For the ideal fluid under discussion, we assume that its internal energy e, regarded as a function $e(\tau, S)$ of τ and S, is strictly convex (see Appendix B for the rationality of this assumption); that is, the matrix

$$\begin{pmatrix} e_{\tau\tau} & e_{\tau S} \\ e_{\tau S} & e_{SS} \end{pmatrix} \quad (2.59)$$

is positive definite. Taking the *Legendre transform* to τ, S, and e (see Appendix B), we know, from the properties of the Legendre transform, that $\tau e_\tau + S e_S - e$ is a strictly convex function of e_τ and e_S. Thus

$$\tau e_\tau + S e_S + \frac{1}{2} u^2 - e (= T v_0) \quad (2.60)$$

is a strictly convex function of e_τ, e_S, u_1, u_2, and u_3 and then a strictly convex function of $-e_\tau$, e_S, u_1, u_2, and u_3.

Lemma 2.1. *Suppose that*

$$L = L(\xi_0, \xi_1, \ldots, \xi_n) \quad (2.61)$$

is a strictly convex function of its arguments and that $L_{\xi_0} < 0$. Then the function

$$\xi_0 = \xi_0(L, \xi_1, \ldots, \xi_n) \quad (2.62)$$

determined by (2.61) is also a strictly convex function of its arguments.

Proof. For simplicity, we give only the proof for the case $n = 1$. Taking the second-order partial derivatives with respect to L and ξ_1 on both sides of $L = L(\xi_0(L, \xi_1), \xi_1)$, we obtain

$$\frac{\partial^2 \xi_0}{\partial L^2} = -L_{\xi_0 \xi_0} \left(\frac{\partial \xi_0}{\partial L} \right)^2 \Big/ L_{\xi_0},$$

$$\frac{\partial^2 \xi_0}{\partial L \partial \xi_1} = -\left(L_{\xi_0 \xi_0} \frac{\partial \xi_0}{\partial L} \frac{\partial \xi_0}{\partial \xi_1} + L_{\xi_0 \xi_1} \frac{\partial \xi_0}{\partial L} \right) \Big/ L_{\xi_0},$$

$$\frac{\partial^2 \xi_0}{\partial \xi_1^2} = -\left(L_{\xi_0 \xi_0} \left(\frac{\partial \xi_0}{\partial L} \right)^2 + 2 L_{\xi_0 \xi_1} \frac{\partial \xi_0}{\partial \xi_1} + L_{\xi_1 \xi_1} \right) \Big/ L_{\xi_0}.$$

From the strict convexity of $L = L(\xi_0, \xi_1)$, we have $L_{\xi_0 \xi_0} > 0$ and $L_{\xi_0 \xi_0} L_{\xi_1 \xi_1} - L_{\xi_0 \xi_1}^2 > 0$. Noting that $L_{\xi_0} < 0$, from the above formulas, we have $\frac{\partial^2 \xi_0}{\partial L^2} > 0$ and

$$\det \begin{pmatrix} \dfrac{\partial^2 \xi_0}{\partial L^2} & \dfrac{\partial^2 \xi_0}{\partial L \partial \xi_1} \\ \dfrac{\partial^2 \xi_0}{\partial \xi_1 \partial L} & \dfrac{\partial^2 \xi_0}{\partial \xi_1^2} \end{pmatrix}$$

$$= L_{\xi_0}^2 \left(\frac{\partial \xi_0}{\partial L} \right)^2 (L_{\xi_0 \xi_0} L_{\xi_1 \xi_1} - L_{\xi_0 \xi_1}^2) > 0. \quad (2.63)$$

This proves the lemma. □

2.1 System of Ideal Fluid Dynamics

Since

$$\frac{\partial}{\partial(-e_\tau)}(Tv_0) = -\tau < 0, \tag{2.64}$$

it is easy to see, from Lemma 2.1, that $-e_\tau = p$ is a strictly convex function of

$$-Tv_0 = e + \rho e_\rho - TS - \frac{1}{2}u^2, \quad u_1, u_2, u_3, e_S = T.$$

Lemma 2.2. *Suppose that function $L(\xi_0, \xi_1, \ldots, \xi_n)$ is strictly convex with respect to its arguments $\xi_0(> 0), \xi_1, \ldots, \xi_n$; then*

$$M = \frac{1}{\xi_0} L(\xi_0, \xi_1, \ldots, \xi_n) \tag{2.65}$$

is strictly convex with respect to the arguments

$$\eta_0 = \frac{1}{\xi_0}, \eta_1 = \frac{\xi_1}{\xi_0}, \ldots, \eta_n = \frac{\xi_n}{\xi_0}. \tag{2.66}$$

The proof of this lemma is left to the reader as an exercise.

It has already been proved that p is a strictly convex function of $-Tv_0, u_1, u_2, u_3$ and T. From Lemma 2.2, we know that $\frac{p}{T}$ is a strictly convex function of the arguments $-v_0, \frac{u_1}{T}, \frac{u_2}{T}, \frac{u_3}{T}$ and $\frac{1}{T}$, i.e., a strictly convex function of the arguments

$$\frac{e + \rho e_\rho - TS - \frac{1}{2}u^2}{T}, \frac{u_1}{T}, \frac{u_2}{T}, \frac{u_3}{T}, \frac{1}{T}, \tag{2.67}$$

and then a strictly convex function of v_0, v_1, v_2, v_3, and v_4 given by (2.56) and (2.57). This reaches our conclusion.

Here we mention that, from the system of conservation laws of the form (2.41), we can obtain an additional conservation law

$$\frac{\partial}{\partial t}\left(\sum_{i=0}^{4} v_i L^0_{v_i} - L^0\right)$$
$$+ \sum_{k=1}^{3} \frac{\partial}{\partial x_k}\left(\sum_{i=0}^{4} v_i L^k_{v_i} - L^k\right) = 0. \tag{2.68}$$

It is easy to verify that the above formula is exactly the conservation equation (2.40) of entropy.

Moreover, from the expressions (2.44) and (2.49) of L^0 and its partial derivatives with respect to the new unknown variable V, together with the already proven strict convexity of L^0 with respect to V, we can see that v_0, v_1, v_2, v_3, v_4, and L^0 are exactly the Legendre transforms of $\rho, \rho u_1, \rho u_2, \rho u_3, \rho e + \frac{1}{2}\rho u^2$, and ρS.

At last, we discuss the general system of conservation laws

$$\frac{\partial}{\partial t} U + \sum_{k=1}^{3} \frac{\partial}{\partial x_k} B^k(U) = 0, \tag{2.69}$$

where $U = (u_1, \ldots, u_n)^T$, $B^k = (b_1^k, \ldots, b_n^k)^T$ ($k = 1, 2, 3$). For this system, we will present a sufficient and necessary condition under which it can be turned into a first-order symmetric hyperbolic system with the form of conservation laws by introducing new unknown variables.

Theorem 2.1. *The first-order system (2.69) of quasi-linear partial differential equations with the form of conservation laws can be turned into a first-order symmetric hyperbolic system with the form of conservation laws through a transformation of unknown variables*

$$U = U(V), \tag{2.70}$$

namely,

$$u_i = u_i(v_1, \ldots, v_n) \quad (i = 1, \ldots, n) \tag{2.71}$$

if and only if there exist a strictly convex scalar function $W(U)$ and a vector function $H = (h_1(U), h_2(U), h_3(U))^T$ such that the additional conservation law

$$\frac{\partial}{\partial t} W(U) + \sum_{k=1}^{3} \frac{\partial}{\partial x_k} h_k(U) = 0 \tag{2.72}$$

holds. Here, the strict convexity of $W(U)$ means the positive definiteness of the Hessian *matrix*

$$\frac{\partial^2 W}{\partial U^2} = \left(\frac{\partial^2 W}{\partial u_i \partial u_j} \right).$$

The scalar function $W(U)$ is also called the *entropy function* of system (2.69) of conservation laws, while $H(U)$ is called the *entropy flux vector*.

Proof. *Necessity.* Suppose that the transformation (2.70) of unknown variables turns the system (2.69) into a first-order symmetric hyperbolic system with the form of conservation laws. Thus, system (2.69) can be written as

$$\frac{\partial U}{\partial V} \frac{\partial V}{\partial t} + \sum_{k=1}^{3} \frac{\partial B^k}{\partial V} \frac{\partial V}{\partial x_k} = 0, \tag{2.73}$$

where

$$\frac{\partial U}{\partial V} = \left(\frac{\partial u_i}{\partial v_j} \right), \quad \left(\frac{\partial B^k}{\partial V} \right) = \left(\frac{\partial b_i^k}{\partial v_j} \right) \quad (k = 1, 2, 3),$$

and $\frac{\partial U}{\partial V}$ is a symmetric and positively definite matrix, and $\frac{\partial B^k}{\partial V}$ ($k = 1, 2, 3$) are symmetric matrices.

From the symmetry of $\frac{\partial U}{\partial V}$, namely,

$$\frac{\partial u_i}{\partial v_j} = \frac{\partial u_j}{\partial v_i}, \quad i, j = 1, \ldots, n,$$

2.1 System of Ideal Fluid Dynamics

we know that there exists a scalar function $L^0(V)$ such that U is the gradient of L^0:

$$\frac{\partial L^0(V)}{\partial v_i} = u_i \quad (i=1,\ldots,n). \tag{2.74}$$

Moreover, $L^0(V)$ is a strictly convex function from the positive definiteness of $\frac{\partial U}{\partial V}$. Similarly, there exist $L^k(V)$ $(k=1,2,3)$ such that

$$\frac{\partial L^k(V)}{\partial v_i} = b_i^k \quad (k=1,2,3;\ i=1,\ldots,n). \tag{2.75}$$

Let

$$W(U) = \sum_{j=1}^{n} u_j v_j - L^0(V), \tag{2.76}$$

$$h_k(U) = \sum_{j=1}^{n} b_j^k v_j - L^k(V) \quad (k=1,2,3). \tag{2.77}$$

$W(U)$, $H(U) = (h_1(U), h_2(U), h_3(U))^{\mathrm{T}}$ are the required entropy function and entropy flux vector, respectively.

To show this, we first prove the strict convexity of $W(U)$. Taking the partial derivatives of (2.76) with respect to u_i and using (2.74), we have

$$\frac{\partial W(U)}{\partial u_i} = v_i + \sum_{j=1}^{n} u_j \frac{\partial v_j}{\partial u_i} - \sum_{j=1}^{n} \frac{\partial L^0}{\partial v_j}\frac{\partial v_j}{\partial u_i} = v_i$$
$$(i=1,\ldots,n). \tag{2.78}$$

Equations (2.76), (2.74), and (2.78) imply that $u_1,\ldots,u_n$ and W are the Legendre transform of $v_1,\ldots,v_n$ and L^0. Then, it follows from the strict convexity of $L^0(V)$ that $W(U)$ is a strictly convex function of U.

Next, we prove that the additional conservation law (2.72) holds. Similarly to how we obtained (2.78), using (2.75), we can easily verify, from (2.77), that

$$\frac{\partial h^k(U)}{\partial u_i} = \sum_{j=1}^{n} \frac{\partial b_j^k}{\partial u_i} v_j. \tag{2.79}$$

From (2.78) and (2.79), together with the original system (2.69), we immediately have

$$\frac{\partial}{\partial t} W(U) + \sum_{k=1}^{3} \frac{\partial}{\partial x_k} h_k(U) = \sum_{j=1}^{n} v_j \frac{\partial u_j}{\partial t} + \sum_{k=1}^{3} \sum_{i,j=1}^{n} \frac{\partial b_j^k}{\partial u_i} v_j \frac{\partial u_i}{\partial x_k}$$
$$= \sum_{j=1}^{n} v_j \left(\frac{\partial u_j}{\partial t} + \sum_{k=1}^{3} \frac{\partial b_j^k}{\partial x_k} \right) = 0.$$

This proves the necessity.

Sufficiency. Suppose that there exist $W(U)$ and $H(U)$ satisfying the conditions in the theorem. First, we show that the entropy function $W(U)$ and the entropy flux vector $H(U)$ satisfy the following relation:

$$\sum_{i=1}^{n} \frac{\partial W}{\partial u_i} \frac{\partial b_i^k}{\partial u_j} - \frac{\partial h_k}{\partial u_j} = 0$$

$$(k = 1,2,3; j = 1,\ldots,n). \tag{2.80}$$

In fact, from the original system (2.69), the additional conservation law (2.72) can be rewritten as

$$-\sum_{i,j=1}^{n}\sum_{k=1}^{3} \frac{\partial W}{\partial u_i}\frac{\partial b_i^k}{\partial u_j}\frac{\partial u_j}{\partial x_k} + \sum_{j=1}^{n}\sum_{k=1}^{3}\frac{\partial h_k}{\partial u_j}\frac{\partial u_j}{\partial x_k} = 0, \tag{2.81}$$

which should be held for any U satisfying the original system (2.69), and then (2.80) follows immediately from (2.81).[1]

From the strict convexity of $W(U)$, we know that

$$v_i = \frac{\partial}{\partial u_i} W(U) \quad (i=1,\ldots,n) \tag{2.82}$$

is an invertible transformation. So U can be defined as function of V from (2.82).

Let

$$L^0(V) = \sum_{i=1}^{n} v_i u_i - W(U), \tag{2.83}$$

$$L^k(V) = \sum_{i=1}^{n} v_i b_i^k - h_k(U) \quad (k=1,2,3). \tag{2.84}$$

Taking the partial derivative of (2.83) with respect to v_j, and using (2.82), we obtain

$$\frac{\partial L^0(V)}{\partial v_j} = u_j \quad (j=1,\ldots,n). \tag{2.85}$$

Similarly, differentiating (2.84) with respect to v_j and using (2.80) and (2.82), it is easy to obtain that

$$\frac{\partial L^k(V)}{\partial v_j} = b_j^k \quad (k=1,2,3; j=1,\ldots,n). \tag{2.86}$$

---

[1] To show this, we need only prove that if

$$\sum_{j=1}^{n}\sum_{k=1}^{3} a_{jk}(U)\frac{\partial u_j}{\partial x_k} = 0$$

holds for any U satisfying the original system (2.69), then $a_{jk}(V^0) = 0$ ($j=1,\ldots,n; k=1,2,3$) hold for any given $V^0 = (v_1^0,\ldots,v_n^0)^T \in \mathbb{R}^n$. To this end, let $u_j^0(x) = v_j^0 + \sum_{k=1}^{3} c_{jk} x_k$ ($j=1,\ldots,n$), where c_{jk} are constants. Then for the solution U to the Cauchy problem of system (2.69) with initial value $U^0 = (u_1^0,\ldots,u_n^0)^T$, it follows from the above formula that $\sum_{j=1}^{n}\sum_{k=1}^{3} a_{jk}(V^0)c_{jk} = 0$ at $t = x_1 = \cdots = x_n = 0$. Suitably choosing c_{jk} can lead to the required result.

2.1 System of Ideal Fluid Dynamics

Equations (2.85) and (2.86) imply that the matrices

$$\frac{\partial U}{\partial V} \quad \text{and} \quad \frac{\partial B^k}{\partial V} \quad (k=1,2,3)$$

are symmetric, while (2.83), (2.82), and (2.85) imply that $v_1,\ldots,v_n$ and L^0 are the Legendre transform of $u_1,\ldots,u_n$ and W. Thus, it yields that $L^0(V)$ is a strictly convex function from the strict convexity of $W(U)$, and then $\frac{\partial U}{\partial V}$ is a positively definite matrix. This finishes the proof of the theorem. □

2.1.4 System of One-Dimensional Ideal Fluid Dynamics

Now we consider a special and important situation: the one-dimensional motion of the ideal fluid. Suppose that the motion occurs in a cylindrical pipeline whose axis is the $x = x_1$ axis; it is also assumed that the direction of the velocity is along the x axis, as is the external force, and the state of the fluid is identical on any cross section vertical to the x axis. Then, we have only the component $u = u_1$ of velocity, while $u_2 = u_3 = 0$, and all state variables are independent of x_2 and x_3.

At this time, if we denote by F the component of the external force along the x direction, then the system (2.10), (2.14) and (2.19) of ideal fluid dynamics can be reduced to

$$\frac{\partial \rho}{\partial t} + \frac{\partial}{\partial x}(\rho u) = 0, \tag{2.87}$$

$$\frac{\partial}{\partial t}(\rho u) + \frac{\partial}{\partial x}(\rho u^2 + p) = \rho F, \tag{2.88}$$

$$\frac{\partial}{\partial t}\left(\rho e + \frac{1}{2}\rho u^2\right) + \frac{\partial}{\partial x}\left(\left(\rho e + \frac{1}{2}\rho u^2 + p\right)u\right) = \rho F u, \tag{2.89}$$

or, from (2.10), (2.15), and (2.25),

$$\frac{\partial \rho}{\partial t} + \frac{\partial}{\partial x}(\rho u) = 0, \tag{2.90}$$

$$\frac{\partial u}{\partial t} + u \frac{\partial u}{\partial x} + \frac{1}{\rho}\frac{\partial p}{\partial x} = F, \tag{2.91}$$

$$\frac{\partial S}{\partial t} + u \frac{\partial S}{\partial x} = 0. \tag{2.92}$$

They are both *first-order quasi-linear hyperbolic systems*.

Since none of the above two systems is a first-order symmetric hyperbolic system, in order to show their hyperbolicity, it is necessary to introduce the definition of a first-order quasi-linear hyperbolic system with one spatial variable. Consider the following first-order quasi-linear system of partial differential equations:

$$A(t,x,U)\frac{\partial U}{\partial t} + B(t,x,U)\frac{\partial U}{\partial x} = F(t,x,U), \tag{2.93}$$

where $U = (u_1,\ldots,u_n)^T$ is the unknown function vector, $A = (a_{ij})_{n \times n}$, $B = (b_{ij})_{n \times n}$, and $F = (f_1,\ldots,f_n)^T$ are suitably smooth matrix functions in the domain under consideration.

Below we always assume that in the domain under consideration,

$$\det A \neq 0. \tag{2.94}$$

Definition 2.1. *Suppose that, for any (t,x,U) in the domain under consideration, we have the following:*
(a) *The characteristic equation*

$$\det(B - \lambda A) = 0 \tag{2.95}$$

has n real roots

$$\lambda_1(t,x,U), \lambda_2(t,x,U), \ldots, \lambda_n(t,x,U). \tag{2.96}$$

(b) *Set $\eta^i = (\eta_1^i, \eta_2^i, \ldots, \eta_n^i)$ to be a generalized left eigenvector corresponding to the above generalized eigenvalues λ_i,*

$$\eta^i B = \lambda_i \eta^i A; \tag{2.97}$$

then η^i ($i = 1, \ldots, n$) constitute a complete set of vectors, i.e.,

$$\det(\eta_j^i) \neq 0. \tag{2.98}$$

Now we call the system (2.93) *a* hyperbolic system.
If the characteristic equation (2.95) *has n distinct real roots,*

$$\lambda_1(t,x,U) < \lambda_2(t,x,U) < \cdots < \lambda_n(t,x,U), \tag{2.99}$$

then hypothesis (b) *is automatically satisfied, and the system* (2.93) *is called a* strictly hyperbolic system.

Suppose that there is a smooth curve C on the (t,x) plane, whose parameterized equation is

$$C: t = t(\sigma), x = x(\sigma), \quad |t'(\sigma)|^2 + |x'(\sigma)|^2 \neq 0.$$

If

$$\det(t'(\sigma)B - x'(\sigma)A) = 0 \tag{2.100}$$

is satisfied on this curve C, then curve C is called the *characteristic curve* of system (2.93). Noting hypothesis (2.94), it is easy to see that when the above formula is satisfied, curve C can be expressed in the form

$$C: x = x(t), \tag{2.101}$$

and (2.100) can be rewritten as

$$\det\left(B - \frac{dx}{dt} A\right) = 0. \tag{2.102}$$

2.1 System of Ideal Fluid Dynamics

Now we explore the hyperbolicity of system (2.87)–(2.89) or system (2.90)–(2.92). Because the invertible transformation of the unknown variables and the invertible linear combinations of the equations in the system do not change the type of the system, we need only deal with system (2.90)–(2.92). Take $U = (\rho, u, S)^T$ to be the unknown function vector, and $p = p(\rho, S)$. Writing this system in the form of (2.93), it is not difficult to see that

$$A = I, \tag{2.103}$$

$$B = \begin{pmatrix} u & \rho & 0 \\ \dfrac{c^2}{\rho} & u & \dfrac{p_S}{\rho} \\ 0 & 0 & u \end{pmatrix}, \tag{2.104}$$

where I is the unit matrix, $c^2 = \dfrac{\partial p}{\partial \rho}$ is the square of the local speed of sound (see section 2.1.3), and $p_S = \dfrac{\partial p}{\partial S}$. So

$$\det(B - \lambda A) = \begin{vmatrix} u - \lambda & \rho & 0 \\ \dfrac{c^2}{\rho} & u - \lambda & \dfrac{p_S}{\rho} \\ 0 & 0 & u - \lambda \end{vmatrix}$$

$$= (u - \lambda)((u - \lambda)^2 - c^2). \tag{2.105}$$

The characteristic equation of the system (2.90)–(2.92) is given by

$$(u - \lambda)((u - \lambda)^2 - c^2) = 0, \tag{2.106}$$

whose roots are

$$\lambda_1 = u - c, \ \lambda_2 = u, \ \lambda_3 = u + c. \tag{2.107}$$

This implies that (2.90)–(2.92) is a strictly hyperbolic system in the region where no vacuum occurs ($c > 0$), whose three families of characteristic curves are given by

$$\frac{dx}{dt} = u - c, \quad \frac{dx}{dt} = u, \quad \frac{dx}{dt} = u + c, \tag{2.108}$$

respectively.

Finally, we consider the special situation which assumes that the entropy is identically equal to a constant in the whole flow field (not only along the streamline), that is, $S \equiv$ constant, and the flow is called an *isentropic flow*. Under this assumption, $p = p(\rho)$ and $c^2 = \dfrac{dp}{d\rho}$. Then, system (2.90)–(2.92) can be written as

$$\frac{\partial \rho}{\partial t} + \frac{\partial}{\partial x}(\rho u) = 0, \tag{2.109}$$

$$\frac{\partial u}{\partial t} + u \frac{\partial u}{\partial x} + \frac{c^2}{\rho} \frac{\partial \rho}{\partial x} = F. \tag{2.110}$$

We can easily verify that it is a strictly hyperbolic system with two unknown functions ρ and u in the region without a vacuum. For the case $F \equiv 0$, there is a series of significant discussions on this kind of system (see [6], [7]). This kind of situation is relatively easy to handle, but there are still many problems for further study.

2.2 System of Viscous Fluid Dynamics

2.2.1 Introduction

The major difference between actual fluid and ideal fluid is that the former has viscosity (internal friction) and heat conductivity. This phenomenon is generated by the molecular structure of the fluid (gas and liquid). Friction stress and heat quantity exchanged from heat conduction, according to the basic laws expressed by the distributions of the fluid velocity and temperature, can be derived from the molecular kinetic theory of the fluid (see Chapter 8) in principle. But as a description of macroscopic phenomena, these laws should be given in advance as some additional physical laws (according to the experiment results). How to give these additional physical laws is the key step of the related mathematical modeling.

First we will make a preliminary investigation.

For the viscosity (internal friction), Newton has set a simple rule: The shear friction stress between two layers of the viscous fluid plane moving along the tangent direction is proportional to the gradient of fluid velocity along the direction vertical to the plane; in short, the friction is proportional to the velocity gradient. For example, suppose that two layers of viscous fluid are parallel to the plane Oxy and move along Ox with velocity u; then the *frictional shear stress* p_{xz} should be

$$p_{xz} = \mu \frac{\partial u}{\partial z}, \tag{2.111}$$

where the subscript z in p_{xz} represents the shear friction stress of the plane vertical to z, and the subscript x indicates that the force is along the x axis. μ is called the *viscosity coefficient*, which is independent of the orientation of the plane under consideration in the isotropic case, and depends only on the fluid under consideration as well as the temperature of the fluid in general (usually regarded as irrelevant to the fluid pressure).

Here we make some remarks on Newton's laws. The friction between two layers of the viscous fluid is an internal interaction force. One layer exerts friction on another, and vice versa, according to Newton's third law of motion; they are of equal value with opposite directions. Therefore, only when a part of the fluid is assumed to be isolated from the whole fluid can the friction on the surface of this part from other parts of the fluid be reflected. Hence, the friction is a kind of internal force, and the internal force per unit area is called *stress*; at the same time, the friction should always be in the direction tangent to the surface, so it is a kind of *shear stress*.

Only when there is a relative motion between two layers of the fluid does there exist friction, and the friction plays a role in impeding movement. Otherwise, the friction is null. Therefore, only when there is a velocity gradient between two layers of the fluid does there appear friction stress. Newton's law says that the friction stress should be proportional to the velocity gradient, and the proportional coefficient depends only on the fluid under consideration and its temperature. This is a kind of linear hypothesis. The fluid meeting this requirement is called *Newtonian fluid*; otherwise it is called *non-Newtonian fluid*.

From (2.111), the dimension of viscosity coefficient is

$$\frac{\text{force}}{\text{length}^2} \cdot \frac{\text{length}}{\text{velocity}} = \frac{\text{force}}{\text{length} \cdot \text{velocity}} = \frac{\text{force} \cdot \text{time}}{\text{length}^2} = \frac{\text{mass}}{\text{length} \cdot \text{time}},$$

2.2 System of Viscous Fluid Dynamics

whose unit in the international system of units is Pa·s:

$$1\,\text{Pa}\cdot\text{s} = 1\frac{\text{N}\cdot\text{s}}{\text{m}^2} = 1\frac{\text{kg}}{\text{m}\cdot\text{s}},$$

and $1\,\text{Pa} = 1\,\text{N/m}^2$. We usually choose mPa·s as the unit:

$$1\,\text{mPa}\cdot\text{s} = 10^{-3}\,\text{Pa}\cdot\text{s},$$

which is equivalent to the viscosity coefficient of water at 20.5°C.

The fluid viscosity has something to do with the temperature. For example, the viscosity coefficient of water decreases when the temperature increases; see Table 2.1. In contrast, the viscosity coefficient of air increases when the temperature increases. See Table 2.2.

Table 2.1. *The viscosity of water changes with temperature.*

Temperature / °C	μ / mPa·s	Temperature / °C	μ / mPa·s
0	1.792	40	0.656
5	1.519	45	0.599
10	1.308	50	0.549
15	1.140	60	0.469
20	1.005	70	0.406
25	0.894	80	0.357
30	0.801	90	0.317
35	0.723	100	0.284

Table 2.2. *The viscosity of air changes with temperature.*

Temperature / °C	μ / 10^{-2} mPa·s	Temperature / °C	μ / 10^{-2} mPa·s
0	1.709	260	2.806
20	1.808	280	2.877
40	1.904	300	2.946
60	1.997	320	3.014
80	2.088	340	3.080
100	2.175	360	3.146
120	2.260	380	3.212
140	2.344	400	3.277
160	2.425	420	3.340
180	2.505	440	3.402
200	2.582	460	3.463
220	2.658	480	3.523
240	2.733	500	3.583

Some liquids have large viscosity, such as glycerol, $\mu = 4.220\text{Pa} \cdot \text{s}$ at 3°C, and engine oil, $\mu = 0.6755\text{Pa} \cdot \text{s}$ at 10°C. For some liquids, its viscosity decreases rapidly as the temperature increases, e.g., the viscosity of glycerol changes with temperature as follows:

Temperature / °C	3	18	21
μ / Pa·s	4.220	1.069	0.778

In cold areas, the engine oil loses utility since its viscosity increases due to cooling; heating methods are usually used to reduce the viscosity.

The relationship between viscosity and temperature is often expressed, for gases, by the following power function:

$$\frac{\mu}{\mu_0} = \left(\frac{T}{T_0}\right)^n, \tag{2.112}$$

where n varies with different gases, and sometimes has a weak dependence relation with temperature. For instance, $n \approx 0.72$ for air, $n \approx 0.64$ for helium, $n \approx 0.69$ for hydrogen, $n \approx 0.95$ for carbon dioxide. In approximate computations, sometimes we take $n = 0.5$ for the gas at a relatively high temperature, while take $n = 1$ in the low temperature case.

Now let us look at the heat conductivity. It has been discussed in the course of mathematical physics equations and known to be described by *Fourier's experimental law*: The heat passing through the infinitesimal surface element dS along the normal direction $\boldsymbol{n}$ in unit time is

$$dq = -\kappa \frac{\partial T}{\partial n} dS, \tag{2.113}$$

where the negative sign indicates the direction in which heat flows from the higher to the lower temperature. Equation (2.113) shows that the heat is proportional to the temperature gradient (this is also a linear relation!), and the proportional coefficient κ is called *heat conductivity*. Heat conductivity is determined by the fluid under consideration and may depend on the temperature. Equation (2.113) can be rewritten as

$$dq = -\kappa \operatorname{grad} T \cdot \boldsymbol{n} dS, \tag{2.114}$$

where $-\kappa \operatorname{grad} T$ is called the *heat flux density vector*.

Next we will focus on extending Newton's law describing viscosity of the fluid to the general situation of the fluid in a state of any movement.

2.2.2 Stress Tensor

In order to extend Newton's law proposed previously in the case of simple motion to the general case of any movement, preparing for the establishment of the system of viscous fluid dynamics, we will first introduce the notion of stress tensor.

Now let us go back to the process of establishing the conservation equations of momentum in section 2.1. In the case of ideal fluid, the impulse term of the surface force

$$-\int_{t_1}^{t_2} \int_{\Omega} \operatorname{div}(p\boldsymbol{I}) dx dt$$

is produced by the pressure of external fluid. Hereinafter, we denote $\boldsymbol{x} = (x_1, x_2, x_3)$ and $dx = dx_1 dx_2 dx_3$. In the case of viscosity, the acting forces on Ω received from the

2.2 System of Viscous Fluid Dynamics

outside include not only the pressure (normal stress) on the surface but also the internal friction (shear stress) on the surface, whose joint effects give rise to the corresponding surface force impulse. Therefore, the general form of the conservation equations of momentum should be kept unchanged, while the second-order tensor $p\boldsymbol{I}$ should be revised to the stress tensor $\{p_{ij}\}$ due to viscosity.

For an area element ΔS at point M in the flow field, assume that its normal direction is $\boldsymbol{n}$, and $\Delta \boldsymbol{p}$ is the surface force acting on ΔS by the fluid on the side to which $\boldsymbol{n}$ is pointing; then

$$\boldsymbol{p}_n = \lim_{\Delta S \to 0} \frac{\Delta \boldsymbol{p}}{\Delta S} \tag{2.115}$$

is the surface force on the unit area with normal direction $\boldsymbol{n}$ at point M, called the *stress vector*. Because there are infinitely many directions through point M, the corresponding stress vectors are generally different. In order to describe the stress state of a point, we need to know the stress on each surface through the point. But the stress vectors at the same point in different directions are not unrelated. In fact, it is enough to know the stress vectors acting on the surfaces parallel to three coordinate planes. Denote by p_{ij} the ith component of the stress vector on the area element with the positive direction of x_j as its normal direction; then $\{p_{ij}\}$ is called the *stress tensor*.

Now we show that $\{p_{ij}\}$ defined above is a second-order tensor. To this end, we will prove that if dS is an area element with unit normal vector $\boldsymbol{n}$, then the stress vector upon it (the force per unit area upon dS by the side to which $\boldsymbol{n}$ is pointing) is

$$\boldsymbol{p}_n = \boldsymbol{P}\boldsymbol{n}, \tag{2.116}$$

where $\boldsymbol{P} = \{p_{ij}\}$, and $\boldsymbol{P}\boldsymbol{n}$ is the product of the matrix $\boldsymbol{P}$ with the vector $\boldsymbol{n}$ in the usual sense, that is, the vector with components $\sum_{j=1}^{3} p_{ij} n_j$ ($i = 1, 2, 3$), where n_j ($j = 1, 2, 3$) are the components of $\boldsymbol{n}$. According to the tensor identification theorem (See Appendix A), it is easy to see from (2.116) that $\boldsymbol{P}$ is a second-order tensor.

Now we prove (2.116). Suppose, without loss of generality, that the point under discussion is in the origin O. Examine the balance of forces on the tetrahedral element shown in Figure 2.1. This tetrahedron receives volume force (including external force and inertial force) and surface force. When the edge length of the tetrahedron tends to zero, the volume force is a third-order infinitesimal of edge length, while the surface force is a second-order infinitesimal. Suppose that the outward normal vector of plane ABC is $\boldsymbol{n}(n_1, n_2, n_3)$; then obviously

the area of $\triangle OBC = n_1 \cdot$ the area of $\triangle ABC$,
the area of $\triangle OAC = n_2 \cdot$ the area of $\triangle ABC$,
the area of $\triangle OAB = n_3 \cdot$ the area of $\triangle ABC$.

Furthermore, from the definition of p_{ij}, we have the following:

The surface force on $\triangle OBC$
$= -(p_{11}, p_{21}, p_{31}) \cdot$ the area of $\triangle OBC$
$= -n_1(p_{11}, p_{21}, p_{31}) \cdot$ the area of $\triangle ABC$;

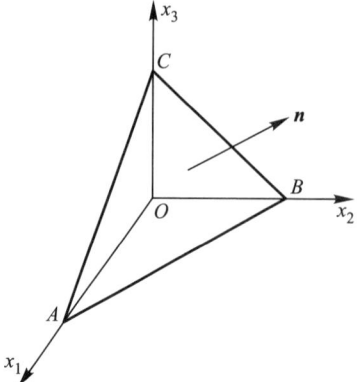

Figure 2.1.

the surface force on $\triangle OAC$

$$= -(p_{12}, p_{22}, p_{32}) \cdot \text{the area of } \triangle OAC$$
$$= -n_2(p_{12}, p_{22}, p_{32}) \cdot \text{the area of } \triangle ABC;$$

the surface force on $\triangle OAB$

$$= -(p_{13}, p_{23}, p_{33}) \cdot \text{the area of } \triangle OAB$$
$$= -n_3(p_{13}, p_{23}, p_{33}) \cdot \text{the area of } \triangle ABC.$$

Suppose that the stress vector on the plane ABC is $\boldsymbol{p}_n$; then from the equilibrium condition of forces on the tetrahedron, we obtain

$$\begin{aligned}(\boldsymbol{p}_n - n_1(p_{11}, p_{21}, p_{31}) &- n_2(p_{12}, p_{22}, p_{32}) \\ -n_3(p_{13}, p_{23}, p_{33})) &\cdot \text{the area of } \triangle ABC \\ +\text{the volume force} &\text{ on the tetrahedron} \\ = 0. & \end{aligned} \qquad (2.117)$$

Dividing both sides of the above formula by the area of $\triangle ABC$, then sending OA, OB, and OC to zero under the condition that $\boldsymbol{n}$ is kept unchanged, we see that (2.116) follows immediately from (2.117).

Next we will prove that $\boldsymbol{P} = \{p_{ij}\}$ is a symmetric tensor, i.e.,

$$p_{ij} = p_{ji} \quad (i, j = 1, 2, 3). \qquad (2.118)$$

Consider the cubic element shown in Figure 2.2, and suppose that its side lengths are dx_1, dx_2, and dx_3, respectively. Now we investigate the moment of the forces on the cube with respect to its center (assumed to be O). From the equilibrium conditions, it should be zero. As before, since the moment of the volume force is a high order infinitesimal, it can be neglected. In addition, the normal component of the stress on the surface passes through

2.2 System of Viscous Fluid Dynamics

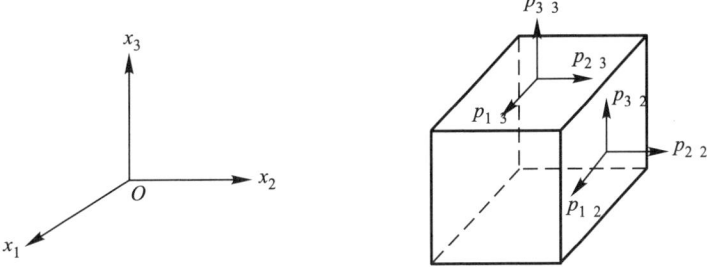

Figure 2.2.

the center of this cube, and its contribution to the moment is zero. Then, by exploring the component of the moment along the x_1 direction, it is easy to get

$$p_{32}\mathrm{d}x_3\mathrm{d}x_1 \cdot \mathrm{d}x_2 - p_{23}\mathrm{d}x_1\mathrm{d}x_2 \cdot \mathrm{d}x_3 = 0,$$

i.e.,

$$p_{32} = p_{23}.$$

Similarly we can prove $p_{13} = p_{31}$ and $p_{12} = p_{21}$.

2.2.3 Generalized Newton's Law: Constitutive Equations

Now the problem is how to specifically represent this second-order tensor $\boldsymbol{P} = \{p_{ij}\}$ according to the characters of viscous flow and, consequently, to obtain the required generalized Newton's law. For this purpose, we write $\boldsymbol{P}$ into the following form:

$$\{p_{ij}\} = -p\{\delta_{ij}\} + \{\tau_{ij}\}, \tag{2.119}$$

where $\{\tau_{ij}\}$ is the part in $\boldsymbol{P}$ corresponding to friction stress, will be discussed below in detail.

First, it is easy to see that, since the internal friction is caused essentially by the relative movement between the layers of the fluid, the viscosity stress should not change when a translation motion is added to the whole fluid. Therefore, $\{\tau_{ij}\}$ should be irrelevant to the fluid velocity, and be only a function of the partial derivatives of velocity components. When the velocity gradient is not considerably large, τ_{ij} can be regarded as a function of only the first-order partial derivatives of velocity components, and thus is a linear function. This fact has been very well confirmed by experiments. Thus, we now assume that τ_{ij} are linear homogeneous functions of the velocity gradient $\left\{\frac{\partial u_k}{\partial x_l}\right\}$:

$$\tau_{ij} = \sum_{k,l=1}^{3} c_{ijkl} \frac{\partial u_k}{\partial x_l}, \tag{2.120}$$

where c_{ijkl} are constants characterizing the viscosity of the fluid, which have a total of $3^4 = 81$. Since $\left\{\frac{\partial u_k}{\partial x_l}\right\}$ and $\{\tau_{ij}\}$ are both second-order tensors, from the tensor identification theorem (see Appendix A), $\{c_{ijkl}\}$ is a fourth-order tensor. Since $\{p_{ij}\}$ is a symmetric tensor, as is $\{\tau_{ij}\}$, we then have

$$c_{ijkl} = c_{jikl}. \tag{2.121}$$

Suppose that the fluid is isotropic, which means that $\{c_{ijkl}\}$ is an isotropic tensor of fourth order. From the general form of fourth-order tensors (see Appendix A), we have

$$c_{ijkl} = \lambda \delta_{ij}\delta_{kl} + \alpha \delta_{ik}\delta_{jl} + \beta \delta_{il}\delta_{jk}. \tag{2.122}$$

Set $\alpha = \mu + \nu$, $\beta = \mu - \nu$; then

$$c_{ijkl} = \lambda \delta_{ij}\delta_{kl} + \mu(\delta_{ik}\delta_{jl} + \delta_{il}\delta_{jk}) + \nu(\delta_{ik}\delta_{jl} - \delta_{il}\delta_{jk}),$$
$$c_{jikl} = \lambda \delta_{ij}\delta_{kl} + \mu(\delta_{jk}\delta_{il} + \delta_{jl}\delta_{ik}) + \nu(\delta_{jk}\delta_{il} - \delta_{jl}\delta_{ik}).$$

Noting (2.121), it is easy to obtain

$$c_{ijkl} = \lambda \delta_{ij}\delta_{kl} + \mu(\delta_{ik}\delta_{jl} + \delta_{il}\delta_{jk}). \tag{2.123}$$

Further, it can be seen from the above formula that $\{c_{ijkl}\}$ is also symmetric with respect to subscripts k and l, that is,

$$c_{ijkl} = c_{ijlk}. \tag{2.124}$$

Thus, in order to determine $\{c_{ijkl}\}$, it suffices to determine only λ and μ. Plugging (2.123) into (2.120), we have

$$\tau_{ij} = \lambda \operatorname{div} \boldsymbol{u} \delta_{ij} + 2\mu s_{ij}, \tag{2.125}$$

where

$$s_{ij} = \frac{1}{2}\left(\frac{\partial u_i}{\partial x_j} + \frac{\partial u_j}{\partial x_i}\right). \tag{2.126}$$

From (2.119) and (2.125), we obtain the expression for the stress tensor,

$$p_{ij} = (-p + \lambda \operatorname{div} \boldsymbol{u})\delta_{ij} + 2\mu s_{ij} \tag{2.127}$$

or

$$\boldsymbol{P} = (-p + \lambda \operatorname{div} \boldsymbol{u})\boldsymbol{I} + 2\mu \boldsymbol{S}, \tag{2.128}$$

where $\boldsymbol{S} = \{s_{ij}\}$.

Next we explain the meaning of λ and μ in the above formula.
First, we consider the shear motion along the x_1 direction:

$$u_1 = u_1(x_3), \quad u_2 = u_3 = 0.$$

From (2.127), for this motion we have

$$p_{13} = \mu \frac{\partial u_1}{\partial x_3}.$$

2.2 System of Viscous Fluid Dynamics

This coincides completely with (2.111) given by Newton's law in section 2.2.1. μ is called the *first coefficient of viscosity* or the *kinetic coefficient of viscosity*.

It can be obtained from (2.125) that

$$\frac{1}{3}\sum_{i=1}^{3}\tau_{ii} = \left(\lambda+\frac{2}{3}\mu\right)\mathrm{div}\,\boldsymbol{u}.$$

The left-hand side of the above formula is the average normal friction stress caused by viscosity, and div $\boldsymbol{u}$ represents the rate of change of the volume (i.e., the amount of change of unit volume per unit time). In fact, from the continuity equation (2.10), it is not hard to get

$$\mathrm{div}\,\boldsymbol{u} = \frac{1}{\tau}\frac{d\tau}{dt},$$

where $\tau = \frac{1}{\rho}$ is the specific volume, and $\frac{d}{dt}$ is defined by (2.17). Then, if we denote

$$\mu' = \lambda + \frac{2}{3}\mu, \tag{2.129}$$

then, as the ratio of the average normal friction stress to the rate of change of the volume, μ' describes the change of the average normal friction stress caused by the expansion or contraction in the motion of the fluid. μ' is called the *second viscosity coefficient* or the *expansive coefficient of viscosity*. For monatomic gases, under the condition that the pressure is not particularly high, we can take $\mu' = 0$. For diatomic gases such as oxygen, we can also regard $\mu' = 0$ when the temperature is not too high. But for the general situation, the affects of μ' have to be taken into consideration.

Using the definition (2.129) of μ', (2.127) can be rewritten as

$$p_{ij} = -p\delta_{ij} + 2\mu\left(s_{ij} - \frac{1}{3}\mathrm{div}\,\boldsymbol{u}\delta_{ij}\right) + \mu'\mathrm{div}\,\boldsymbol{u}\delta_{ij}, \tag{2.130}$$

and so the friction stress tensor is

$$\tau_{ij} = 2\mu\left(s_{ij} - \frac{1}{3}\mathrm{div}\,\boldsymbol{u}\delta_{ij}\right) + \mu'\mathrm{div}\,\boldsymbol{u}\delta_{ij}. \tag{2.131}$$

This is exactly the *generalized Newton's law*, where $\mu > 0$, $\mu' \geq 0$. Equations (2.130) are also called the *constitutive equations* of the fluid.

2.2.4 System of Viscous Heat-Conducting Fluid Dynamics

Now we consider the dynamical system of the fluid with viscosity and heat conduction. The conservation law of mass still takes its original form (see (2.10)):

$$\frac{\partial \rho}{\partial t} + \mathrm{div}(\rho\boldsymbol{u}) = 0. \tag{2.132}$$

In the conservation law of momentum, we replace the original pressure tensor $-p\boldsymbol{I}$ with the stress tensor $\boldsymbol{P} = \{p_{ij}\}$. Similarly to (2.13), we have

$$\frac{\partial}{\partial t}(\rho\boldsymbol{u}) + \mathrm{div}(\rho\boldsymbol{u}\otimes\boldsymbol{u} - \boldsymbol{P}) = \rho\boldsymbol{F}, \tag{2.133}$$

where
$$p_{ij} = -p\delta_{ij} + \mu\left(\frac{\partial u_i}{\partial x_j} + \frac{\partial u_j}{\partial x_i} - \frac{2}{3}\operatorname{div}\boldsymbol{u}\delta_{ij}\right) + \mu'\operatorname{div}\boldsymbol{u}\,\delta_{ij}. \tag{2.134}$$

Using the continuity equation (2.132), the momentum equation (2.133) can be rewritten as

$$\rho\frac{du_i}{dt} + \frac{\partial p}{\partial x_i} - \frac{\partial}{\partial x_i}\left(\left(\mu' - \frac{2}{3}\mu\right)\operatorname{div}\boldsymbol{u}\right)$$
$$-\sum_{j=1}^{3}\frac{\partial}{\partial x_j}\left(\mu\left(\frac{\partial u_i}{\partial x_j} + \frac{\partial u_j}{\partial x_i}\right)\right) = \rho F_i \quad (i = 1, 2, 3) \tag{2.135}$$

or as the corresponding vector form

$$\rho\frac{d\boldsymbol{u}}{dt} + \operatorname{grad} p - \operatorname{grad}\left(\left(\mu' - \frac{2}{3}\mu\right)\operatorname{div}\boldsymbol{u}\right) - 2\operatorname{div}(\mu S) = \rho\boldsymbol{F}, \tag{2.136}$$

where $\frac{d}{dt}$ is defined as in (2.17), and both μ and μ' can be the functions of temperature T. Equations (2.135) or (2.136) are also called *Euler equations*.

In the conservation law of energy, we replace the original work done by the pressure tensor with the work done by the stress tensor. From (2.116), in the time interval $[t_1, t_2]$, the work done by the stress tensor on the fluid in Ω is

$$\int_{t_1}^{t_2}\int_{\Gamma}\boldsymbol{u}\cdot(\boldsymbol{Pn})\mathrm{d}S\mathrm{d}t = \int_{t_1}^{t_2}\int_{\Omega}\operatorname{div}(\boldsymbol{Pu})\mathrm{d}x\mathrm{d}t,$$

where Γ is the boundary of Ω. Thus, the term $-\operatorname{div}(p\boldsymbol{u})$ in equation (2.19) for ideal fluids should be revised to

$$\operatorname{div}(\boldsymbol{Pu}) = \sum_{i,j=1}^{3}\frac{\partial}{\partial x_i}(p_{ij}u_j)$$
$$= -\operatorname{div}(p\boldsymbol{u}) + \sum_{i=1}^{3}\frac{\partial}{\partial x_i}\left(\mu\sum_{j=1}^{3}\left(\frac{\partial u_i}{\partial x_j} + \frac{\partial u_j}{\partial x_i}\right)u_j\right.$$
$$\left. + \left(\mu' - \frac{2}{3}\mu\right)u_i\operatorname{div}\boldsymbol{u}\right). \tag{2.137}$$

Besides, considering the heat conduction, in the time interval $[t_1, t_2]$, the increment of the fluid energy in Ω has to contain the work (or energy) converted by the heat flowing into Ω through the boundary Γ during this period of time. According to Fourier's law (2.114), the contribution of this term should be

$$\int_{t_1}^{t_2}\int_{\Gamma}\kappa\operatorname{grad} T\cdot\boldsymbol{n}\mathrm{d}S\mathrm{d}t = \int_{t_1}^{t_2}\int_{\Omega}\operatorname{div}(\kappa\operatorname{grad} T)\mathrm{d}x\mathrm{d}t. \tag{2.138}$$

Then, the conservation equation of energy takes the form (see (2.19))

$$\frac{\partial}{\partial t}\left(\rho e + \frac{1}{2}\rho u^2\right) + \operatorname{div}\left(\left(\rho e + \frac{1}{2}\rho u^2\right)\boldsymbol{u} - \boldsymbol{Pu}\right)$$
$$= \operatorname{div}(\kappa\operatorname{grad} T) + \rho\boldsymbol{F}\cdot\boldsymbol{u}, \tag{2.139}$$

that is,

$$\frac{\partial}{\partial t}\left(\rho e + \frac{1}{2}\rho u^2\right) + \operatorname{div}\left(\left(\rho e + \frac{1}{2}\rho u^2 + p\right)u\right)$$
$$- \sum_{i=1}^{3} \frac{\partial}{\partial x_i}\left(\mu \sum_{j=1}^{3}\left(\frac{\partial u_i}{\partial x_j} + \frac{\partial u_j}{\partial x_i}\right)u_j + \left(\mu' - \frac{2}{3}\mu\right)u_i \operatorname{div} \boldsymbol{u}\right)$$
$$= \operatorname{div}(\kappa \operatorname{grad} T) + \rho \boldsymbol{F} \cdot \boldsymbol{u}. \tag{2.140}$$

Using the continuity equation (2.132), the above formula can be rewritten as

$$\rho \frac{d}{dt}\left(e + \frac{u^2}{2}\right) + \operatorname{div}(p\boldsymbol{u})$$
$$- \sum_{i=1}^{3} \frac{\partial}{\partial x_i}\left(\mu \sum_{j=1}^{3}\left(\frac{\partial u_i}{\partial x_j} + \frac{\partial u_j}{\partial x_i}\right)u_j + \left(\mu' - \frac{2}{3}\mu\right)u_i \operatorname{div} \boldsymbol{u}\right)$$
$$= \operatorname{div}(\kappa \operatorname{grad} T) + \rho \boldsymbol{F} \cdot \boldsymbol{u}, \tag{2.141}$$

where $\frac{d}{dt}$ is defined as in (2.17).

Noting $\operatorname{div}(p\boldsymbol{u}) = p \operatorname{div} \boldsymbol{u} + \boldsymbol{u} \cdot \operatorname{grad} p$ and $\frac{d}{dt}\left(\frac{u^2}{2}\right) = \boldsymbol{u} \cdot \frac{d\boldsymbol{u}}{dt}$ and using the Euler equations (2.135), equation (2.141) can also be reduced to

$$\rho \frac{de}{dt} + p \operatorname{div} \boldsymbol{u} - \mu \sum_{i,j=1}^{3}\left(\frac{\partial u_i}{\partial x_j} + \frac{\partial u_j}{\partial x_i}\right)\frac{\partial u_j}{\partial x_i}$$
$$- \left(\mu' - \frac{2}{3}\mu\right)(\operatorname{div} \boldsymbol{u})^2 = \operatorname{div}(\kappa \operatorname{grad} T). \tag{2.142}$$

Similarly to the case of the system of ideal fluid dynamics, after being matched with appropriate equations of state, for instance, $p = p(\rho, T)$, $e = e(\rho, T)$, and relations of μ and μ' depending on T, formulas (2.132), (2.133), and (2.139), or formulas (2.132), (2.135), and (2.142), constitute a closed system of partial differential equations with five unknown functions.

2.2.5 Mathematical Structure of the System of Viscous Heat-Conducting Fluid Dynamics

Now we claim that the system of viscous heat-conducting fluid dynamics can be written in the form of a quasi-linear symmetric hyperbolic-parabolic coupled system, whose significance will be illustrated in what follows.

The conservation equation (2.132) of mass can be written in the following form:

$$\frac{\partial \rho}{\partial t} + u_1 \frac{\partial \rho}{\partial x_1} + u_2 \frac{\partial \rho}{\partial x_2} + u_3 \frac{\partial \rho}{\partial x_3} = f_0(\rho, \operatorname{grad} \boldsymbol{u}), \tag{2.143}$$

where

$$f_0(\rho, \operatorname{grad} \boldsymbol{u}) = -\rho \operatorname{div} \boldsymbol{u}. \tag{2.144}$$

The conservation equation (2.135) of momentum can be written as

$$\rho \frac{\partial u_i}{\partial t} - \mu \Delta u_i - \left(\mu' + \frac{1}{3}\mu\right) \frac{\partial}{\partial x_i}(\operatorname{div} \boldsymbol{u})$$
$$= \rho F_i + \widetilde{f}_i(\rho, T, \boldsymbol{u}, \operatorname{grad} \rho, \operatorname{grad} \boldsymbol{u}, \operatorname{grad} T) \quad (i = 1, 2, 3), \tag{2.145}$$

where

$$\widetilde{f}_i(\rho, T, \boldsymbol{u}, \operatorname{grad} \rho, \operatorname{grad} \boldsymbol{u}, \operatorname{grad} T)$$
$$= -c^2 \frac{\partial \rho}{\partial x_i} - \frac{\partial p}{\partial T} \frac{\partial T}{\partial x_i} - \rho \sum_{k=1}^{3} u_k \frac{\partial u_i}{\partial x_k}$$
$$+ \sum_{j=1}^{3} \frac{d\mu}{dT} \frac{\partial T}{\partial x_j} \left(\frac{\partial u_i}{\partial x_j} + \frac{\partial u_j}{\partial x_i}\right)$$
$$+ \frac{d(\mu' - \frac{2}{3}\mu)}{dT} \frac{\partial T}{\partial x_i} \operatorname{div} \boldsymbol{u} \quad (i = 1, 2, 3) \tag{2.146}$$

and $c^2 = \frac{\partial p}{\partial \rho}$. Equation (2.145) can be written in the form of components, that is,

$$\frac{\partial u_1}{\partial t} - \frac{1}{\rho}\left(\left(\mu' + \frac{4}{3}\mu\right)\frac{\partial^2 u_1}{\partial x_1^2} + \mu \frac{\partial^2 u_1}{\partial x_2^2} + \mu \frac{\partial^2 u_1}{\partial x_3^2}\right)$$
$$- \frac{1}{\rho}\left(\mu' + \frac{1}{3}\mu\right)\frac{\partial^2 u_2}{\partial x_1 \partial x_2} - \frac{1}{\rho}\left(\mu' + \frac{1}{3}\mu\right)\frac{\partial^2 u_3}{\partial x_1 \partial x_3}$$
$$= F_1 + f_1(\rho, T, \boldsymbol{u}, \operatorname{grad} \rho, \operatorname{grad} \boldsymbol{u}, \operatorname{grad} T), \tag{2.147}$$

$$\frac{\partial u_2}{\partial t} - \frac{1}{\rho}\left(\mu' + \frac{1}{3}\mu\right)\frac{\partial^2 u_1}{\partial x_1 \partial x_2}$$
$$- \frac{1}{\rho}\left(\mu \frac{\partial^2 u_2}{\partial x_1^2} + \left(\mu' + \frac{4}{3}\mu\right)\frac{\partial^2 u_2}{\partial x_2^2} + \mu \frac{\partial^2 u_2}{\partial x_3^2}\right)$$
$$- \frac{1}{\rho}\left(\mu' + \frac{1}{3}\mu\right)\frac{\partial^2 u_3}{\partial x_2 \partial x_3}$$
$$= F_2 + f_2(\rho, T, \boldsymbol{u}, \operatorname{grad} \rho, \operatorname{grad} \boldsymbol{u}, \operatorname{grad} T), \tag{2.148}$$

$$\frac{\partial u_3}{\partial t} - \frac{1}{\rho}\left(\mu' + \frac{1}{3}\mu\right)\frac{\partial^2 u_1}{\partial x_1 \partial x_3} - \frac{1}{\rho}\left(\mu' + \frac{1}{3}\mu\right)\frac{\partial^2 u_2}{\partial x_2 \partial x_3}$$
$$- \frac{1}{\rho}\left(\mu \frac{\partial^2 u_3}{\partial x_1^2} + \mu \frac{\partial^2 u_3}{\partial x_2^2} + \left(\mu' + \frac{4}{3}\mu\right)\frac{\partial^2 u_3}{\partial x_3^2}\right)$$
$$= F_3 + f_3(\rho, T, \boldsymbol{u}, \operatorname{grad} \rho, \operatorname{grad} \boldsymbol{u}, \operatorname{grad} T), \tag{2.149}$$

where $f_i = \frac{\widetilde{f}_i}{\rho}$ ($i = 1, 2, 3$), and $\operatorname{grad} \boldsymbol{u}$ denotes the second-order tensor $\left(\frac{\partial u_i}{\partial x_j}\right)$.

2.2 System of Viscous Fluid Dynamics

Finally, let us look at the conservation equation (2.142) of energy. From the equation of state $e = e(\rho, T)$ (in particular, $e = e(T)$ for ideal gases) and using the continuity equation (2.132), we have

$$\frac{de}{dt} = \frac{\partial e}{\partial \rho}\frac{d\rho}{dt} + \frac{\partial e}{\partial T}\frac{dT}{dt}$$

$$= -\frac{\partial e}{\partial \rho}\rho \operatorname{div} \boldsymbol{u} + \frac{\partial e}{\partial T}\frac{dT}{dt}.$$

Plugging this into (2.142), we obtain

$$\frac{\partial T}{\partial t} + \sum_{k=1}^{3} u_k \frac{\partial T}{\partial x_k} - \frac{1}{\rho \frac{\partial e}{\partial T}} \operatorname{div}(\kappa \operatorname{grad} T) = \widetilde{f}(\rho, T, \operatorname{grad} \boldsymbol{u}), \tag{2.150}$$

where

$$\widetilde{f}(\rho, T, \operatorname{grad} \boldsymbol{u}) = \left\{ \left(\rho \frac{\partial e}{\partial \rho} - \frac{p}{\rho} \right) \operatorname{div} \boldsymbol{u} \right.$$

$$+ \frac{\mu}{\rho} \sum_{i,j=1}^{3} \left(\frac{\partial u_i}{\partial x_j} + \frac{\partial u_j}{\partial x_i} \right) \frac{\partial u_j}{\partial x_i}$$

$$\left. + \frac{1}{\rho} \left(\mu' - \frac{2}{3}\mu \right) (\operatorname{div} \boldsymbol{u})^2 \right\} \Big/ \frac{\partial e}{\partial T}. \tag{2.151}$$

Equation (2.150) can also be written as

$$\frac{\partial T}{\partial t} - \frac{\kappa}{\rho \frac{\partial e}{\partial T}} \Delta T = f_4(\rho, T, \boldsymbol{u}, \operatorname{grad} \boldsymbol{u}, \operatorname{grad} T), \tag{2.152}$$

where

$$f_4(\rho, T, \boldsymbol{u}, \operatorname{grad} \boldsymbol{u}, \operatorname{grad} T) = \widetilde{f}(\rho, T, \operatorname{grad} \boldsymbol{u})$$

$$- \sum_{k=1}^{3} u_k \frac{\partial T}{\partial x_k} + \frac{1}{\rho \frac{\partial e}{\partial T}} \operatorname{grad} \kappa \cdot \operatorname{grad} T, \tag{2.153}$$

and $\frac{\partial e}{\partial T} > 0$.

In the system consisting of the above five equations (2.143), (2.147)–(2.149), and (2.152), the mass conservation equation (2.143) is a first-order partial differential equation. Regarding it as an equation for ρ, it is a special case of a symmetric hyperbolic system of equations, and thus a first-order hyperbolic equation, while its coefficients and right-hand side depend also on $\boldsymbol{u}$.

Now we consider the latter four equations. Setting $U = (u_1, u_2, u_3, T)^T$, the left-hand sides of equations (2.147)–(2.149) and (2.152) include only the first-order partial derivatives of U with respect to t and the second-order derivatives with respect to x. The system formed by these four equations can be written in the following matrix form:

$$\frac{\partial U}{\partial t} - \sum_{i,j=1}^{3} B_{ij}(\rho, U) \frac{\partial^2 U}{\partial x_i \partial x_j} = C(\rho, \operatorname{grad} \rho, U, \operatorname{grad} U), \tag{2.154}$$

where

$$B_{11} = \begin{pmatrix} \frac{\mu' + \frac{4}{3}\mu}{\rho} & 0 & 0 & 0 \\ 0 & \frac{\mu}{\rho} & 0 & 0 \\ 0 & 0 & \frac{\mu}{\rho} & 0 \\ 0 & 0 & 0 & \frac{\kappa}{\rho \frac{\partial e}{\partial T}} \end{pmatrix},$$

$$B_{22} = \begin{pmatrix} \frac{\mu}{\rho} & 0 & 0 & 0 \\ 0 & \frac{\mu' + \frac{4}{3}\mu}{\rho} & 0 & 0 \\ 0 & 0 & \frac{\mu}{\rho} & 0 \\ 0 & 0 & 0 & \frac{\kappa}{\rho \frac{\partial e}{\partial T}} \end{pmatrix},$$

$$B_{33} = \begin{pmatrix} \frac{\mu}{\rho} & 0 & 0 & 0 \\ 0 & \frac{\mu}{\rho} & 0 & 0 \\ 0 & 0 & \frac{\mu' + \frac{4}{3}\mu}{\rho} & 0 \\ 0 & 0 & 0 & \frac{\kappa}{\rho \frac{\partial e}{\partial T}} \end{pmatrix},$$

$$B_{12} = B_{21} = \frac{1}{2} \begin{pmatrix} 0 & \frac{\mu' + \frac{1}{3}\mu}{\rho} & 0 & 0 \\ \frac{\mu' + \frac{1}{3}\mu}{\rho} & 0 & 0 & 0 \\ 0 & 0 & 0 & 0 \\ 0 & 0 & 0 & 0 \end{pmatrix},$$

$$B_{23} = B_{32} = \frac{1}{2} \begin{pmatrix} 0 & 0 & 0 & 0 \\ 0 & 0 & \frac{\mu' + \frac{1}{3}\mu}{\rho} & 0 \\ 0 & \frac{\mu' + \frac{1}{3}\mu}{\rho} & 0 & 0 \\ 0 & 0 & 0 & 0 \end{pmatrix},$$

$$B_{13} = B_{31} = \frac{1}{2} \begin{pmatrix} 0 & 0 & \frac{\mu' + \frac{1}{3}\mu}{\rho} & 0 \\ 0 & 0 & 0 & 0 \\ \frac{\mu' + \frac{1}{3}\mu}{\rho} & 0 & 0 & 0 \\ 0 & 0 & 0 & 0 \end{pmatrix},$$

which are all symmetric matrices. Besides, for any given $\boldsymbol{\xi} = (\xi_1, \xi_2, \xi_3) \in \mathbb{R}^3$, $|\boldsymbol{\xi}| = 1$,

$$\sum_{i,j=1}^{3} B_{ij} \xi_i \xi_j$$

2.2 System of Viscous Fluid Dynamics

is always a positively definite matrix. In fact,

$$\sum_{i,j=1} B_{ij}\xi_i\xi_j = \begin{pmatrix} a\xi_1^2+b & a\xi_1\xi_2 & a\xi_1\xi_3 & 0 \\ a\xi_1\xi_2 & a\xi_2^2+b & a\xi_2\xi_3 & 0 \\ a\xi_1\xi_3 & a\xi_2\xi_3 & a\xi_3^2+b & 0 \\ 0 & 0 & 0 & \frac{\kappa}{\rho\frac{\partial e}{\partial T}} \end{pmatrix},$$

where

$$a = \frac{\mu' + \frac{1}{3}\mu}{\rho}, \quad b = \frac{\mu}{\rho}.$$

Since $\mu > 0$, $\mu' \geq 0$, when $\rho > 0$ (a vacuum does not appear), a and b are both positive. Then the principal subdeterminant of the above matrix is

$$a\xi_1^2 + b > 0,$$

$$\begin{vmatrix} a\xi_1^2+b & a\xi_1\xi_2 \\ a\xi_1\xi_2 & a\xi_2^2+b \end{vmatrix} = ab(\xi_1^2+\xi_2^2) + b^2 > 0,$$

$$\begin{vmatrix} a\xi_1^2+b & a\xi_1\xi_2 & a\xi_1\xi_3 \\ a\xi_1\xi_2 & a\xi_2^2+b & a\xi_2\xi_3 \\ a\xi_1\xi_3 & a\xi_2\xi_3 & a\xi_3^2+b \end{vmatrix} = ab^2 > 0.$$

So this matrix is positive definite.

Now we call the system (2.154) a *symmetric parabolic system*.

Generally speaking, **if a system**

$$\frac{\partial U}{\partial t} - \sum_{i,j=1}^{n} B_{ij}\frac{\partial^2 U}{\partial x_i \partial x_j} = C,$$

where $U = (u_1,\ldots,u_m)^{\mathrm{T}}$ and B_{ij} are $m \times m$ **matrices, satisfying that**

(1°) B_{ij} ($i = 1,\ldots,n$) are symmetric matrices;
(2°) for any given $\boldsymbol{\xi} = (\xi_1,\ldots,\xi_n) \in \mathbb{R}^n$, $|\boldsymbol{\xi}| = 1$,

$$\sum_{i,j=1}^{n} B_{ij}\xi_i\xi_j \text{ is always a symmetric positively definite matrix,}$$

then we say that this system is a *symmetric parabolic system in the sense of Petrovsky*.

Therefore, the latter four equations in the system of viscous fluid dynamics form a second-order symmetric parabolic system for unknown functions (u_1,u_2,u_3,T), while the first equation is a first-order symmetric hyperbolic equation for unknown function ρ, and they are coupled with each other to constitute a *quasi-linear symmetric hyperbolic-parabolic coupled system*. This is exactly the mathematical structure of the system of equations.

We can usually formulate the Cauchy problem for this kind of system, that is, give initial states

$$t = 0 : (\rho, u_1, u_2, u_3, T) \\ = (\rho^0(x), u_1^0(x), u_2^0(x), u_3^0(x), T^0(x)), \quad (2.155)$$

and then find the states in later time (i.e., $t > 0$). When the initial values are suitably smooth, the existence and uniqueness of classical solutions to this problem can be easily proved in the local time period $0 \leq t \leq \delta$ ($\delta > 0$) by using the methods of energy integral and iteration. This type of solution is call the local classical solution. Different from the case of ideal fluids without viscosity and heat conduction, discontinuity of shock-type does not occur mechanically in the current situation; it still remains open whether there exist global classical solutions (i.e., classical solutions for any time $t \geq 0$) for any given smooth initial values. For the time being, what can be proved is that if the initial values are sufficiently small, then there must exist global classical solutions to this initial value problem (for example, see [12]).

Besides initial conditions, sometimes there are possibly boundary conditions. When studying the problem of the flow past a body for viscous fluids, on the surface Γ of the body, the velocity of the fluid is zero not only along the normal direction (since the fluid does not permeate), but also along the tangential direction due to viscosity, so the boundary condition for the velocity is

$$\mathbf{u}|_\Gamma = \mathbf{0}, \quad (2.156)$$

while for temperature T, we can take one of the usual three kinds of boundary conditions (i.e., to give one of T, $\frac{\partial T}{\partial n}$, or $\frac{\partial T}{\partial n} + \alpha T$ ($\alpha > 0$) on boundary Γ, where $\mathbf{n}$ is the unit outward normal vector on Γ). Here, we do not need any boundary conditions for the unknown function ρ.

2.2.6 System of One-Dimensional Viscous Heat-Conducting Fluid Dynamics

Now we consider an important special case of one-dimensional motion. We assume that the flow proceeds along a cylindrical tube whose axis is $x = x_1$, the motion velocity has only the component of the x direction, and the state variables on any section vertical to the x axis are always equal, that is, the states variables depend only on t and x. Therefore, all the partial derivatives with respect to x_2 and x_3 should be eliminated in the equations stated in section 2.2.4, and $u_2 = u_3 = 0$. The external force has only the component of $x = x_1$ direction. Then, the system corresponding to (2.132), (2.133), and (2.139) is

$$\frac{\partial \rho}{\partial t} + \frac{\partial}{\partial x}(\rho u) = 0, \quad (2.157)$$

$$\frac{\partial}{\partial t}(\rho u) + \frac{\partial}{\partial x}\left(\rho u^2 + p - \left(\frac{4}{3}\mu + \mu'\right)\frac{\partial u}{\partial x}\right) = \rho F, \quad (2.158)$$

$$\frac{\partial}{\partial t}\left(\rho e + \frac{1}{2}\rho u^2\right) + \frac{\partial}{\partial x}\left(\left(\rho e + \frac{1}{2}\rho u^2 + p\right)u \\ - \left(\frac{4}{3}\mu + \mu'\right)u\frac{\partial u}{\partial x}\right) = \frac{\partial}{\partial x}\left(\kappa\frac{\partial T}{\partial x}\right) + \rho F u. \quad (2.159)$$

2.3 Navier–Stokes Equations

Corresponding to (2.132), (2.135), and (2.142), similarly we have

$$\frac{\partial \rho}{\partial t} + \frac{\partial}{\partial x}(\rho u) = 0, \tag{2.160}$$

$$\frac{\partial u}{\partial t} + u\frac{\partial u}{\partial x} + \frac{1}{\rho}\frac{\partial p}{\partial x} - \frac{1}{\rho}\frac{\partial}{\partial x}\left(\left(\frac{4}{3}\mu + \mu'\right)\frac{\partial u}{\partial x}\right) = F, \tag{2.161}$$

$$\rho\frac{\partial e}{\partial t} + \rho u\frac{\partial e}{\partial x} + p\frac{\partial u}{\partial x} - \left(\frac{4}{3}\mu + \mu'\right)\left(\frac{\partial u}{\partial x}\right)^2 = \frac{\partial}{\partial x}\left(\kappa\frac{\partial T}{\partial x}\right), \tag{2.162}$$

where the last equation can also be written as

$$\rho\frac{\partial e}{\partial T}\left(\frac{\partial T}{\partial t} + u\frac{\partial T}{\partial x}\right) - \rho^2\frac{\partial e}{\partial \rho}\frac{\partial u}{\partial x} + p\frac{\partial u}{\partial x}$$
$$- \left(\frac{4}{3}\mu + \mu'\right)\left(\frac{\partial u}{\partial x}\right)^2 = \frac{\partial}{\partial x}\left(\kappa\frac{\partial T}{\partial x}\right). \tag{2.163}$$

Here, the first equation (2.160) is still a first-order hyperbolic equation for unknown function ρ, while the latter two equations (2.161) and (2.162) form a second-order parabolic system for unknown functions u and T; the above system of equations is still a quasi-linear hyperbolic-parabolic coupled system. We can formulate a Cauchy problem for this system,

$$t = 0 : (\rho, u, T) = (\rho^0(x), u^0(x), T^0(x)), \tag{2.164}$$

or an initial-boundary value problem with the following boundary conditions:

$x = 0 : u = 0$, T satisfies one of the usual three kinds of boundary conditions,

$x = 1 :$ similar boundary conditions.

Here we assume that the fluid under discussion is bounded between $x = 0$ and $x = 1$. For the above problem, it is not necessary to give boundary conditions for ρ.

The local solvability of these problems has been completely settled, while the global solvability is still an issue for research.

2.3 Navier–Stokes Equations

Now we consider the motions of incompressible viscous fluids. This is a situation with practical significance. A typical incompressible fluid is water under normal temperature and pressure. Since the density hardly changes with pressure and temperature, it can be approximately regarded to be incompressible. For *incompressible fluid*, its equation of state is $\rho \equiv$ constant. We assume without loss of generality that

$$\rho \equiv 1. \tag{2.165}$$

For the motion of this kind of fluid, due to (2.165), the mass conservation equation (2.132) is reduced to

$$\text{div}\, \boldsymbol{u} = 0. \tag{2.166}$$

From (2.166), now μ' does not appear in the constitutive equation (2.130), and then not in the system of fluid dynamics. Besides, we can assume μ to be a constant. Then, noting (2.165), (2.166), Euler equations (2.136) can be reduced to

$$\frac{d\boldsymbol{u}}{dt} - \mu \Delta \boldsymbol{u} + \operatorname{grad} p = \boldsymbol{F}, \tag{2.167}$$

where p is a function of t and $\boldsymbol{x} = (x_1, x_2, x_3)$. It is necessary to mention that, even under normal temperature, it cannot be inferred from the equation of state that p is a constant, since the equation of state is now (2.165) and does not take the form of $p = p(\rho, T)$. In fact, there does not exist an absolutely incompressible fluid. The so-called incompressibility is a simplification of the phenomenon that the density ρ of the fluid changes little when the pressure p and temperature T change. In other words, for this kind of fluid, the tiny change of density ρ may cause a remarkable change in pressure p, etc. Therefore, if we take the equation of state to be in the form of (2.165), we have to regard p as a function of t and $\boldsymbol{x}$.

Under the current situation, equations (2.166) and (2.167), regarded as a system for unknown functions $\boldsymbol{u} = (u_1, u_2, u_3)^T$ and p, is a closed system. Then, we obtain the fundamental system for incompressible fluid dynamics,

$$\frac{\partial \boldsymbol{u}}{\partial t} - \mu \Delta \boldsymbol{u} + \sum_{k=1}^{3} u_k \frac{\partial \boldsymbol{u}}{\partial x_k} + \operatorname{grad} p = \boldsymbol{F}, \tag{2.168}$$

$$\operatorname{div} \boldsymbol{u} = 0. \tag{2.169}$$

This system is usually called the *Navier–Stokes equations* or, more precisely, *three-dimensional Navier–Stokes equations*.

For Navier–Stokes equations (2.168), (2.169), we usually consider the following Cauchy problem: Given initial velocity

$$t = 0 : \boldsymbol{u} = \boldsymbol{u}^0(x_1, x_2, x_3), \tag{2.170}$$

determine the velocity $\boldsymbol{u}(t, x_1, x_2, x_3)$ and pressure $p(t_1, x_1, x_2, x_3)$ in later time $(t > 0)$. Here it is natural to assume that the initial velocity $\boldsymbol{u}^0$ satisfies the condition $\operatorname{div} \boldsymbol{u}^0 = 0$, while accordingly p can be determined up to an additive arbitrary function of t (independent of $\boldsymbol{x}$).

Besides initial value problems, we can also consider initial-boundary value problems for this system; that is, in addition to the initial condition (2.170), the solution of this system is required to satisfy some given boundary conditions on the boundary Γ of the domain Ω occupied by the fluid, for example

$$\boldsymbol{u}|_\Gamma = \boldsymbol{0}. \tag{2.171}$$

This condition says that the velocity of the fluid should be zero on the boundary due to viscosity.

Although the system of Navier–Stokes equations (2.168), (2.169) is formally simpler than the general system of viscous fluid dynamics, it is still a rather complicated nonlinear system. Since the system of Navier–Stokes equations can be used not only to describe the motion of the viscous fluid (say, water), but also, according to what researchers tend to think today, to interpret the mechanism for the occurrence of the turbulence, it is of great importance, and it is the most important research target of the infinite-dimensional

2.3 Navier–Stokes Equations

dynamical system (see [14]). Until now, some fundamental problems, such as the existence and uniqueness of solutions to its initial-boundary value problems in three space dimensions, the existence of its global attractors and inertial manifolds, etc., still remains to be solved and is expected to be a very hot topic of study in theory as well as in numerical methods for a rather long period of time from now.

The unknown function p and equation div $\boldsymbol{u} = 0$ do not look so natural mathematically in system (2.168)–(2.169). The most commonly used method to solve this system, according to the initiative work of French academician J. Leray (see [13]), is to chose a suitable function space and attract those unnatural things into the space so that the problem looks "neat and tidy" mathematically. For instance, consider the initial-boundary problem (2.168)–(2.171) in L^2 space. Set $H = (L^2(\Omega))^3$, and make an orthogonal decomposition of H. In Lemma 1.3 of Chapter 1, we proved that any vector field can be decomposed into the superposition of a longitudinal field and a transverse field. Actually, we can demand more on the transverse field in the above decomposition.

Lemma 2.3. *Suppose that $\boldsymbol{u}$ is a suitably smooth vector field in Ω; then $\boldsymbol{u}$ can be represented uniquely in the following form:*

$$\boldsymbol{u} = \boldsymbol{w} + \mathrm{grad}\, p, \tag{2.172}$$

where $\boldsymbol{w}$ satisfies

$$\begin{cases} \mathrm{div}\,\boldsymbol{w} = 0 & in \quad \Omega, \\ \boldsymbol{w} \cdot \boldsymbol{n} = 0 & on \quad \Gamma, \end{cases} \tag{2.173}$$

with $\boldsymbol{n}$ being the unit outward normal vector on Γ.

Proof. First we notice that, if (2.173) holds, then $\boldsymbol{w}$ must be orthogonal to any grad p in H. In fact, using Green's formula, we have

$$\int_\Omega \boldsymbol{w} \cdot \mathrm{grad}\, p\, dx = \int_\Gamma (\boldsymbol{w}\cdot\boldsymbol{n}) p\, dS - \int_\Omega (\mathrm{div}\,\boldsymbol{w}) p\, dx = 0.$$

Now we prove the uniqueness of the decomposition (2.172). If there exist $\boldsymbol{w}_1$, p_1 and $\boldsymbol{w}_2$, p_2 such that both satisfy (2.172), then

$$\boldsymbol{w}_1 - \boldsymbol{w}_2 + \mathrm{grad}(p_1 - p_2) = \boldsymbol{0}.$$

Taking the scalar product of the above formula with $\boldsymbol{w}_1 - \boldsymbol{w}_2$ and integrating over Ω, we obtain

$$\int_\Omega |\boldsymbol{w}_1 - \boldsymbol{w}_2|^2\, dx + \int_\Omega (\boldsymbol{w}_1 - \boldsymbol{w}_2) \cdot \mathrm{grad}(p_1 - p_2) dx = 0;$$

then from the orthogonality of $\boldsymbol{w}_1 - \boldsymbol{w}_2$ and $\mathrm{grad}(p_1 - p_2)$ in H, we have

$$\|\boldsymbol{w}_1 - \boldsymbol{w}_2\|_H = 0,$$

i.e., $\boldsymbol{w}_1 = \boldsymbol{w}_2$. Regarding the determination of p, it is certainly allowed a difference up to a constant, which will not affect the result of the decomposition (2.172).

Consider the following Neumann problem:

$$\begin{cases} \Delta p = \operatorname{div} \boldsymbol{u} & \text{in } \Omega, \\ \dfrac{\partial p}{\partial n} = \boldsymbol{u} \cdot \boldsymbol{n} & \text{on } \Gamma. \end{cases}$$

The above problem has a unique solution up to an additive arbitrary constant (see [1]). For the obtained p, take $\boldsymbol{w} = \boldsymbol{u} - \operatorname{grad} p$, which can be easily seen to satisfy (2.173). Thus we obtain the required decomposition (2.172). The proof of Lemma 2.3 is completed. □

For general $\boldsymbol{u} \in H$, it can be proved (see [5]) that $\boldsymbol{u}$ can still be represented uniquely in the form of (2.172), where $\boldsymbol{w} \in H_\sigma$,

$$H_\sigma = \left\{ \boldsymbol{w} \mid \boldsymbol{w} \in H, \int_\Omega \boldsymbol{w} \cdot \operatorname{grad} \phi \, \mathrm{d}x = 0, \, \forall \phi \in H^1(\Omega) \right\},$$

and $p \in H^1(\Omega)$, here $H^1(\Omega)$ is the Sobolev space in the usual sense. This indicates that H can be decomposed into the direct sum of the following two spaces:

$$H = H_\sigma \oplus H_\sigma^\perp,$$

where

$$H_\sigma^\perp = \{ \boldsymbol{f} \mid \boldsymbol{f} = \operatorname{grad} p, p \in H^1(\Omega) \}.$$

Obviously, H_σ and $H_\sigma^\perp$, the two subspaces of H, are orthogonal to each other.

Denote by P the projection operator from H to H_σ:

$$P\boldsymbol{u} = \boldsymbol{w}, \quad \forall \boldsymbol{u} \in H,$$

where $\boldsymbol{w} \in H_\sigma$ is given by the decomposition formula (2.172) of $\boldsymbol{u}$. Using the projection operator P on both sides of equations (2.168) (for simplicity and without loss of generality, suppose that $\boldsymbol{F} = \boldsymbol{0}$) and noting that (2.169) implies $\boldsymbol{u} \in H_\sigma$, we get

$$P \frac{\partial \boldsymbol{u}}{\partial t} = \frac{\partial \boldsymbol{u}}{\partial t}, \quad P \operatorname{grad} p = \boldsymbol{0},$$

and thus we have

$$\frac{\partial \boldsymbol{u}}{\partial t} = P \left(\mu \Delta \boldsymbol{u} - \sum_{k=1}^3 u_k \frac{\partial \boldsymbol{u}}{\partial x_k} \right). \tag{2.174}$$

For system (2.174) together with initial condition (2.170) and the boundary conditions, we can use the usual Galerkin's method or the method of operator semigroup to discuss the existence of solutions.

If the flow is restricted to the two-dimensional case, that is, $u_3 \equiv 0$, while u_1, u_2, and p depend only on t, x_1, and x_2, then we have the *two-dimensional Navier–Stokes equations*

$$\begin{cases} \dfrac{\partial \boldsymbol{u}}{\partial t} - \mu \Delta \boldsymbol{u} + \sum_{k=1}^2 u_k \dfrac{\partial \boldsymbol{u}}{\partial x_k} + \operatorname{grad} p = \boldsymbol{F}, \\ \operatorname{div} \boldsymbol{u} = 0, \end{cases}$$

where $\boldsymbol{u} = (u_1, u_2)^\mathrm{T}$. For the above mixed initial-boundary value problem of this two-dimensional system, the existence and uniqueness of global solutions have to be proved, and the spaces of existence and uniqueness are the same. While, for the three-dimensional case, there are already results on existence and uniqueness, however, the spaces of these two do not coincide with each other, so the problem still needs a perfect solution.

There are already quite a few books on Navier–Stokes equations; we refer the interested reader to, for instance, [5], [15], and [16], etc.

2.4 Shock Waves

In this section we consider the strongly discontinuous solutions, i.e., shock waves, to the one-dimensional system (2.87)–(2.89) of ideal fluid dynamics. For narrative simplicity, let $F = 0$.

2.4.1 Jump Conditions

The equations (2.87)–(2.89) of differential form hold only for solutions in smooth regions. They are no longer valid on the discontinuous curve of solutions, while the conservation laws (2.9), (2.11), and (2.18) (which should be reduced to the form of one-dimensional flow!) of integral form, from which the differential form is inferred, still hold. From such integral conservation laws, we can directly derive the connection conditions on the discontinuous curve of piecewise smooth solutions.

Let us first make a general investigation.

Any one-dimensional conservation law of integral form can be written in the following form:

$$\int_{x_1}^{x_2} f(t_2, x) \mathrm{d}x - \int_{x_1}^{x_2} f(t_1, x) \mathrm{d}x$$
$$= \int_{t_1}^{t_2} q(t, x_1) \mathrm{d}t - \int_{t_1}^{t_2} q(t, x_2) \mathrm{d}t,$$
$$\forall (x_1, x_2), \forall t_1 < t_2. \tag{2.175}$$

Equations (2.175) can also be written as

$$\oint_{\Gamma} f \mathrm{d}x - q \mathrm{d}t = 0, \tag{2.176}$$

where Γ is the rectangular contour given in Figure 2.3. It is not hard to deduce that (2.176) holds for any closed curve Γ in the (t, x) plane.

From Green's formula, if f and q are smooth, then (2.176) leads to

$$\int_{\Omega} \left(\frac{\partial f}{\partial t} + \frac{\partial q}{\partial x} \right) \mathrm{d}x \mathrm{d}t = 0, \tag{2.177}$$

where Ω is the domain bounded by Γ. From the arbitrariness of Ω, we get

$$\frac{\partial f}{\partial t} + \frac{\partial q}{\partial x} = 0. \tag{2.178}$$

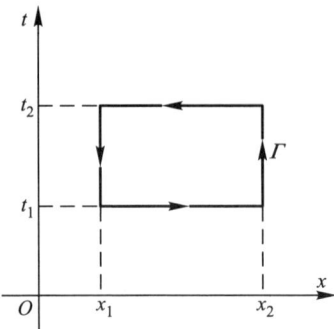

Figure 2.3.

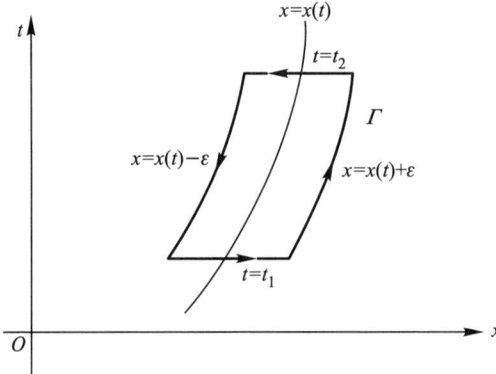

Figure 2.4.

This is the corresponding conservation law of differential form. In the domain where f and q are smooth, it is equivalent to (2.175) or (2.176).

Let $x = x(t)$ be a discontinuous curve, beside which f and q are continuous, and on which we have discontinuity of the first class. To obtain the connection conditions satisfied by f and q on the discontinuous curve, we take a contour Γ as shown in Figure 2.4 and use formula (2.176) on it; then we have

$$\int_{x(t_1)-\varepsilon}^{x(t_1)+\varepsilon} f(t_1,x)\mathrm{d}x - \int_{x(t_2)-\varepsilon}^{x(t_2)+\varepsilon} f(t_2,x)\mathrm{d}x$$
$$+ \int_{t_1}^{t_2} (f(t,x(t)+\varepsilon)\frac{\mathrm{d}x}{\mathrm{d}t} - q(t,x(t)+\varepsilon))\mathrm{d}t$$
$$- \int_{t_1}^{t_2} (f(t,x(t)-\varepsilon)\frac{\mathrm{d}x}{\mathrm{d}t} - q(t,x(t)-\varepsilon))\mathrm{d}t = 0.$$

2.4 Shock Waves

Setting $\varepsilon \to 0$, it follows from the above formula that

$$\int_{t_1}^{t_2} \left([f]\frac{dx}{dt} - [q]\right) dt = 0, \quad \forall t_1 < t_2;$$

then, noting again the arbitrariness of t_1 and t_2, we obtain on $x = x(t)$ that

$$[f]\frac{dx}{dt} - [q] = 0, \tag{2.179}$$

where $[f] = f_+ - f_-$ is the jump of f across the discontinuity $x = x(t)$, and $f_+ = f(t, x(t) + 0)$, $f_- = f(t, x(t) - 0)$, $[q]$ means the same. Condition (2.179) is called a *jump condition*. It is matched with the differential conservation law (2.178).

Applying (2.179) to equations (2.87)–(2.89), we get

$$[\rho]\frac{dx}{dt} = [\rho u], \tag{2.180}$$

$$[\rho u]\frac{dx}{dt} = [\rho u^2 + p], \tag{2.181}$$

$$\left[\rho e + \frac{1}{2}\rho u^2\right]\frac{dx}{dt} = \left[\left(\rho e + \frac{1}{2}\rho u^2 + p\right)u\right]. \tag{2.182}$$

These jump conditions are called *Rankine–Hugoniot conditions*.

Denoting by $U = \frac{dx}{dt}$ the propagation speed of the discontinuity, and in addition with

$$v_- = u_- - U, \quad v_+ = u_+ - U,$$

(2.180)–(2.182) can be rewritten as

$$\rho_- v_- = \rho_+ v_+, \tag{2.183}$$

$$\rho_- v_-^2 + p_- = \rho_+ v_+^2 + p_+, \tag{2.184}$$

$$\left(\rho_- e_- + \frac{1}{2}\rho_- v_-^2 + p_-\right) v_- = \left(\rho_+ e_+ + \frac{1}{2}\rho_+ v_+^2 + p_+\right) v_+. \tag{2.185}$$

Let $m = \rho_- v_- = \rho_+ v_+$.

If $m = 0$, we call the discontinuity $x = x(t)$ a *contact discontinuity*. For contact discontinuity, $v_- = v_+ = 0$, i.e., $u_+ = u_- = U$, so this discontinuity moves with the fluid at the same speed, and no fluid crosses the discontinuity. At this moment, due to (2.184), $p_- = p_+$, but $\rho_- \neq \rho_+$ (otherwise no discontinuity occurs).

If $m \neq 0$, we call the discontinuity $x = x(t)$ a *shock wave*. For a shock wave, $v_- \neq 0$, $v_+ \neq 0$, so the fluid crosses the discontinuity. We denote by subscript "0" the states before crossing the discontinuity, and by subscript "1" the states after crossing the discontinuity. From (2.183)–(2.185), we have

$$\rho_0 v_0 = \rho_1 v_1 = m, \tag{2.186}$$

$$\rho_0 v_0^2 + p_0 = \rho_1 v_1^2 + p_1, \tag{2.187}$$

$$\left(\rho_0 e_0 + \frac{1}{2}\rho_0 v_0^2 + p_0\right) v_0 = \left(\rho_1 e_1 + \frac{1}{2}\rho_1 v_1^2 + p_1\right) v_1. \tag{2.188}$$

Now we eliminate v_0 and v_1 from (2.186)–(2.188) to obtain the equations satisfied by the thermodynamical variables when crossing the shock wave. From (2.186) and (2.187), we have

$$m = -\frac{p_0 - p_1}{v_0 - v_1}. \tag{2.189}$$

From $\tau = \frac{1}{\rho}$, and substituting $v_0 = m\tau_0$, $v_1 = m\tau_1$ into the above formula, we get

$$m^2 = -\frac{p_0 - p_1}{\tau_0 - \tau_1}. \tag{2.190}$$

Noting $m^2 = \rho_0 v_0 \rho_1 v_1$, the above formula can also be written as

$$v_0 v_1 = \frac{p_0 - p_1}{\rho_0 - \rho_1}. \tag{2.191}$$

Using (2.186) and noting $\tau = \frac{1}{\rho}$, (2.188) can be rewritten as

$$\left(\rho_0 e_0 + \frac{1}{2}\rho_0 v_0^2\right)\tau_0 - \left(\rho_1 e_1 + \frac{1}{2}\rho_1 v_1^2\right)\tau_1 = p_1 \tau_1 - p_0 \tau_0. \tag{2.192}$$

But

$$\text{left-hand side of (2.192)} = \frac{1}{2}(v_0 - v_1)(v_0 + v_1) + e_0 - e_1$$
$$= (e_0 - e_1) - \frac{1}{2}(p_0 - p_1)(\tau_0 + \tau_1),$$

where (2.186) and (2.189) are used in the last step. Hence, (2.192) is rewritten as

$$e_1 - e_0 + \frac{1}{2}(p_0 + p_1)(\tau_1 - \tau_0) = 0. \tag{2.193}$$

Equation (2.193) is called the *Hugoniot equation* or the *thermodynamical shock condition on a shock wave*. It depends only on thermodynamical variables τ and p and is independent of v (and hence of u).

Denoting

$$H(\tau, p; \tau_0, p_0) = e(\tau, p) - e(\tau_0, p_0) + \frac{1}{2}(p_0 + p)(\tau - \tau_0), \tag{2.194}$$

and calling it the *Hugoniot function*, Hugoniot equation (2.193) can be simplified to

$$H(\tau_1, p_1; \tau_0, p_0) = 0. \tag{2.195}$$

For polytropic gases, the equation of state takes the following form:

$$e = \frac{1}{\gamma - 1} p\tau. \tag{2.196}$$

Then, from (2.194), its Hugoniot function satisfies

$$2\mu^2 H(\tau, p; \tau_0, p_0) = (\tau - \mu^2 \tau_0)p - (\tau_0 - \mu^2 \tau)p_0, \tag{2.197}$$

where $\mu^2 = (\gamma - 1)/(\gamma + 1)$. The above formula implies that, for polytropic gases, curve $H(\tau, p; \tau_0, p_0) = 0$ is a hyperbola across point (τ_0, p_0) (see Figure 2.5).

2.4 Shock Waves

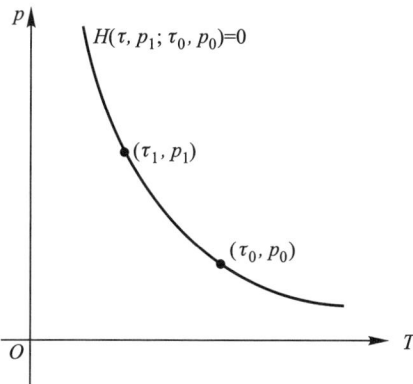

Figure 2.5.

2.4.2 Entropy Conditions

In order to ensure that the established theories are enough to depict the flow field including shock waves, the uniqueness of solutions has to be taken into consideration. For the one-dimensional system (2.87)–(2.89) of ideal fluid dynamics, can the jump connection conditions (2.180)–(2.182) guarantee the uniqueness of discontinuous solutions (or more generally, weak solutions) to the given problem (say, the initial value problem)? The answer is negative. We will explain this in the following Example 2.1 by taking a simple equation as an example.

Now we consider this problem from the physical point of view. Equations (2.87)–(2.89) and the corresponding jump connection conditions (2.180)–(2.182) describe the conservation laws of mass, momentum, and energy, respectively, satisfied by the fluid in the flow. The conservation law of energy is exactly the first law of thermodynamics. But as an actual process of fluid flow, it has to satisfy the second law of thermodynamics, which determines whether a thermodynamical process can proceed. We know from section 2.1 that, in a smooth flow field, the conservation law of energy is equivalent to the conservation law of entropy for ideal fluids. Is the entropy of the fluid still conserved across a shock? Let us look at the case of polytropic gases. If the entropy is still conserved across the shock, then we have

$$p_1 \rho_1^{-\gamma} = p_0 \rho_0^{-\gamma} \quad \text{or} \quad p_1 \tau_1^{\gamma} = p_0 \tau_0^{\gamma}.$$

But generally speaking, curve $p\tau^{\gamma} = p_0 \tau_0^{\gamma}$ does not coincide with the Hugoniot curve

$$H(\tau, p; \tau_0, p_0) = 0$$

formed by $H(\tau, p; \tau_0, p_0)$ given by (2.197), and therefore it is not an isotropic process for the fluid crossing a shock. Regarding the actual fluid, the so-called shock implies a sharp change of pressure, velocity, and density, etc., of the fluid in a small range. It can be seen from the investigation on viscous fluids in section 2.2, that the effects caused by the viscous friction cannot be neglected any longer even if the viscosity coefficient is very small, so this is analogous to an irreversible process, and the entropy should increase when the fluid crosses the shock (see Appendix B).

Now we discuss how to conveniently describe in mathematics the thermodynamical condition that the entropy increases across a shock. First, we consider the following simplest nonlinear equation of conservation law.

Example 2.1. For equation

$$\frac{\partial u}{\partial t} + \frac{\partial}{\partial x}\left(\frac{u^2}{2}\right) = 0, \quad x \in \mathbb{R}, \, t > 0, \tag{2.198}$$

its jump condition (4.5) is

$$[u]\frac{dx}{dt} - \left[\frac{u^2}{2}\right] = 0,$$

i.e.,

$$\frac{dx}{dt} = \frac{u_- + u_+}{2}. \tag{2.199}$$

Corresponding to the following two initial conditions,

$$u(0,x) = \begin{cases} 0, & x \leq 0, \\ 1, & x \geq 0 \end{cases} \tag{2.200}$$

and

$$u(0,x) = \begin{cases} 1, & x \leq 0, \\ 0, & x \geq 0, \end{cases} \tag{2.201}$$

we can construct, respectively, two discontinuous solutions of equation (2.198) satisfying the jump condition (2.199):

$$u(t,x) = \begin{cases} 0, & x \leq \frac{1}{2}t, \\ 1, & x \geq \frac{1}{2}t \end{cases} \tag{2.202}$$

and

$$u(t,x) = \begin{cases} 1, & x \leq \frac{1}{2}t, \\ 0, & x \geq \frac{1}{2}t. \end{cases} \tag{2.203}$$

For solution (2.202), the characteristic curves (defined by $\frac{dx}{dt} = u$) on both sides of the discontinuity $x = \frac{1}{2}t$ all spread upward, namely, in the direction along which t increases (see Figure 2.6), while for solution (2.203), the characteristic curves on both sides of the discontinuity $x = \frac{1}{2}t$ all spread downward, namely, in the direction along which t decreases till the initial axis $t = 0$ (see Figure 2.7). Since any solution of (2.198) takes a constant value along every characteristic curve $\frac{dx}{dt} = u$, the initial value (2.201) given at $t = 0$ can

2.4 Shock Waves

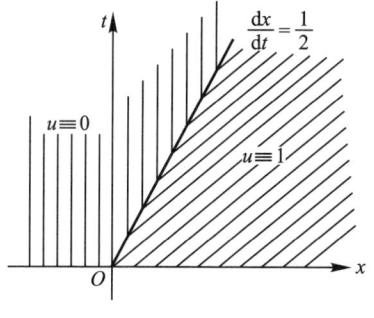

Figure 2.6.

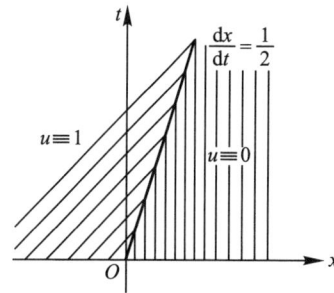

Figure 2.7.

determine the solution (2.203) for $t > 0$ along characteristic curves, while the solution (2.202) cannot be completely determined by the initial value (2.200) along characteristic curves. According to the law of causality (it is essentially the invertibility with respect to time), the solution at $t > 0$ and the shock wave should be determined by the initial data and not determined by the "future." Thus, we have every reason to believe that the solution given by (2.203) is reasonable, while the solution given by (2.202) to problem (2.198) and (2.200) does not fit the law of causality and should be discarded. In fact, for initial condition (2.200), we can construct another solution of equation (2.198):

$$u(t,x) = \begin{cases} 1, & x \geq t, \\ x/t, & 0 \leq x \leq t, \\ 0, & x \leq 0. \end{cases} \qquad (2.204)$$

We usually call this a *centered rarefaction wave* (see Figure 2.8). Thus, it is only when giving up the solution given by (2.202) that it possible to guarantee the uniqueness of discontinuous solutions satisfying the jump conditions. ∎

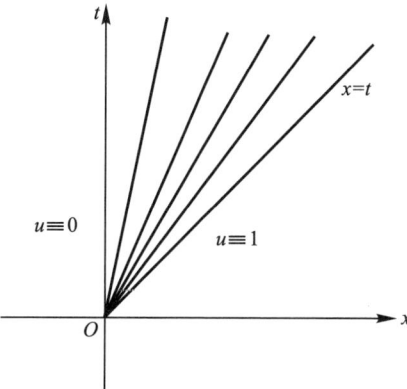

Figure 2.8.

Based on the above discussion, we now consider the system (2.87)–(2.89) of one-dimensional gas dynamics. The roots of its characteristic equation are

$$\lambda_1 = u - c, \quad \lambda_2 = u, \quad \lambda_3 = u + c. \tag{2.205}$$

The characteristic curves corresponding to the above eigenvalues are determined, respectively, by the following three formulas:

$$\frac{dx}{dt} = u - c, \quad \frac{dx}{dt} = u, \quad \frac{dx}{dt} = u + c. \tag{2.206}$$

As a mathematical description of the condition that the entropy increases across a shock, P. D. Lax proposed the following shock inequalities: On both sides of the shock, for a certain $k \in \{1, 2, 3\}$, there hold

$$\lambda_k(u_+, c_+) < U < \lambda_k(u_-, c_-), \tag{2.207}$$

$$\lambda_{k-1}(u_-, c_-) < U < \lambda_{k+1}(u_+, c_+), \tag{2.208}$$

where U is the propagation speed of shock. Neither the inequality on the left-hand side of (2.208) for $k = 1$ nor the inequality on the right-hand side of (2.208) for $k = 3$ are needed. The shock jump satisfying this group of inequalities is called a *k-shock*. Equations (2.207) and (2.208) are sometimes called *entropy inequalities* or *entropy conditions*.

For shock waves, it is known from (2.186) that v_0 and v_1 have the same sign. If both v_0 and v_1 are negative, it means that the fluid crosses the shock from right to left, that is, the shock wave propagates to the right relative to the fluid. Then we call the shock wave a *rightward shock wave*. Similarly, if both v_0 and v_1 are positive, it is then called a *leftward shock wave*.

Now we claim that, for the system of one-dimensional gas dynamics, there does not exist any 2-shock wave; i.e., there does not exist a shock jump such that (2.207), (2.208) hold for $k = 2$. In fact, if the shock wave propagates to the right, then from the definition of v, we have

$$u_0, u_1 < U;$$

if the shock wave propagates to the left, then

$$u_0, u_1 > U.$$

So no matter what the situation, it is always impossible for (2.207) to hold for $k = 2$. Similarly, it is easy to prove that there do not exist a rightward 1-shock wave and a leftward 3-shock wave. Hence, **for the system of one-dimensional gas dynamics, the only possible shock waves are the leftward 1-shock wave and the rightward 3-shock wave**.

We call a shock wave *compressive* if the fluid possesses larger pressure when crossing the shock wave, i.e.,

$$p_0 < p_1. \tag{2.209}$$

For the entropy conditions (2.207), (2.208) and the shock compressive condition (2.209), we have the following results.

Theorem 2.2. *For polytropic gases, a shock wave is compressive if and only if it satisfies the entropy conditions (2.207), (2.208).*

2.4 Shock Waves

Proof. The theorem can be proved in the following steps.

(1°) **The propagation speeds v_0 and v_1 of the gas, relative to the shock wave, on both sides of the shock wave are subsonic and supersonic, respectively.**

In fact, from (2.186) and (2.188) we have

$$\frac{1}{2}v_0^2 + e_0 + p_0\tau_0 = \frac{1}{2}v_1^2 + e_1 + p_1\tau_1. \tag{2.210}$$

Substituting $\mu^2 = (\gamma-1)/(\gamma+1)$, $e = p\tau/(\gamma-1)$ and $c^2 = \gamma p\tau$ into the above formula, we get

$$\mu^2 v_0^2 + (1-\mu^2)c_0^2 = \mu^2 v_1^2 + (1-\mu^2)c_1^2. \tag{2.211}$$

Denote by c_*^2 the common value on both sides of the above formula; then

$$(1-\mu^2)(v_0^2 - c_0^2) = v_0^2 - c_*^2,$$
$$(1-\mu^2)(v_1^2 - c_1^2) = v_1^2 - c_*^2.$$

The above two formulas imply that

$$|v_0| > c_0 \text{ is equivalent to } |v_0| > c_*, \tag{2.212}$$
$$|v_1| > c_1 \text{ is equivalent to } |v_1| > c_*. \tag{2.213}$$

From the right-hand side of (2.211) which defines c_*^2, noting $c_1^2 = \gamma p_1 \tau_1$ and $(1-\mu^2)\gamma = 1+\mu^2$, we have

$$\rho_1 c_*^2 = \mu^2(\rho_1 v_1^2 + p_1) + p_1.$$

Similarly, we get

$$\rho_0 c_*^2 = \mu^2(\rho_0 v_0^2 + p_0) + p_0.$$

Subtracting the above two formulas and using (2.187), we obtain

$$c_*^2 = \frac{p_1 - p_0}{\rho_1 - \rho_0}. \tag{2.214}$$

Comparing (2.214) with (2.191), we have

$$c_*^2 = v_0 v_1. \tag{2.215}$$

Equation (2.215) is called *Prandtl's relation*. The required conclusion follows from Prandtl's relation and (2.212), (2.213).

(2°) **A shock wave is compressive if and only if**

$$\rho_0 < \rho_1. \tag{2.216}$$

This conclusion follows immediately from the definition of compressive shock wave and (2.214).

(3°) A noncompressive shock wave does not satisfy the entropy conditions (2.207), (2.208).

We know from (2°) that $\rho_1 < \rho_0$ for a noncompressive shock wave. Suppose that the shock wave propagates toward the right, namely, $v_0, v_1 < 0$. Then, we know from the previous discussion that the entropy conditions (2.207), (2.208) cannot possibly hold for $k = 1, 2$. Now we prove that it is also impossible for (2.207) to hold for $k = 3$. It is not difficult to verify directly that, for the function $p = p(\tau)$ determined by the Hugoniot equation $H(\tau, p; \tau_0, p_0) = 0$, $c = \sqrt{\gamma p(\tau) \tau}$ is a strictly decreasing function of τ, and then $c(\rho_1) < c(\rho_0)$. In addition, (4.12) yields $v_1 < v_0 < 0$. Therefore,

$$v_1 + c(\rho_1) < v_0 + c(\rho_0).$$

But from (1°), we have

$$|v_1| > c(\rho_1), \qquad |v_0| < c(\rho_0).$$

This implies

$$v_1 + c(\rho_1) < 0 < v_0 + c(\rho_0),$$

i.e.,

$$u_1 + c(\rho_1) < U < u_0 + c(\rho_0). \tag{2.217}$$

For the rightward shock wave, "0" means the right state "+", "1" means the left state "−". Thus (2.217) contradicts the inequality (2.207) for $k = 3$. A similar conclusion can be made for the leftward shock wave.

(4°) A compressive shock wave satisfies the entropy conditions (2.207), (2.208).

From (2°), $\rho_1 > \rho_0$ for compressive shock waves. We still assume that the shock wave propagates toward the right. Then $v_0 < v_1 < 0$ and $c(\rho_1) > c(\rho_0)$. Thus

$$v_1 + c(\rho_1) > v_0 + c(\rho_0).$$

From (1°), we have

$$|v_0| > c(\rho_0), \qquad |v_1| < c(\rho_1).$$

This implies

$$v_1 + c(\rho_1) > 0 > v_0 + c(\rho_0),$$

i.e.,

$$u_1 + c(\rho_1) > U > u_0 + c(\rho_0). \tag{2.218}$$

This is exactly the inequality (2.207) for $k = 3$. Since the shock wave propagates toward the right, $v_1 < 0$, i.e., $u_1 < U$, which is nothing but the inequality (2.208) for $k = 3$. So the entropy conditions (2.207), (2.208) hold. Similar conclusions can be done for the leftward shock wave. The proof of Theorem 2.2 is finished. □

The relation between the entropy conditions (2.207), (2.208) and the physical entropy is given in the following theorem.

2.5 System of One-Dimensional Fluid Dynamics in Lagrangian Representation 119

Theorem 2.3. *For polytropic gases, a shock wave satisfies the entropy conditions* (2.207), (2.208) *if and only if the entropy S increases after crossing the shock wave, i.e.,*

$$S_1 > S_0. \qquad (2.219)$$

Proof. From Theorem 2.2 we know that the shock wave satisfies the entropy conditions (2.207), (2.208) if and only if

$$\tau_1 < \tau_0. \qquad (2.220)$$

So, from the expression (2.27) of the entropy for the polytropic gas, in order to prove (2.219), we need only prove that $p(\tau)\tau^\gamma$ is a strictly decreasing function of τ, where the function $p = p(\tau)$ is determined by the Hugoniot equation $H(\tau, p; \tau_0, p_0) = 0$. Similarly to the proof of Theorem 2.2, this can be directly verified from the expression (2.197) of $H(\tau, p; \tau_0, p_0)$. The proof is completed. □

For polytropic gases, Theorem 2.3 shows that the entropy inequalities (2.207), (2.208) satisfied on the shock wave are equivalent to the physical fact that the entropy increases when the gas crosses the shock wave. When our discussion is not restricted to polytropic gases, it can be proved that this result also holds for weak shock waves (see [4]). The so-called weak shock wave means a shock wave with sufficiently small intensity which is the jump of physical variables across the shock.

At last, to explain that the shock wave is a kind of compressive phenomenon, we give the following result.

Theorem 2.4. *For polytropic gases, the shock wave satisfies the entropy conditions* (2.207), (2.208) *if and only if*

$$u_- > u_+ \qquad (2.221)$$

on both sides of the shock.

Proof. From Theorem 2.2, the shock wave satisfies the entropy conditions (2.207)–(2.208) if and only if

$$\rho_1 > \rho_0 \qquad (2.222)$$

across the shock.

If the shock wave propagates toward the right, i.e., $v_0, v_1 < 0$, then, from (2.186), condition (2.222) is equivalent to

$$v_0 < v_1, \qquad (2.223)$$

i.e.,

$$u_0 < u_1. \qquad (2.224)$$

But for the rightward shock wave, the state marked with "0" is the right state "+", while the state marked with "1" is the left state "−". So (2.224) is exactly (2.221).

The case of leftward shock wave can be discussed similarly. The proof of Theorem 2.4 is finished. □

2.5 System of One-Dimensional Fluid Dynamics in Lagrangian Representation

2.5.1 Introduction

In the previous derivation of equations of fluid dynamics for the ideal fluid and the viscous heat-conducting fluid, we represented the variables describing the states of the fluid motion as functions of time t and space Cartesian coordinates $\boldsymbol{x} = (x_1, x_2, x_3)$. The established equations are called *equations in Eulerian representation*, where we explore the law of the fluid state changing with time t at any fixed position (location). There is also another way to look at this; in other words, we can also observe the law of the fluid state changing with time t at any fixed (but in motion!) fluid particle. The established equations are called *equations in Lagrangian representation*. In short, we discuss the former under the coordinate system fixed in space, and discuss the latter under the coordinate system fixed on the particle. In the Lagrangian representation, the variables describing the fluid states will be denoted by functions of time t and *material coordinates (Lagrangian coordinates)*, while the position of the particle in space, namely, its *Eulerian coordinates* $\boldsymbol{x}$, will be determined by functions of time and particle coordinates. Employing the Lagrangian coordinates will bring us significant convenience in describing the equations in the case of one-dimensional motion, so the following discussion is restricted to this case.

2.5.2 Lagrangian Coordinates

Now we show the method for introducing the Lagrangian coordinates, explain its physical meaning, and give the transform relation between Lagrangian coordinates and Euler coordinates.

From the continuity equation

$$\frac{\partial \rho}{\partial t} + \frac{\partial}{\partial x}(\rho u) = 0,$$

we know that

$$\rho \mathrm{d}x - \rho u \mathrm{d}t$$

is a total differential. So there exists a function $m = m(t, x)$ such that

$$\mathrm{d}m = \rho \mathrm{d}x - \rho u \mathrm{d}t. \tag{2.225}$$

Taking m as a new argument instead of x (the other argument is still t), we get the Lagrangian coordinates (t', m), while

$$t' = t, \tag{2.226}$$
$$m = m(t, x) \tag{2.227}$$

is the transform formula between Euler coordinates and Lagrangian coordinates. Since $\rho > 0$ and $\frac{\partial m}{\partial x} = \rho$, the above transform is globally invertible.

2.5 System of One-Dimensional Fluid Dynamics in Lagrangian Representation

The physical meaning of the introduced Lagrangian coordinate m can be explained as follows. From (2.225), we can choose

$$m(t,x) = \int_{(0,0)}^{(t,x)} \rho dx - \rho u dt, \tag{2.228}$$

and the integral value is independent of the path of integration from $(0,0)$ to (t,x). Then we can take a particular path OAB as shown in Figure 2.9, where OA is the graph in the (t,x) plane of the motion law $x = x(t)$ of the particle which is at the origin at $t = 0$, while AB is a line segment through $B(t,x)$ parallel to the x axis. Thus

$$m(t,x) = \int_{OA+AB} \rho dx - \rho u dt. \tag{2.229}$$

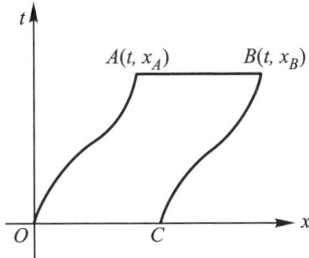

Figure 2.9.

Since $\frac{dx}{dt} = u$ on OA, $\int_{OA} \rho dx - \rho u dt = 0$, while on AB, $dt = 0$, and then (2.229) is reduced to

$$m(t,x) = \int_{x_A}^{x_B} \rho dx, \tag{2.230}$$

where x_A and x_B denote the Eulerian coordinates of points A and B, respectively. The right-hand side of (2.230) represents the fluid mass between AB (If B lies to the right of A, it is positive; if B lies to the left of A, it is negative). Suppose that the particle which is at point B at time t lies at point C at time $t = 0$. Since the particle which is at point A at time t lies at the origin at time $t = 0$, from the conservation law of mass, $m(t,x)$ should be equal to the fluid mass between OC at time $t = 0$. Therefore, the Lagrangian coordinate m represents the mass, while $m = $ constant indicates the same particle, so m is the particle coordinate. If we draw through point C in the (t,x) plane the curve CB of the motion law of the particle which is located at point C at time $t = 0$, then $m = $ constant on the whole curve. Therefore, the transition from Eulerian coordinates (t,x) to Lagrangian coordinates $(t',m) = (t,m)$ is essentially that the curves of the motion law of fluid particles at the (t,x) plane are taken as coordinate curves.

Now we illustrate some transform rules between Eulerian coordinates and Lagrangian coordinates. From

$$dm = -\rho u dt + \rho dx, \tag{2.231}$$
$$dt' = dt, \tag{2.232}$$

we have

$$\frac{\partial}{\partial t} = \frac{\partial}{\partial t'} - \rho u \frac{\partial}{\partial m}, \tag{2.233}$$

$$\frac{\partial}{\partial x} = \rho \frac{\partial}{\partial m}. \tag{2.234}$$

Together with

$$\mathrm{d}x = u\mathrm{d}t' + \tau \mathrm{d}m, \tag{2.235}$$

$$\mathrm{d}t = \mathrm{d}t', \tag{2.236}$$

we have

$$\frac{\partial}{\partial t'} = \frac{\partial}{\partial t} + u \frac{\partial}{\partial x}, \tag{2.237}$$

$$\frac{\partial}{\partial m} = \tau \frac{\partial}{\partial x}. \tag{2.238}$$

The transform formulas (2.233), (2.234) and (2.237), (2.238) give the relations of partial derivatives with respect to (t,x) and to (t',m).

Furthermore, when Lagrangian coordinates (t',m) are known, then from (2.235), Eulerian coordinate x can be obtained by the following formula:

$$x = \int_{(0,0)}^{(t',m)} u\mathrm{d}t' + \tau \mathrm{d}m. \tag{2.239}$$

When Eulerian coordinates (t,x) are known, the Lagrangian coordinate m can be found from (2.228).

2.5.3 System of One-Dimensional Ideal Fluid Dynamics in Lagrangian Representation

The system of one-dimensional ideal fluid dynamics in Eulerian coordinates can be written as

$$\frac{\partial \rho}{\partial t} + \frac{\partial}{\partial x}(\rho u) = 0, \tag{2.240}$$

$$\frac{\partial u}{\partial t} + u \frac{\partial u}{\partial x} + \frac{1}{\rho} \frac{\partial p}{\partial x} = F, \tag{2.241}$$

$$\frac{\partial S}{\partial t} + u \frac{\partial S}{\partial x} = 0 \tag{2.242}$$

(see (2.90)–(2.92)). By introducing Lagrangian coordinates (t',m), it can be reduced to the corresponding system in Lagrangian coordinates according to the aforementioned transform formulas of coordinates. Here note particularly that

$$\frac{\partial}{\partial t'} = \frac{\partial}{\partial t} + u \frac{\partial}{\partial x} = \frac{\mathrm{d}}{\mathrm{d}t}$$

2.5 System of One-Dimensional Fluid Dynamics in Lagrangian Representation

represents exactly taking derivative with respect to t when fixing the particle. This is identical to the nature of Lagrangian coordinates.

The transition of equation (2.240) can be done using the above transform formulas (2.233), (2.234). The simpler method is to notice that (2.240) is already used to define Lagrangian coordinate m (in other words, $\rho dx - \rho u dt = dm$ is a total differential, which is going to be replaced by the condition that $dx = u dt' + \tau dm$ is a total differential), and thus there should be

$$\frac{\partial \tau}{\partial t'} - \frac{\partial u}{\partial m} = 0, \tag{2.243}$$

while from (2.233), (2.234), it is easy to reduce (2.241) and (2.242) to

$$\frac{\partial u}{\partial t'} + \frac{\partial p}{\partial m} = F, \tag{2.244}$$

$$\frac{\partial S}{\partial t'} = 0. \tag{2.245}$$

Therefore, the system of one-dimensional ideal fluid dynamics in Lagrangian coordinates takes the form (for continuously differentiable flow)

$$\frac{\partial \tau}{\partial t} - \frac{\partial u}{\partial x} = 0, \tag{2.246}$$

$$\frac{\partial u}{\partial t} + \frac{\partial p}{\partial x} = F, \tag{2.247}$$

$$\frac{\partial S}{\partial t} = 0. \tag{2.248}$$

Here, for convenience Lagrangian coordinates (t', m) are still denoted by (t, x).

The above system has a simpler form than the original system (2.240)–(2.242): (2.246) and (2.248) are linear equations with constant coefficients, and the nonlinearity of the whole system is reflected only in the term $\frac{\partial p}{\partial x}$ ($p = p(\tau, S)$). In particular, equation (2.248) is very simple. For the Cauchy problem, as long as the initial value of S,

$$t = 0 : S = S_0(x), \tag{2.249}$$

is given, we then have $S = S_0(x)$ from equation (2.248). Thus, system (2.246)–(2.248) can be reduced to a system composed of two equations:

$$\frac{\partial \tau}{\partial t} - \frac{\partial u}{\partial x} = 0, \tag{2.250}$$

$$\frac{\partial u}{\partial t} + \frac{\partial}{\partial x} p(\tau, S_0(x)) = F. \tag{2.251}$$

Especially the system for isentropic flow ($S \equiv$ constant) in Lagrangian coordinates has a much simpler form:

$$\frac{\partial \tau}{\partial t} - \frac{\partial u}{\partial x} = 0, \tag{2.252}$$

$$\frac{\partial u}{\partial t} + \frac{\partial}{\partial x} p(\tau) = F. \tag{2.253}$$

When $F \equiv 0$ (the external force is zero), this is a quasi-linear hyperbolic system with simple form and typical significance, sometimes called the *p-system*, and is still being studied by many researchers.

2.5.4 System of One-Dimensional Viscous Heat-Conducting Fluid Dynamics in Lagrangian Representation

The system of one-dimensional viscous heat-conducting fluid dynamics in Eulerian coordinates takes the form

$$\frac{\partial \rho}{\partial t} + \frac{\partial}{\partial x}(\rho u) = 0, \tag{2.254}$$

$$\frac{\partial u}{\partial t} + u\frac{\partial u}{\partial x} + \frac{1}{\rho}\frac{\partial p}{\partial x} - \frac{1}{\rho}\frac{\partial}{\partial x}\left(\left(\frac{4}{3}\mu + \mu'\right)\frac{\partial u}{\partial x}\right) = F, \tag{2.255}$$

$$\rho\frac{\partial e}{\partial t} + \rho u\frac{\partial e}{\partial x} + p\frac{\partial u}{\partial x} - \left(\frac{4}{3}\mu + \mu'\right)\left(\frac{\partial u}{\partial x}\right)^2 = \frac{\partial}{\partial x}\left(\kappa\frac{\partial T}{\partial x}\right) \tag{2.256}$$

(see (2.160)–(2.162)). For the same reason, the corresponding system in Lagrangian coordinates can be written as

$$\frac{\partial \tau}{\partial t'} - \frac{\partial u}{\partial m} = 0, \tag{2.257}$$

$$\frac{\partial u}{\partial t'} + \frac{\partial p}{\partial m} - \frac{\partial}{\partial m}\left(\left(\frac{4}{3}\mu + \mu'\right)\rho\frac{\partial u}{\partial m}\right) = F, \tag{2.258}$$

$$\frac{\partial e}{\partial t'} + p\frac{\partial u}{\partial m} - \rho\left(\frac{4}{3}\mu + \mu'\right)\left(\frac{\partial u}{\partial x}\right)^2 = \frac{\partial}{\partial m}\left(\kappa\rho\frac{\partial T}{\partial m}\right). \tag{2.259}$$

Using (2.257), the third equation (2.259) can also be written as

$$\frac{\partial e}{\partial T}\frac{\partial T}{\partial t'} + \left(p - \frac{\partial e}{\partial \tau}\right)\frac{\partial u}{\partial m} - \rho\left(\frac{4}{3}\mu + \mu'\right)\left(\frac{\partial u}{\partial m}\right)^2 = \frac{\partial}{\partial m}\left(\kappa\rho\frac{\partial T}{\partial m}\right). \tag{2.260}$$

Again denoting (t', m) by (t, x), we have

$$\frac{\partial \tau}{\partial t} - \frac{\partial u}{\partial x} = 0, \tag{2.261}$$

$$\frac{\partial u}{\partial t} + \frac{\partial p}{\partial x} - \frac{\partial}{\partial x}\left(\left(\frac{4}{3}\mu + \mu'\right)\rho\frac{\partial u}{\partial x}\right) = F, \tag{2.262}$$

$$\frac{\partial e}{\partial t} + p\frac{\partial u}{\partial x} - \rho\left(\frac{4}{3}\mu + \mu'\right)\left(\frac{\partial u}{\partial x}\right)^2 = \frac{\partial}{\partial x}\left(\kappa\rho\frac{\partial T}{\partial x}\right) \tag{2.263}$$

or

$$\frac{\partial e}{\partial T}\frac{\partial T}{\partial t} + \left(p - \frac{\partial e}{\partial \tau}\right)\frac{\partial u}{\partial x} - \rho\left(\frac{4}{3}\mu + \mu'\right)\left(\frac{\partial u}{\partial x}\right)^2 = \frac{\partial}{\partial x}\left(\kappa\rho\frac{\partial T}{\partial x}\right). \tag{2.263}'$$

This possesses a simpler form than the original system (2.254)–(2.256) in Euler coordinates.

2.5 System of One-Dimensional Fluid Dynamics in Lagrangian Representation

Finally, we point out that the Lagrangian coordinate transform can simplify the form of the original nonlinear system since it is a transform relevant to the solution—taking the curves of the motion law as the coordinate curves. It is impossible to give Lagrangian coordinates specifically before the solution is found. However, the above arguments tell us that, as long as the solution exists, a Lagrangian coordinate transform can always be conducted, and it reduces the system to the form in Lagrangian coordinates. As long as the system in Lagrangian coordinates is solved, the corresponding solution in Eulerian coordinates can be found with the help of the inverse transform, and therefore the whole argument is valid.

In order to solve the system in Lagrangian coordinates, it should also be explained how to transfer those conditions originally given in Eulerian coordinates to the conditions in Lagrangian coordinates. For instance, if the initial conditions are given in Eulerian coordinates as follows:

$$t = 0: u = u_0(x),\ \rho = \rho_0(x),\ T = T_0(x)\ (\text{or } S = S_0(x)), \tag{2.264}$$

then from (2.228) we have

$$m(0,x) = \int_0^x \rho_0(x)\mathrm{d}x. \tag{2.265}$$

The value of m corresponding to x at initial time $t = t' = 0$ can be obtained from the above formula, and then x can be solved as a function of m: $x = x(m)$. Thus the initial conditions in Lagrangian coordinates are

$$t' = 0: u = \bar{u}_0(m),\ \rho = \bar{\rho}_0(m),\ T = \bar{T}_0(m)\ (\text{or } S = \bar{S}_0(m)), \tag{2.266}$$

where $\bar{u}_0(m) = u_0(x(m))$, etc.

The boundary conditions given along the curve of the motion law of a particle in Eulerian coordinates possess very simple forms in Lagrangian coordinates. For instance, for system (2.257)–(2.259), if boundary conditions on the curve $x = x_0(t)$ ($x_0(0) = 0$) passing through the origin are given (see Figure 2.10) as

$$u = u_0(t), \qquad T = T_0(t), \tag{2.267}$$

and

$$\dot{x}_0(t) = u_0(t), \tag{2.268}$$

then $x = x_0(t)$ turns to $m = 0$ when transformed into Lagrangian coordinates, and so the boundary reduces to a straight line; while boundary conditions (2.267) are turned into

$$m = 0: u = u_0(t),\ T = T_0(t). \tag{2.269}$$

This greatly simplifies the problem, which is also one of the benefits of introducing Lagrangian coordinates.

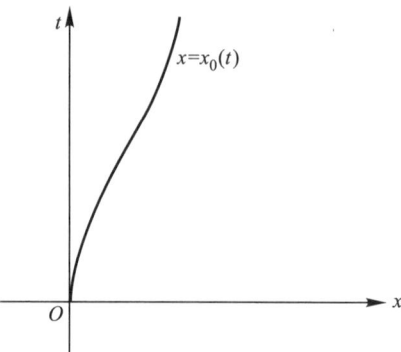

Figure 2.10.

Exercises

1. Prove that when the flow field is irrotational, i.e., rot $u = 0$, the Euler equations for the ideal fluid can be written in the following form:

$$\frac{\partial u}{\partial t} + \text{grad}\, \frac{u^2}{2} + \frac{1}{\rho}\, \text{grad}\, p = F.$$

2. Prove that if the mass force F has potential, that is, there exists ϕ such that $F = -\text{grad}\, \phi$, then the conservation equation of energy in differential form for the ideal fluid can be written as

$$\frac{d}{dt}\left(e + \frac{u^2}{2} + \frac{p}{\rho} + \phi\right) = \frac{1}{\rho}\frac{\partial p}{\partial t} + \frac{\partial \phi}{\partial t}.$$

3. Suppose that Ω is a simply connected domain and that the vector field u_B is given on the boundary Γ of Ω. Prove that there exists a velocity field u in $\overline{\Omega}$ such that div $u = 0$ holds in Ω, and this velocity field has potential (which means that there exists a scalar field ϕ such that $u = -\text{grad}\, \phi$), and its normal component on Γ is $u \cdot n = u_B \cdot n$, if and only if

$$\int_\Gamma u_B \cdot n\, dS = 0,$$

where n is the unit outward normal vector.

4. Let u be the solution satisfying the conditions given in exercise 3. Prove that u is the solution of the variational problem

$$\min_{w \in A} \frac{1}{2} \int_\Omega \rho |w|^2 dx,$$

where

$$A = \{w \in C^1(\Omega) \cap C^0(\overline{\Omega});\quad \text{div}\, w = 0 \text{ in } \Omega;$$
$$w \cdot n = u_B \cdot n,\, \text{on}\, \Gamma\}.$$

5. Let $\boldsymbol{P} = \{p_{ij}\}$ be the stress tensor of the fluid in the flow field. Prove that the average of the normal components of the stress on the sphere S_r with radius r centered at a certain point tends to the average of the normal stresses at this point as $r \to 0$, that is,

$$\lim_{r \to 0} \frac{1}{4\pi r^2} \int_{S_r} \boldsymbol{p}_n \cdot \boldsymbol{n} \, \mathrm{d}S = \frac{1}{3}(p_{11} + p_{22} + p_{33}),$$

where $\boldsymbol{p}_n$ is defined by (2.115) or (2.116).

6. Prove that the fluid motion governed by Navier–Stokes equations is always rotational, that is, if $\mathrm{rot}\, \boldsymbol{u} \equiv \boldsymbol{0}$, then Navier–Stokes equations (2.168), (2.169) are reduced to Euler equations (2.15).

7. Suppose that there is a pipeline with axis x and with a constant cross section, which is fully occupied with incompressible ideal fluid flowing along the x direction, the states of fluid are the same on each cross section, and $p = p(x)$. If $p(0) = p_1$ and $p(L) = p_2$ are given and $p_1 > p_2$, determine the fluid velocity inside the pipeline (neglecting the volume force).

8. Consider the x directional flow of a steady-state viscous incompressible fluid fixed between the two planes located at $y = 0$ and $y = 1$. Suppose that $p = p(x)$, $p(0) = p_1$, and $p(L) = p_2$ are given and that $p_1 > p_2$. Try to find the velocity $u(x,y)$ and the pressure $p(x)$ of this flow field (neglecting the volume force).

9. Suppose that $\Omega \subset \mathbb{R}^3$ is a bounded domain and that $\boldsymbol{u}$ is the solution of Navier–Stokes equations (2.168), (2.169) satisfying boundary conditions (2.171), where the volume force $\boldsymbol{F} \equiv \boldsymbol{0}$. Prove that the kinetic energy of the fluid decreases when the time increases, i.e.,

$$\frac{\mathrm{d}}{\mathrm{d}t} \int_\Omega \frac{1}{2} |\boldsymbol{u}|^2 \mathrm{d}x \le 0.$$

10. Prove that system (2.246)–(2.248) of one-dimensional ideal fluid dynamics in Lagrangian coordinates can also be written in the following form:

$$\begin{cases} \dfrac{\partial \tau}{\partial t} - \dfrac{\partial u}{\partial x} = 0, \\ \dfrac{\partial u}{\partial t} + \dfrac{\partial p}{\partial x} = F, \\ \dfrac{\partial}{\partial t}\left(e + \dfrac{u^2}{2}\right) + \dfrac{\partial}{\partial x}(pu) = Fu. \end{cases}$$

11. For the system of one-dimensional ideal fluid dynamics in Lagrangian form given in exercise 10, give the jump conditions satisfied by the solution on a strong discontinuity (suppose that the volume force $F \equiv 0$).

12. Let function $L(\xi_0, \xi_1, \ldots, \xi_n)$ be strictly convex with respect to the variables $\xi_0 > 0$, $\xi_1, \ldots, \xi_n$. Prove that function

$$M = \frac{1}{\xi_0} L(\xi_0, \xi_1, \ldots, \xi_n)$$

is strictly convex with respect to the variables

$$\eta_0 = \frac{1}{\xi_0}, \quad \eta_1 = \frac{\xi_1}{\xi_0}, \ldots, \eta_n = \frac{\xi_n}{\xi_0}.$$

13. Introduce a new unknown function to reduce the p system

$$\frac{\partial \tau}{\partial t} - \frac{\partial u}{\partial x} = 0,$$
$$\frac{\partial u}{\partial t} + \frac{\partial}{\partial x} p(\tau) = 0$$

into a first-order quasi-linear symmetric hyperbolic system of conservation laws. Here we assume $p'(\tau) < 0$.

Bibliography

[1] Courant R., Hilbert D. *Methods of Mathematical Physics*, Volume II. New York: John Wiley & Sons, 1989.

[2] Wu W. *Fluid Dynamics*, Volume I (in Chinese). Beijing: Peking University Press, 1982.

[3] Chorin A. J., Marsden J. E. *A Mathematical Introduction to Fluid Mechanics*, Second Edition. New York: Springer-Verlag, 1990.

[4] Smoller J. *Shock Waves and Reaction–Diffusion Equations*. New York: Springer–Verlag, 1983.

[5] Temam R. *Navier–Stokes Equations, Theory and Numerical Analysis*. Dordrecht, The Netherlands: North–Holland, 1984.

[6] Courant R, Friedrichs K. O. *Supersonic Flow and Shock Waves*. New York: Interscience Publishers, 1948.

[7] Li T. *Global Classical Solutions for Quasilinear Hyperbolic Systems*. Research in Applied Mathematics 32. Paris/New York: Masson/John Wiley, 1994.

[8] Harten A., Lax P. D. *A random choice finite difference scheme for hyperbolic conservation laws*. SIAM J. Numer. Anal., 18 (1981), 289–315.

[9] Lax P. D. *Symmetrizing hyperbolic differential equations*. In Nonlinear Hyperbolic Problems, Eds. C. Carasso, P.-A. Raviart, and D. Serre, Lecture Notes in Mathematics 1270, New York: Springer–Verlag, 1987, 150–151.

[10] Godunov S. K. *Lois de conservation et intégrales d'énergie des équations hyperboliques*. In Nonlinear Hyperbolic Problems, Eds. C. Carasso, P.-A. Raviart, and D. Serre, Lecture Notes in Mathematics 1270, New York: Springer–Verlag, 1987, 136–149.

Bibliography

[11] Panel on Mathematical Sciences, Board on Mathematical Sciences, Commission on Physical Sciences, Mathematics, and Resources, National Research Council. *Mathematical Sciences: A Unifying and Dynamic Resource*. Washington, D.C.: National Academy Press, 1986.

[12] Matsumura A., Nishida T. *The initial value problem for the equations of motion of viscous and heat-conductive gases*. J. Math. Kyoto Univ., 20 (1980), 67–104.

[13] Leray J. *Essai sur les mouvements plans d'un liquide visqueux que limitent des parois*. J. Math. Pures Appl., 13 (1934), 331–418.

[14] Temam R. *Infinite-Dimensional Dynamical Systems in Mechanics and Physics*. Applied Mathematical Sciences 68, New York: Springer–Verlag, 1988.

[15] Ladyzhenskaya O. A. *The Mathematical Theory of Viscous Incompressible Flow*, Second Edition. New York: Gordon and Breach, 1969.

[16] Constantin P., Foias C. *Navier–Stokes Equations*. Chicago Lectures in Mathematics, Chicago: The University of Chicago Press, 1988.

Chapter 3
Magnetohydrodynamics

3.1 Plasma

Magnetohydrodynamics is the study of the motion of a conductive fluid such as plasma in the electromagnetic field. First, we give a preliminary introduction to plasma, the fourth state of matter, and display some of its properties to be used hereafter.

We already know that any matter may exist in different states such as solid, liquid, or gas under different conditions on temperature, pressure, and so on. These are three common aggregative states of matter. They can transform into each other under certain conditions. If the temperature is increased continuously, the state of matter may still change. Under high temperature (tens of thousands or hundreds of thousands of degrees), the movement between molecules and atoms of the matter will be very intense. Far from being mutually bound with each other, some outer electrons in the atom can break away from their nucleus to become free electrons since they have considerably large kinetic energies, while the atom turns into a positively charged ion when it loses those free electrons. In this way, the matter becomes a mixture of electrons and ions. It is not only different from solid and liquid but also essentially different from ordinary gas. It is a brand-new aggregative state of matter called *plasma* or the fourth state of matter.

To be specific, as the ionized gas (whether partially or completely ionized), the plasma has something in common with the ordinary gas, for example, some macroscopic physical quantities describing the ordinary gas, such as density, temperature, pressure, and so on, are still applicable to the plasma (the ionized gas), but an essential change has taken place in its principal properties. It is a kind of electrically conductive fluid with high conductivity, and the influence from the electromagnetic field can be obviously seen on it. And as long as more than a thousandth of the gas is ionized, its behavior is then dominated by the Coulomb force between the ions and electrons, while the interaction between neutral particles takes second place. Since the total number of negative charges of electrons is equal to that of positive charges of ions, macroscopically, this state of aggregation presents electrically neutral, and so there comes the word "plasma."

From the above we know that the plasma is exactly the ionized gas, which is made up of electrons, ions, and neutral particles. Because the energy of the thermal motion of the gas at normal temperature is not large and no spontaneous ionization occurs, the matter in our environment usually exists in three states as solid, liquid, and gas, instead of in the fourth state as plasma. However, in the endless universe, more than 99% of matters are plasmas. In fact, the temperature in the core of the sun is up to tens of millions of degrees, and the matter there exists in the form of plasma. Some sun-like stars, galaxies, and matter in boundless interstellar space are all plasmas. However, there are very few "cold planets" such as earth in the universe.

At the beginning of 19th century, physicists began to consider the existence of the fourth state of matter that is essentially different from the three known aggregative states of matter. In 1879, British physicist Crookes studied the discharge process of the vacuum tube and pointed out for the first time the existence of the fourth state of matter. In 1929, Langmuir and Tanks first introduced the term *plasma* to denote the fourth state of matter, i.e., a completely or partially ionized physical state. The ionosphere was studied from the 1930s to the end of World War II and was shown to be composed of plasmas. Magnetohydrodynamics was established in the 1940s by H. Alfvén and was successfully applied to space physics and astrophysics, for instance, to the study of the galaxy, nebulae, and the sun, which are all composed of plasmas at extremely high temperatures and can always be handled as continuum media in the research on massive (large scale) processes no matter how rarefied. The objects of these studies are all limited to outer space and the earth's upper atmosphere and seem too far away from us. If restricted to only these applications, the study of plasma physics and magnetohydrodynamics may not be greatly pushed forward.

Later, in the summer of 1945, two atomic bombs exploded over Hiroshima and Nagasaki in Japan. Over the explosion for the first time on earth there formed artificial plasmas with temperature up to millions of degrees on the surface of the earth! In 1952, a hydrogen bomb was successfully exploded in the United States, and it caused the same effect over the earth—an uncontrollable nuclear fusion reaction. Then in the early 1950s, the former Soviet Union, the United States, and Britain each independently and secretly launched controllable nuclear fusion research, and a new climax was reached in the research on plasma physics and magnetohydrodynamics.

The research on controllable nuclear fusion is very intriguing. Seawater is taken as raw materials to extract deuterium (a stable isotope of hydrogen), and at extremely high temperatures of nearly 100 million degrees, a thermonuclear reaction occurs and releases vast amounts of energy. This energy has inexhaustible raw materials with little threat of radioactive pollution, and hence it is a very ideal new energy. If such research is successful, the energy crisis faced by human beings will be ultimately solved. It depends on the intensive study of high-temperature plasma physics. At the same time, along with the launch of man-made satellites and space rockets and the invention of radio astronomical telescopes, observational data have been greatly enriched in space physics and astrophysics, which further promotes the study of plasma physics. It can be said that astrophysics and controllable nuclear fusion research are the two driving forces in the new and developing discipline of magnetohydrodynamics.

A fundamental characteristic of plasma is *electric neutrality*. From what we discussed above, the plasma made up of electrons, ions, and neutral particles is macroscopically electrically neutral; that is, the total number of negative charges of electrons is equal to that of positive charges of ions. Moreover, the electric neutrality of plasma assumed to hold in any infinitesimal element, namely, the numbers of positive and negative charges are equal everywhere, and the algebraic sum of the positive and negative charges is zero in any infinitesimal element. This is because the plasma is very sensitive to breaks in electric neutrality. Once there appears a charge separation somewhere in the plasma, it immediately generates a tremendously large electric field to recover the electric neutrality. Take hydrogen plasma, for example. We denote by n_e and n_i, respectively, the concentrations of electrons and ions (the number of electrons and ions per unit volume), and the electric neutrality condition is $n_e \approx n_i$. Even if there is a very slight charge separation in plasma, i.e., the electric neutrality is slightly broken, it will produce a huge electric field. We consider, for instance, a ball with a radius of 10 mm, in which the concentration of plasma is $n_e \approx n_i \approx 10^{20}/\text{m}^3$. Suppose that one percent of electrons move out of the ball due to a certain perturbation; then the positive charges carried by ions become more than negative charges. These excess positive charges produce a potential distribution. Computation tells us that the potential difference between the spherical surface and center (only 10 mm apart!) will be up to 640k volts, which is incredible and impossible. Therefore, the numbers of positive and negative charges in the plasma have to be equal everywhere, and no divergence is allowed. Of course, this is not absolute, as the thermal motion of charged particles will always slightly affect electric neutrality. Nevertheless, as long as the electric field caused by the excessive charge is too weak to affect most particles in the plasma, the plasma can still be regarded as meeting the electric neutrality condition. Henceforth, we will follow this fact.

From (1.180) in Chapter 1, we know that for the medium there holds

$$\boldsymbol{B} = \mu \boldsymbol{H},$$

where $\mu = \mu_r \mu_0$ is the permeability, μ_0 is the permeability in a vacuum, and $\mu_r = 1 + \chi_m$ is the relative permeability. For the plasma, the susceptibility $\chi_m \approx 0$ due to the extremely high temperature, so the relative permeability $\mu_r \approx 1$. For simplicity, we will always assume $\mu_r = 1$ hereinafter; then

$$\boldsymbol{B} = \mu_0 \boldsymbol{H}. \tag{3.1}$$

In addition, it is difficult to set up a strong electric field $\boldsymbol{E}$ in the plasma since it is a good conductor. If there is a strong electric field, it will immediately lead to the motion and redistribution of electrons and charged particles, which will offset the electric field. Hence, $\boldsymbol{E}$ is a small quantity compared to $\boldsymbol{H}$. Hereinafter, $\boldsymbol{H}$ will be the focus of discussion. Therefore, this kind of mechanics for electromagnetic fluids is generally called *magnetohydrodynamics*.

3.2 System of Magnetohydrodynamics

We consider a continuous fluid medium such as plasma, which is a kind of fluid with good conductivity because of the presence of large amounts of charged particles. Its movement in the electromagnetic field generates electric current and then induces the electromagnetic field to change the original electromagnetic field; on the other hand, the existence of the electromagnetic field produces force (Lorentz force) on the moving conductive plasma, which will affect the motion of the conductive plasma. Therefore, it gives rise to a complicated interaction between the electromagnetic phenomenon and the fluid dynamical phenomenon. Hence, the motion of the plasma in the electromagnetic field should abide by both the basic law of motion in the electromagnetic field and law of motion in fluid dynamics. On one hand, the impact of electromagnetic force must be added into the fluid dynamics equations; on the other hand, the electric current produced by the motion of the conductive fluid (plasma) must be added into Maxwell's equations describing the electromagnetic field. Then we have to appropriately couple fluid dynamics equations and Maxwell's equations. Such a joint system is the fundamental system of magnetohydrodynamics, called the *magnetohydrodynamics system*.

3.2.1 Amended Form of Maxwell's Equations Considering the Motion of Conductive Medium (Plasma)

Maxwell's equations in the medium are (see (1.183)–(1.186) in Chapter 1)

$$\text{div } \boldsymbol{D} = \rho_f, \tag{3.2}$$

$$\text{rot } \boldsymbol{E} = -\frac{\partial \boldsymbol{B}}{\partial t}, \tag{3.3}$$

$$\text{div } \boldsymbol{B} = 0, \tag{3.4}$$

$$\text{rot } \boldsymbol{H} = \frac{\partial \boldsymbol{D}}{\partial t} + \boldsymbol{j}_f, \tag{3.5}$$

where $\boldsymbol{D} = \varepsilon \boldsymbol{E}$, $\boldsymbol{B} = \mu \boldsymbol{H}$.

From (3.1), (3.3) and (3.4) can be written, respectively, as

$$\text{rot } \boldsymbol{E} = -\mu_0 \frac{\partial \boldsymbol{H}}{\partial t}, \tag{3.6}$$

$$\text{div } \boldsymbol{H} = 0. \tag{3.7}$$

In the case of static medium, we have Ohm's law in differential form (see (1.243) in Chapter 1)

$$\boldsymbol{j}_f = \sigma \boldsymbol{E},$$

where σ is the electric conductivity of the medium. Here $\boldsymbol{E}$ is the force acting on the unit positive charge at this point, which determines the corresponding free current density vector $\boldsymbol{j}_f$. Now the conductive medium itself is in motion with velocity $\boldsymbol{u}$. From the Lorentz force formula (see (1.64) in Chapter 1), the force acting on the unit positive charge attached to this moving medium is

$$\boldsymbol{E} + \boldsymbol{u} \times \boldsymbol{B} = \boldsymbol{E} + \mu_0 \boldsymbol{u} \times \boldsymbol{H}.$$

Then it can be assumed that

$$\boldsymbol{j}_f = \sigma(\boldsymbol{E} + \mu_0 \boldsymbol{u} \times \boldsymbol{H}), \tag{3.8}$$

where σ is assumed to be a constant, and $\sigma \gg 1$ since the plasma is a good conductor.

Now consider (3.5). Plugging (3.8) into it yields

$$\text{rot } \boldsymbol{H} = \frac{\partial \boldsymbol{D}}{\partial t} + \sigma(\boldsymbol{E} + \mu_0 \boldsymbol{u} \times \boldsymbol{H}). \tag{3.9}$$

3.2 System of Magnetohydrodynamics

Take the electric field of oscillating form $\boldsymbol{E} = \mathrm{e}^{\mathrm{i}\omega t}\boldsymbol{E}_0$, where $\boldsymbol{E}_0$ is independent of t. Then the displacement current term (ε is assumed to be constant) is

$$\frac{\partial \boldsymbol{D}}{\partial t} = \mathrm{i}\varepsilon\omega\mathrm{e}^{\mathrm{i}\omega t}\boldsymbol{E}_0,$$

while

$$\sigma\boldsymbol{E} = \sigma\mathrm{e}^{\mathrm{i}\omega t}\boldsymbol{E}_0.$$

It can be seen from the above two formulas that, if $\sigma \gg \varepsilon\omega$, i.e., $\frac{\sigma}{\omega} \gg \varepsilon$, the displacement current term does not take effect. Since $\sigma \gg 1$, the above assumption always holds except for oscillation with extremely high frequency. Therefore, we can ignore the displacement current term in (3.5) (i.e., (3.9)) and write it as

$$\operatorname{rot}\boldsymbol{H} = \sigma(\boldsymbol{E} + \mu_0 \boldsymbol{u} \times \boldsymbol{H}). \tag{3.10}$$

From the process of deriving Maxwell's equations, the introduction of the displacement current is to match the charge conservation equation $\frac{\partial \rho_f}{\partial t} + \operatorname{div}\boldsymbol{j}_f = 0$. Now this assumption means that $\operatorname{div}\boldsymbol{j}_f = 0$, which can be obtained by using the divergence operator div on both sides of $\operatorname{rot}\boldsymbol{H} = \boldsymbol{j}_f$. This implies that charges are impossible to accumulate (that is, for any region, in any fixed time period, the inflow charges are equal to the outflow charges), which is consistent with the electric neutrality of the plasma in any infinitesimal element.

The purpose of the following arguments is to eliminate $\boldsymbol{E}$ in the equations in order to highlight $\boldsymbol{H}$, which plays a crucial role in magnetohydrodynamics, and to obtain the equations involving only $\boldsymbol{H}$. To this end, it comes from (3.10) that

$$\boldsymbol{E} = \frac{1}{\sigma}\operatorname{rot}\boldsymbol{H} - \mu_0 \boldsymbol{u} \times \boldsymbol{H}. \tag{3.11}$$

Plugging this into (3.6), we obtain

$$\frac{\partial \boldsymbol{H}}{\partial t} = -\frac{1}{\sigma\mu_0}\operatorname{rot}\operatorname{rot}\boldsymbol{H} + \operatorname{rot}(\boldsymbol{u} \times \boldsymbol{H}). \tag{3.12}$$

Since $\operatorname{rot}\operatorname{rot}\boldsymbol{H} = \operatorname{grad}\operatorname{div}\boldsymbol{H} - \Delta\boldsymbol{H}$, and noting (3.7), we have

$$\frac{\partial \boldsymbol{H}}{\partial t} - \operatorname{rot}(\boldsymbol{u} \times \boldsymbol{H}) = \frac{1}{\sigma\mu_0}\Delta\boldsymbol{H} \tag{3.13}$$

and

$$\operatorname{div}\boldsymbol{H} = 0. \tag{3.14}$$

These are the equations that the magnetic field intensity $\boldsymbol{H}$ should satisfy, but the motion velocity $\boldsymbol{u}$ of the medium is included. They cannot be solved independently and have to be solved jointly with the equations of fluid dynamics.

Here we point out that (3.14) can actually be reduced to the requirement of the initial values. In fact, if we have the initial conditions

$$t = 0: \boldsymbol{H} = \boldsymbol{H}_0(x_1, x_2, x_3)$$

with div $\boldsymbol{H}_0 = 0$, then by using the divergence operator div on both sides of (3.13), and noting div rot $\equiv 0$, we have

$$\frac{\partial}{\partial t}(\text{div}\,\boldsymbol{H}) = \frac{1}{\sigma\mu_0}\Delta(\text{div}\,\boldsymbol{H}),$$
$$t = 0 : \text{div}\,\boldsymbol{H}(= \text{div}\,\boldsymbol{H}_0) = 0.$$

Equation (3.14) follows immediately from the uniqueness of solutions to the Cauchy problem of the heat equation.

3.2.2 Amended Form of Fluid Dynamics System Considering the Existence of Electromagnetic Field

For the plasma, the physical quantities such as density, temperature, pressure, and velocity still make sense. The conservation law of mass is still the *continuity equation* (see formula (2.132) in Chapter 2)

$$\frac{\partial \rho}{\partial t} + \text{div}(\rho \boldsymbol{u}) = 0. \tag{3.15}$$

But when considering the conservation laws of momentum and energy, in order to regard the fluid and the electromagnetic filed as a whole, we should put the momentum and energy and their corresponding momentum flux density tensor and energy flux density vector of the electromagnetic field in the equations of momentum and energy for the fluid.

For this purpose, we first give some simplification and an explanation for the above electromagnetic quantities, combining them with the features of the plasma.

From section 1.7 of Chapter 1, we know that, for the case of $\mu_r = 1$, the electromagnetic momentum density vector is

$$\frac{1}{c^2}\boldsymbol{S} = \frac{1}{c^2}\boldsymbol{E} \times \boldsymbol{H},$$

the electromagnetic momentum flux density tensor is

$$\frac{1}{2}(\varepsilon E^2 + \mu_0 H^2)\boldsymbol{I} - \varepsilon \boldsymbol{E} \otimes \boldsymbol{E} - \mu_0 \boldsymbol{H} \otimes \boldsymbol{H},$$

the electromagnetic energy density is

$$\frac{1}{2}(\varepsilon E^2 + \mu_0 H^2),$$

and the electromagnetic flux density vector is $\boldsymbol{S} = \boldsymbol{E} \times \boldsymbol{H}$.

In the current situation, combining with the features of the plasma, we will give some simplification to the above electromagnetic quantities just as we have done to the Maxwell's equations (and to match this simplification).

Just as was stated in the end of the last section, since the plasma is a good conductor, the electric filed $\boldsymbol{E}$ (compared with $\boldsymbol{H}$) is a small quantity, and thus the term of E^2 can be neglected compared with H^2. Furthermore, the electromagnetic momentum density vector

3.2 System of Magnetohydrodynamics

possesses an E and a c^2 in the denominator, and its order of magnitude is very small and can be neglected, too. Therefore, we have that

- the electromagnetic momentum density vector is $\mathbf{0}$,
- the electromagnetic momentum flux density tensor is $\mu_0\left(\frac{1}{2}H^2 I - H \otimes H\right)$,
- the electromagnetic energy density is $\frac{1}{2}\mu_0 H^2$,
- the electromagnetic flux density vector is $S = E \times H$.

Now we establish the conservation equation of momentum. From what we stated above, we need only add the momentum term, entering the domain under consideration due to the electromagnetic momentum flux, into the conservation equation of the fluid dynamics. Then the equation can be written as (see formula (2.133) in Chapter 2)

$$\frac{\partial}{\partial t}(\rho \mathbf{u}) + \mathrm{div}\,(\rho \mathbf{u} \otimes \mathbf{u} - P) - \mu_0 \mathrm{div}\left(H \otimes H - \frac{1}{2}H^2 I\right) = \rho F. \qquad (3.16)$$

Denoting

$$\mathbf{\Pi} = \{\pi_{ij}\} = \rho \mathbf{u} \otimes \mathbf{u} - P - \mu_0\left(H \otimes H - \frac{1}{2}H^2 I\right), \qquad (3.17)$$

where $P = \{p_{ij}\}$ is determined by the generalized Newton's law in Chapter 2 (see formula (2.130) in that chapter), (3.16) can be written as

$$\frac{\partial}{\partial t}(\rho \mathbf{u}) + \mathrm{div}\,\mathbf{\Pi} = \rho F. \qquad (3.18)$$

It can be proved from direct calculation that

$$\mathrm{div}\left(H \otimes H - \frac{1}{2}H^2 I\right) = \mathrm{rot}\, H \times H. \qquad (3.19)$$

In fact, denoting $H = (H_1, H_2, H_3)$ and noting (3.14), we have

the first component on the left-hand side of (3.19)

$$= \frac{\partial}{\partial x_1}\left(H_1^2 - \frac{1}{2}H^2\right) + \frac{\partial}{\partial x_2}(H_1 H_2) + \frac{\partial}{\partial x_3}(H_1 H_3)$$

$$= H_2\left(\frac{\partial H_1}{\partial x_2} - \frac{\partial H_2}{\partial x_1}\right) + H_3\left(\frac{\partial H_1}{\partial x_3} - \frac{\partial H_3}{\partial x_1}\right)$$

$$= \text{the first component on the right-hand side of (3.19).}$$

Similarly, we can prove that the second and third components on both sides of (3.19) are equal. Thus, substituting (3.19) into (3.16), we get

$$\frac{\partial}{\partial t}(\rho \mathbf{u}) + \mathrm{div}(\rho \mathbf{u}\mathbf{u} - P) - \mu_0 \mathrm{rot}\, H \times H = \rho F. \qquad (3.20)$$

This equation can also be obtained by regarding the force acting on the conductive fluid by the electromagnetic field as the external force, and by bringing its impulse into the equilibrium equation of momentum when considering the system of fluid independently. In fact, for any infinitesimal element dx of the fluid, from the Lorentz force formula, the force received from the electromagnetic field should be

$$(\widetilde{\rho}\boldsymbol{E} + \boldsymbol{j}_f \times \boldsymbol{B})dx,$$

where $\widetilde{\rho}$ represents the density of volume charges, and $\boldsymbol{j}_f$ is the density of conductive currents. Noting that the plasma is electrically neutral in any volume element (see section 3.1), we have $\widetilde{\rho} = 0$; noting (3.1) as well, this force is then

$$\mu_0 \boldsymbol{j}_f \times \boldsymbol{H} dx.$$

Together with (3.5) (i.e., (3.8) and (3.10)), ignoring the displacement current

$$\operatorname{rot} \boldsymbol{H} = \boldsymbol{j}_f,$$

this force can also be written as

$$\mu_0 \operatorname{rot} \boldsymbol{H} \times \boldsymbol{H} dx,$$

where no term involving $\boldsymbol{E}$ appears. Thus, in domain Ω and in time period $[t_1, t_2]$, the impulse of the electromagnetic field force is

$$\int_{t_1}^{t_2} \int_{\Omega} \mu_0 \operatorname{rot} \boldsymbol{H} \times \boldsymbol{H} dx dt.$$

Adding this term into the conservation equation (see (2.133) in Chapter 2) leads to (3.20).

By using continuity equation (3.15) and plugging in the following expression of $\boldsymbol{P}$,

$$p_{ij} = -p\delta_{ij} + \overline{\mu}\left(\frac{\partial u_i}{\partial x_j} + \frac{\partial u_j}{\partial x_i} - \frac{2}{3}\operatorname{div}\boldsymbol{u}\delta_{ij}\right)$$
$$+ \overline{\mu}'\operatorname{div}\boldsymbol{u}\delta_{ij}, \tag{3.21}$$

(see (2.130) in Chapter 2; hereinafter, to avoid confusion with the magnetic permeability, the viscosity coefficients μ and μ' are denoted by $\overline{\mu}$ and $\overline{\mu}'$, respectively), the momentum equation (3.20) can be written as

$$\frac{du_i}{dt} + \frac{1}{\rho}\frac{\partial p}{\partial x_i} = \frac{1}{\rho}\sum_{j=1}^{3}\frac{\partial}{\partial x_j}\left(\overline{\mu}\left(\frac{\partial u_i}{\partial x_j} + \frac{\partial u_j}{\partial x_i}\right)\right)$$
$$+ \frac{1}{\rho}\frac{\partial}{\partial x_i}\left(\left(\overline{\mu}' - \frac{2}{3}\overline{\mu}\right)\sum_{l=1}^{3}\frac{\partial u_l}{\partial x_l}\right) - \frac{\mu_0}{\rho}(\boldsymbol{H} \times \operatorname{rot}\boldsymbol{H})_i + F_i$$
$$(i = 1, 2, 3) \tag{3.22}$$

or

$$\rho\frac{d\boldsymbol{u}}{dt} - \operatorname{div}\boldsymbol{P} - \mu_0 \operatorname{rot}\boldsymbol{H} \times \boldsymbol{H} = \rho\boldsymbol{F}, \tag{3.23}$$

which is exactly the present *Euler equation*, and $\frac{d}{dt}$ is defined as in (2.17) in Chapter 2.

3.2 System of Magnetohydrodynamics

Now consider the energy equation. Here we need to add the contributions of the electromagnetic energy density and the electromagnetic energy flux density vector into the original energy equilibrium equation of the fluid. Plugging the expression of E solved from (3.11) into the electromagnetic energy flux density vector, we have

$$S = E \times H$$
$$= \frac{1}{\sigma} \operatorname{rot} H \times H - \mu_0 (u \times H) \times H.$$

Then, we can obtain the following energy equation in the form of a conservation law (see (2.139) in Chapter 2):

$$\frac{\partial}{\partial t}\left(\rho e + \frac{1}{2}\rho u^2 + \frac{1}{2}\mu_0 H^2\right) + \operatorname{div}\left(\left(\rho e + \frac{1}{2}\rho u^2\right)u - Pu\right)$$
$$+ \operatorname{div}\left(\frac{1}{\sigma}\operatorname{rot} H \times H - \mu_0(u \times H) \times H\right)$$
$$= \operatorname{div}(\kappa \operatorname{grad} T) + \rho F \cdot u. \tag{3.24}$$

By using the continuity equation (3.15) and the expression (3.21) of the tensor P, the above formula can be reduced into (see (2.141) in Chapter 2)

$$\rho \frac{d(e + \frac{1}{2}u^2)}{dt} + \operatorname{div}(p u)$$
$$- \sum_{i=1}^{3} \frac{\partial}{\partial x_i}\left(\overline{\mu} \sum_{j=1}^{3}\left(\frac{\partial u_i}{\partial x_j} + \frac{\partial u_j}{\partial x_i}\right)u_j + \left(\overline{\mu}' - \frac{2}{3}\overline{\mu}\right)u_i \operatorname{div} u\right)$$
$$+ \mu_0 H \cdot \frac{\partial H}{\partial t} + \operatorname{div}\left(\frac{1}{\sigma}\operatorname{rot} H \times H - \mu_0(u \times H) \times H\right)$$
$$= \operatorname{div}(\kappa \operatorname{grad} T) + \rho F \cdot u. \tag{3.25}$$

By using Euler equation (3.22), together with (3.12), (3.25) can be rewritten as

$$\rho \frac{de}{dt} + p \operatorname{div} u - \overline{\mu} \sum_{i,j=1}^{3} \left(\frac{\partial u_i}{\partial x_j} + \frac{\partial u_j}{\partial x_i}\right)\frac{\partial u_j}{\partial x_i}$$
$$- \left(\overline{\mu}' - \frac{2}{3}\overline{\mu}\right)(\operatorname{div} u)^2 - \frac{1}{\sigma} H \cdot \operatorname{rot} \operatorname{rot} H$$
$$+ \mu_0 H \cdot \operatorname{rot}(u \times H) - \mu_0 H \times \operatorname{rot} H \cdot u$$
$$+ \operatorname{div}\left(\frac{1}{\sigma}\operatorname{rot} H \times H - \mu_0(u \times H) \times H\right)$$
$$= \operatorname{div}(\kappa \operatorname{grad} T). \tag{3.26}$$

To further simplify the above formula, we first prove the following two formulas:

$$\boldsymbol{H} \cdot \text{rot rot } \boldsymbol{H} - \text{div}(\text{rot } \boldsymbol{H} \times \boldsymbol{H}) = |\text{rot } \boldsymbol{H}|^2, \tag{3.27}$$

$$\boldsymbol{H} \cdot \text{rot}(\boldsymbol{u} \times \boldsymbol{H}) - \text{div}((\boldsymbol{u} \times \boldsymbol{H}) \times \boldsymbol{H})$$
$$- \boldsymbol{H} \times \text{rot } \boldsymbol{H} \cdot \boldsymbol{u} = 0. \tag{3.28}$$

First, using the formula $\text{div}(\boldsymbol{a} \times \boldsymbol{b}) = \boldsymbol{b} \cdot \text{rot } \boldsymbol{a} - \boldsymbol{a} \cdot \text{rot } \boldsymbol{b}$ in vector analysis, we can immediately obtain (3.27), and the left-hand side of (3.28) can be reduced to

$$(\boldsymbol{u} \times \boldsymbol{H}) \cdot \text{rot } \boldsymbol{H} - \boldsymbol{H} \times \text{rot } \boldsymbol{H} \cdot \boldsymbol{u}.$$

From the property $(\boldsymbol{a} \times \boldsymbol{b}) \cdot \boldsymbol{c} = (\boldsymbol{b} \times \boldsymbol{c}) \cdot \boldsymbol{a}$ of the mixed product of vectors, the above quantity is zero, which proves (3.28).

Noting (3.27) and (3.28), the energy conservation equation (3.26) can also be written as

$$\rho \frac{de}{dt} + p \, \text{div } \boldsymbol{u} - \overline{\mu} \sum_{i,j=1}^{3} \left(\frac{\partial u_i}{\partial x_j} + \frac{\partial u_j}{\partial x_i} \right) \frac{\partial u_j}{\partial x_i}$$
$$- \left(\overline{\mu}' - \frac{2}{3}\overline{\mu} \right) (\text{div } \boldsymbol{u})^2 - \frac{1}{\sigma} |\text{rot } \boldsymbol{H}|^2 = \text{div}(\kappa \, \text{grad } T). \tag{3.29}$$

Just as in the fluid dynamics equations, the internal energy e can be replaced by the entropy S in (3.29) by means of the thermodynamics relation $de = T\,dS - p\,d\tau$. In fact,

$$\rho \frac{de}{dt} + p \, \text{div } \boldsymbol{u} = \rho \frac{de}{dt} - \frac{p}{\rho} \frac{d\rho}{dt}$$
$$= \rho \left(\frac{de}{dt} + p \frac{d\tau}{dt} \right) = \rho T \frac{dS}{dt}. \tag{3.30}$$

In the derivation process of the above formulas, we used the continuity equation (3.15). Plugging (3.30) into (3.29), we obtain

$$\rho T \frac{dS}{dt} - \overline{\mu} \sum_{i,j=1}^{3} \left(\frac{\partial u_i}{\partial x_j} + \frac{\partial u_j}{\partial x_i} \right) \frac{\partial u_j}{\partial x_i}$$
$$- \left(\overline{\mu}' - \frac{2}{3}\overline{\mu} \right) (\text{div } \boldsymbol{u})^2 - \frac{1}{\sigma} |\text{rot } \boldsymbol{H}|^2$$
$$= \text{div}(\kappa \, \text{grad } T). \tag{3.31}$$

This is exactly the corresponding heat transmission equation.

It can be noticed that (3.31) can be obtained by the heat transmission equation (see (2.142) and (2.22)–(2.24) in Chapter 2)

$$\rho T \frac{dS}{dt} - \overline{\mu} \sum_{i,j=1}^{3} \left(\frac{\partial u_i}{\partial x_j} + \frac{\partial u_j}{\partial x_i} \right) \frac{\partial u_j}{\partial x_i}$$
$$- \left(\overline{\mu}' - \frac{2}{3}\overline{\mu} \right) (\text{div } \boldsymbol{u})^2 = \text{div}(\kappa \, \text{grad } T) \tag{3.32}$$

3.2 System of Magnetohydrodynamics

in fluid dynamics plus the Joule heat term caused by the electric current. To this end, we first derive the expression of *Joule heat*. First, we assume that the conductive medium is static. There is a steady current in it, with current density j_f and electric field intensity E. Then, we have

$$j_f = \sigma E,$$

where σ is the electric conductivity. Next we will compute the Joule heat of unit volume of medium in unit time produced by the current. It is transferred from the mechanical work done to the moving charges in the conductor by the electric field maintaining the current. We can easily see from section 1.4 in Chapter 1 that it is equal to $j_f \cdot E$ (this is left to the reader as an exercise). Then the heat issued by per unit volume in unit time is

$$j_f \cdot E = j_f^2/\sigma.$$

This is called the *Joule–Lenz law*. In the case when the conductive medium is in motion, the energy dissipation (say, giving out heat) generated by the current in the conductor should be independent of the motion of the conductor and depends only on the current j_f. Therefore, the Joule heat density issued by the moving conductor in unit time has the same expression as above, i.e., j_f^2/σ, when expressed by the current density. But from (3.5) and ignoring the displacement current, we have $j_f = \mathrm{rot}\, H$, so

$$\frac{j_f^2}{\sigma} = \frac{1}{\sigma}|\mathrm{rot}\, H|^2.$$

Adding this term into the heat transmission equation (3.32) yields the required equation (3.31).

3.2.3 System of Magnetohydrodynamics

Combining equations from the above two sections, i.e., (3.13)–(3.14), (3.15), (3.20), and (3.24), we obtain the *magnetohydrodynamics system* for the general cases

$$\frac{\partial H}{\partial t} - \mathrm{rot}(u \times H) = \frac{1}{\sigma \mu_0} \Delta H, \tag{3.33}$$

$$\mathrm{div}\, H = 0, \tag{3.34}$$

$$\frac{\partial \rho}{\partial t} + \mathrm{div}(\rho u) = 0, \tag{3.35}$$

$$\frac{\partial}{\partial t}(\rho u) + \mathrm{div}(\rho u u - P) - \mu_0 \mathrm{rot}\, H \times H = \rho F, \tag{3.36}$$

$$\frac{\partial}{\partial t}\left(\rho e + \frac{1}{2}\rho u^2 + \frac{1}{2}\mu_0 H^2\right) + \mathrm{div}\left(\left(\rho e + \frac{1}{2}\rho u^2\right)u - Pu\right)$$

$$+ \mathrm{div}\left(\frac{1}{\sigma}\mathrm{rot}\, H \times H - \mu_0(u \times H) \times H\right)$$

$$= \mathrm{div}(\kappa\, \mathrm{grad}\, T) + \rho F \cdot u, \tag{3.37}$$

where (3.34) can be reduced to the requirement of the initial values.

Replacing (3.36) and (3.37) with (3.22) and (3.31), respectively, the above system can also be written as

$$\frac{\partial \boldsymbol{H}}{\partial t} - \text{rot}(\boldsymbol{u} \times \boldsymbol{H}) = \frac{1}{\sigma \mu_0} \Delta \boldsymbol{H}, \tag{3.38}$$

$$\text{div}\, \boldsymbol{H} = 0, \tag{3.39}$$

$$\frac{\partial \rho}{\partial t} + \text{div}(\rho \boldsymbol{u}) = 0, \tag{3.40}$$

$$\frac{du_i}{dt} + \frac{1}{\rho}\frac{\partial p}{\partial x_i} = \frac{1}{\rho}\sum_{j=1}^{3}\frac{\partial}{\partial x_j}\left(\overline{\mu}\left(\frac{\partial u_i}{\partial x_j} + \frac{\partial u_j}{\partial x_i}\right)\right)$$

$$+ \frac{1}{\rho}\frac{\partial}{\partial x_i}\left(\left(\overline{\mu}' - \frac{2}{3}\overline{\mu}\right)\text{div}\,\boldsymbol{u}\right)$$

$$- \frac{\mu_0}{\rho}(\boldsymbol{H} \times \text{rot}\,\boldsymbol{H})_i + F_i \quad (i = 1, 2, 3), \tag{3.41}$$

$$\rho T \frac{dS}{dt} - \overline{\mu}\sum_{i,j=1}^{3}\left(\frac{\partial u_i}{\partial x_j} + \frac{\partial u_j}{\partial x_i}\right)\frac{\partial u_j}{\partial x_i}$$

$$- \left(\overline{\mu}' - \frac{2}{3}\overline{\mu}\right)(\text{div}\,\boldsymbol{u})^2 - \frac{1}{\sigma}|\text{rot}\,\boldsymbol{H}|^2$$

$$= \text{div}(\kappa\,\text{grad}\,T), \tag{3.42}$$

where $\frac{d}{dt}$ is defined as in (2.17) in Chapter 2.

In the above equations, the unknown variables can be taken as (say) $\boldsymbol{H} = (H_1, H_2, H_3)$, ρ, $\boldsymbol{u} = (u_1, u_2, u_3)$, and T, together with the equation of state $p = p(\rho, T)$, and thus we get a closed system of equations.

Equation (3.38) implies that the magnetic field is diffusive because of the existence of conductivity σ (whose value is finite). In particular, for the static medium, (3.38) turns into the *diffusion equation*

$$\frac{\partial \boldsymbol{H}}{\partial t} = \frac{1}{\sigma \mu_0} \Delta \boldsymbol{H}. \tag{3.43}$$

From the expression of the solution to the diffusion (heat conduction) equation, if the characteristic scale of the space change of the magnetic field $\boldsymbol{H}$ is L_0, then the characteristic time of the diffusion of the magnetic field is

$$\tau = \sigma \mu_0 L_0^2.$$

For instance, for the magnetic field of the earth, $\frac{1}{\sigma \mu_0} \approx 1\text{m}^2/\text{s}$ corresponding to the conductive fluid in the earth's core, the radius of the earth is $L_0 \approx 6 \times 10^6$m, the characteristic diffusion time of the earth's magnetic field is $\tau \approx 4 \times 10^{12}\text{s} \approx 10^6$ years. The age of the earth is approximately 10^9 years, which is much bigger than the characteristic diffusion time of the earth's magnetic field. That is to say, if there existed an original magnetic field when the earth was formed, it diffused into outer space during the first 10^6 years. Therefore, it

3.2 System of Magnetohydrodynamics

is impossible that the current earth magnetic field is the original magnetic field that existed when the earth was formed, and there must exist some motion process in the interior of the earth to generate and maintain the current earth magnetic field.

3.2.4 System of Incompressible Magnetohydrodynamics

Now we assume that the moving fluid is incompressible. Then the continuity equation (3.40) is reduced to

$$\text{div}\, \boldsymbol{u} = 0. \tag{3.44}$$

Using the above formula, (3.38) can be reduced to

$$\frac{d\boldsymbol{H}}{dt} - (\boldsymbol{H} \cdot \nabla)\boldsymbol{u} = \frac{1}{\sigma \mu_0} \Delta \boldsymbol{H}, \tag{3.45}$$

where $\frac{d}{dt}$ is defined as in (2.17) in Chapter 2, and $\nabla = (\frac{\partial}{\partial x_1}, \frac{\partial}{\partial x_2}, \frac{\partial}{\partial x_3}) = \text{grad}$ is the gradient operator. In fact, using the following formula about the rotation of vector:

$$\text{rot}(\boldsymbol{a} \times \boldsymbol{b}) = (\boldsymbol{b} \cdot \nabla)\boldsymbol{a} - (\boldsymbol{a} \cdot \nabla)\boldsymbol{b}$$
$$+ \boldsymbol{a}\, \text{div}\, \boldsymbol{b} - \boldsymbol{b}\, \text{div}\, \boldsymbol{a},$$

where

$$\boldsymbol{b} \cdot \nabla = b_1 \frac{\partial}{\partial x_1} + b_2 \frac{\partial}{\partial x_2} + b_3 \frac{\partial}{\partial x_3},$$

and noticing (3.39), we have

$$\text{rot}(\boldsymbol{u} \times \boldsymbol{H}) = (\boldsymbol{H} \cdot \nabla)\boldsymbol{u} - (\boldsymbol{u} \cdot \nabla)\boldsymbol{H} - \boldsymbol{H}\, \text{div}\, \boldsymbol{u}.$$

Then (3.38) can be written as

$$\frac{\partial \boldsymbol{H}}{\partial t} - (\boldsymbol{H} \cdot \nabla)\boldsymbol{u} + (\boldsymbol{u} \cdot \nabla)\boldsymbol{H} + \boldsymbol{H}\, \text{div}\, \boldsymbol{u} = \frac{1}{\sigma \mu_0} \Delta \boldsymbol{H},$$

i.e.,

$$\frac{d\boldsymbol{H}}{dt} - (\boldsymbol{H} \cdot \nabla)\boldsymbol{u} + \boldsymbol{H}\, \text{div}\, \boldsymbol{u} = \frac{1}{\sigma \mu_0} \Delta \boldsymbol{H}. \tag{3.46}$$

Equation (3.45) follows immediately from (3.44) and the above formula.

Similarly to the case of Navier–Stokes equations, we assume $\overline{\mu}$ to be a constant, and assume, without loss of generality, that $\rho = 1$, together with (3.44), and we can reduce the momentum equation (3.41) to

$$\frac{d\boldsymbol{u}}{dt} + \mathrm{grad}\, p = \overline{\mu}\Delta\boldsymbol{u} - \mu_0 \boldsymbol{H} \times \mathrm{rot}\, \boldsymbol{H} + \boldsymbol{F}. \tag{3.47}$$

Noting (3.19) and

$$\mathrm{div}\left(\frac{1}{2}H^2 \boldsymbol{I}\right) = \frac{1}{2}\mathrm{grad}\, H^2 = \frac{1}{2}\nabla H^2,$$

$$\mathrm{div}(\boldsymbol{H} \otimes \boldsymbol{H}) = (\boldsymbol{H}\cdot\nabla)\boldsymbol{H},$$

(3.47) can also be written as

$$\frac{d\boldsymbol{u}}{dt} + \nabla p = \overline{\mu}\Delta\boldsymbol{u} - \frac{1}{2}\mu_0 \nabla H^2 + \mu_0(\boldsymbol{H}\cdot\nabla)\boldsymbol{H} + \boldsymbol{F},$$

i.e.,

$$\frac{d\boldsymbol{u}}{dt} = -\nabla\left(p + \frac{1}{2}\mu_0 H^2\right) + \mu_0(\boldsymbol{H}\cdot\nabla)\boldsymbol{H} + \overline{\mu}\Delta\boldsymbol{u} + \boldsymbol{F}. \tag{3.48}$$

Now, just as in the case of Navier–Stokes equations, if we do not need to know the temperature distribution of the fluid, we can neglect the energy equation and obtain the incompressible magnetohydrodynamics system

$$\mathrm{div}\, \boldsymbol{H} = 0, \tag{3.49}$$

$$\frac{d\boldsymbol{H}}{dt} - (\boldsymbol{H}\cdot\nabla)\boldsymbol{u} = \frac{1}{\sigma\mu_0}\Delta\boldsymbol{H}, \tag{3.50}$$

$$\mathrm{div}\, \boldsymbol{u} = 0, \tag{3.51}$$

$$\frac{d\boldsymbol{u}}{dt} = -\nabla\left(p + \frac{1}{2}\mu_0 H^2\right) + \mu_0(\boldsymbol{H}\cdot\nabla)\boldsymbol{H} + \overline{\mu}\Delta\boldsymbol{u} + \boldsymbol{F}, \tag{3.52}$$

where (3.49) can be reduced to the requirement of the initial values, and p can be solved up to an arbitrarily given function of t.

This is the system of equations corresponding to the Navier–Stokes equations in the case of magnetohydrodynamics.

3.3 System of Magnetohydrodynamics When the Conductivity σ Is Infinite

3.3.1 System of Magnetohydrodynamics When the Conductivity σ Is Infinite

Since plasma is a good conductor, σ is usually taken to be considerably large. Now we consider its limit case: $\sigma = +\infty$. Then the diffusion phenomenon of the magnetic field will

3.3 System of Magnetohydrodynamics When the Conductivity σ Is Infinite

not appear, and the magnetohydrodynamics system (3.38)–(3.42) can be reduced to

$$\frac{\partial \boldsymbol{H}}{\partial t} - \operatorname{rot}(\boldsymbol{u} \times \boldsymbol{H}) = \boldsymbol{0}, \tag{3.53}$$

$$\operatorname{div} \boldsymbol{H} = 0, \tag{3.54}$$

$$\frac{\partial \rho}{\partial t} + \operatorname{div}(\rho \boldsymbol{u}) = 0, \tag{3.55}$$

$$\frac{du_i}{dt} + \frac{1}{\rho} \frac{\partial p}{\partial x_i} = \frac{1}{\rho} \sum_{j=1}^{3} \frac{\partial}{\partial x_j} \left(\overline{\mu} \left(\frac{\partial u_i}{\partial x_j} + \frac{\partial u_j}{\partial x_i} \right) \right)$$

$$+ \frac{1}{\rho} \frac{\partial}{\partial x_i} \left(\left(\overline{\mu}' - \frac{2}{3}\overline{\mu} \right) \operatorname{div} \boldsymbol{u} \right)$$

$$- \frac{\mu_0}{\rho} (\boldsymbol{H} \times \operatorname{rot} \boldsymbol{H})_i + F_i \quad (i = 1, 2, 3), \tag{3.56}$$

$$\rho T \frac{dS}{dt} - \overline{\mu} \sum_{i,j=1}^{3} \left(\frac{\partial u_i}{\partial x_j} + \frac{\partial u_j}{\partial x_i} \right) \frac{\partial u_j}{\partial x_i}$$

$$- \left(\overline{\mu}' - \frac{2}{3}\overline{\mu} \right) (\operatorname{div} \boldsymbol{u})^2 = \operatorname{div}(\kappa \operatorname{grad} T). \tag{3.57}$$

By using (3.46), equation (3.53) can also be written as

$$\frac{d\boldsymbol{H}}{dt} - (\boldsymbol{H} \cdot \nabla)\boldsymbol{u} + \boldsymbol{H} \operatorname{div} \boldsymbol{u} = \boldsymbol{0}. \tag{3.58}$$

Now we illustrate two important properties for the electromagnetic fluid with $\sigma = +\infty$.

3.3.2 Differential Expression with Respect to the Time of the Flux of a Vector Field Across Any Surface Moving with the Fluid and Its Applications

Suppose that there is a vector field $\boldsymbol{a}(t, x_1, x_2, x_3)$ and that surface S moves with the fluid velocity $\boldsymbol{u}$ (at each point). We are going to discuss the rate of change with respect to the time

$$\frac{d}{dt} \int_S \boldsymbol{a} \cdot \boldsymbol{n} dS$$

of the flux

$$\int_S \boldsymbol{a} \cdot \boldsymbol{n} dS$$

of $\boldsymbol{a}$ across S along its normal direction $\boldsymbol{n}$.

The change with respect to time t of the above flux is caused by two factors. One is that $\boldsymbol{a}$ itself changes with t, while the other is that the surface S changes with t. The contribution of the first part is obviously

$$\int_S \frac{\partial \boldsymbol{a}}{\partial t} \cdot \boldsymbol{n} dS. \tag{3.59}$$

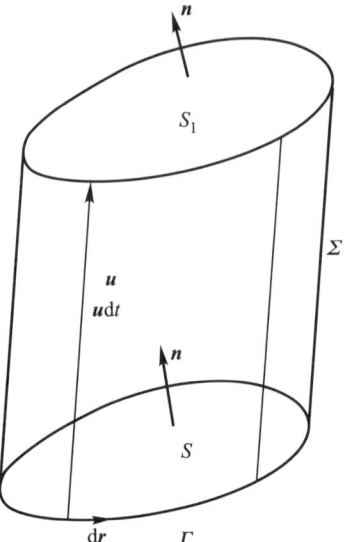

Figure 3.1.

Now we compute the contribution of the second part to the total rate of change. At this moment we assume that a does not change (i.e., freezes the time), and S changes to the position of S_1 after time dt. What we are going to compute is

$$\frac{1}{dt}\left(\int_{S_1} a \cdot n dS - \int_S a \cdot n dS\right),$$

where n takes the same direction for both S and S_1.

Suppose Γ is the boundary curve of S. After time dt, it sweeps a surface Σ, which, together with S and S_1, constitutes a closed surface (see Figure 3.1). We use Green's formula in the space region V surrounded by this closed surface to obtain

$$\int_V \text{div}\, a\, dx = \int_{S_1} a \cdot n dS - \int_S a \cdot n dS + \int_\Sigma a \cdot d\Sigma, \qquad (3.60)$$

where

$$d\Sigma = dr \times u dt,$$

and dr is the directional curve element on Γ. Obviously, we have

$$\int_V \text{div}\, a\, dx = \int_S (\text{div}\, a) u \cdot n dS dt, \qquad (3.61)$$

while

$$\int_\Sigma a \cdot d\Sigma = \int_\Gamma a \cdot (dr \times u dt)$$
$$= \int_\Gamma (u \times a) \cdot dr dt = \int_S \text{rot}(u \times a) \cdot n dS dt. \qquad (3.62)$$

3.3 System of Magnetohydrodynamics When the Conductivity σ Is Infinite 147

Here in the last step we used the Stokes formula. From (3.60)–(3.62), we have

$$\frac{1}{dt}\left(\int_{S_1} \boldsymbol{a}\cdot\boldsymbol{n}\mathrm{d}S - \int_S \boldsymbol{a}\cdot\boldsymbol{n}\mathrm{d}S\right)$$
$$= \int_S ((\mathrm{div}\,\boldsymbol{a})\boldsymbol{u} - \mathrm{rot}(\boldsymbol{u}\times\boldsymbol{a}))\cdot\boldsymbol{n}\mathrm{d}S.$$

Combining (3.59) with the contribution given by the above formula, we get

$$\frac{\mathrm{d}}{\mathrm{d}t}\int_S \boldsymbol{a}\cdot\boldsymbol{n}\mathrm{d}S = \int_S \left(\frac{\partial \boldsymbol{a}}{\partial t} + (\mathrm{div}\,\boldsymbol{a})\boldsymbol{u} - \mathrm{rot}(\boldsymbol{u}\times\boldsymbol{a})\right)\cdot\boldsymbol{n}\mathrm{d}S. \qquad (3.63)$$

This is exactly the required formula.

Now we apply the above formula to the magnetic field $\boldsymbol{H}$ with $\sigma = +\infty$. Noting (3.53), (3.54) and taking $\boldsymbol{a} = \boldsymbol{H}$ in (3.63), we have

$$\frac{\mathrm{d}}{\mathrm{d}t}\int_S \boldsymbol{H}\cdot\boldsymbol{n}\mathrm{d}S = 0. \qquad (3.64)$$

Therefore, in the case of $\sigma = +\infty$, we obtain the following *conservation law of magnetic field*: **When the conductivity is infinite, the magnetic flux across any surface S moving with the medium is conserved, i.e., independent of time t**. This conclusion is called the *Alfvén theorem*.

3.3.3 "Frozen" Principle of Magnetic Field Lines

One of the fundamental problems in magnetohydrodynamics is determining the change of the magnetic field intensity when the conductive medium is in motion. This problem can usually be solved with the aid of the following "frozen" principle of magnetic field lines. This principle claims that **in the case of $\sigma = +\infty$ (infinitely conductive), particles distributed on any given magnetic field line at the initial time will stay on the same magnetic field line in the process of motion, i.e., the magnetic field lines seem to be "frozen" on the matter.**

To prove this, it is easy to see, from equation (3.58) and the continuity equation (3.55), that

$$\frac{1}{\rho}\frac{\mathrm{d}\boldsymbol{H}}{\mathrm{d}t} - \frac{1}{\rho^2}\boldsymbol{H}\frac{\mathrm{d}\rho}{\mathrm{d}t} = \left(\frac{\boldsymbol{H}}{\rho}\cdot\nabla\right)\boldsymbol{u},$$

i.e.,

$$\frac{\mathrm{d}}{\mathrm{d}t}\left(\frac{\boldsymbol{H}}{\rho}\right) = \left(\frac{\boldsymbol{H}}{\rho}\cdot\nabla\right)\boldsymbol{u}. \qquad (3.65)$$

On the other hand, let us look at any given curve moving with the fluid. We assume that there is an infinitesimal element vector $\mathrm{d}\boldsymbol{r}$ on this curve at time t, and now we inspect its change with time. If we suppose that $\boldsymbol{u}$ is the velocity of the fluid at the beginning of $\mathrm{d}\boldsymbol{r}$, then the velocity of the fluid at the end of $\mathrm{d}\boldsymbol{r}$ is $\boldsymbol{u} + (\mathrm{d}\boldsymbol{r}\cdot\nabla)\boldsymbol{u}$. So in the time period $\mathrm{d}t$, $\mathrm{d}\boldsymbol{r}$ will turn into

$$\mathrm{d}\boldsymbol{r} + (\mathrm{d}\boldsymbol{r}\cdot\nabla)\boldsymbol{u}\mathrm{d}t$$

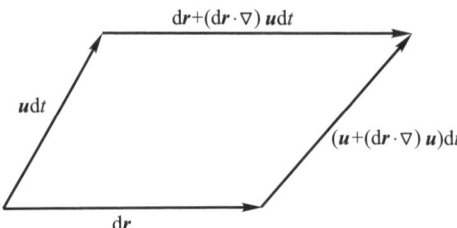

Figure 3.2.

(see Figure 3.2). Thus, the rate of change of $d\boldsymbol{r}$ with respect to t is

$$\frac{d}{dt}(d\boldsymbol{r}) = (d\boldsymbol{r} \cdot \nabla)\boldsymbol{u}. \tag{3.66}$$

It can be seen from (3.65) and (3.66) that $\frac{\boldsymbol{H}}{\rho}$ and $d\boldsymbol{r}$ (for any fixed particle) satisfy a linear homogeneous system of ordinary differential equations of the same form. Therefore, if $\frac{\boldsymbol{H}}{\rho}$ and $d\boldsymbol{r}$ are parallel at the initial time so that, for any fixed particle, we have

$$\frac{\boldsymbol{H}}{\rho} = k d\boldsymbol{r},$$

where k is a constant when the particle is fixed, then the same relation must hold in later time. Hence, $\frac{\boldsymbol{H}}{\rho}$ and $d\boldsymbol{r}$ are parallel as well in later time, and the ratio of their lengths keeps the same value. In other words, if a curve is a magnetic field line at the initial time, then it will remain a magnetic field line in the process of motion in later time, and the ratio between $\frac{\boldsymbol{H}}{\rho}$ and $d\boldsymbol{r}$ will remain unchanged.

Therefore, each magnetic field line moves with the particles on this line, and magnetic field lines seem to be "frozen" on the fluid moving with them. On each point, $\frac{\boldsymbol{H}}{\rho}$ changes in proportion to the elongation of the magnetic field line. In particular, for the incompressible fluid, ρ is a constant, and $\boldsymbol{H}$ itself changes in proportion to the elongation of the magnetic field line; namely, the magnetic field intensity strengthens with the stretch of the magnetic field line. In summary, we can see that the change of the magnetic field can be depicted according to the motion of the fluid.

3.4 Mathematical Structure of Magnetohydrodynamics System

In the case that there exist viscosity and heat conduction ($\overline{\mu} > 0$, $\overline{\mu}' \geq 0$, $\kappa > 0$) in the fluid along with $\sigma \neq +\infty$, it can be seen from (3.41) and (3.42) that the corresponding magnetohydrodynamics equations can be obtained as long as the terms regarding $\boldsymbol{H}$ and the first-order partial derivatives of $\boldsymbol{H}$ with respect to spatial variables are added in the momentum and energy equations in fluid dynamics, while (3.38) is the diffusion (heat conduction) equation of $\boldsymbol{H}$, which has the form of a decoupled principal part with (3.41), (3.42) (not involving partial derivatives like $\frac{\partial}{\partial t}$ and $\frac{\partial^2}{\partial x_k^2}$ of $\boldsymbol{u}$ and T). It is not difficult

3.4 Mathematical Structure of Magnetohydrodynamics System

to see that the system of magnetohydrodynamics composed of (3.38) and (3.40)–(3.42) is a system with a first-order (symmetric) hyperbolic equation coupled with a symmetric system of parabolic equations, and is quasi-linear, so it is still a quasi-linear symmetric hyperbolic-parabolic coupled system.

In the case that there exist viscosity and heat conduction, while $\sigma = +\infty$, the energy equation (3.42) does not involve $\boldsymbol{H}$, and the momentum equation (3.41) contains the term $\frac{\mu_0}{\rho} \boldsymbol{H} \times \mathrm{rot}\, \boldsymbol{H}$ which involves $\boldsymbol{H}$ and the first-order partial derivatives of $\boldsymbol{H}$ with respect to spatial variables. Then, (3.38) is reduced into

$$\frac{\partial \boldsymbol{H}}{\partial t} - \mathrm{rot}(\boldsymbol{u} \times \boldsymbol{H}) = \boldsymbol{0}, \tag{3.67}$$

i.e.,

$$\frac{\mathrm{d}\boldsymbol{H}}{\mathrm{d}t} - (\boldsymbol{H} \cdot \nabla)\boldsymbol{u} + \boldsymbol{H}\,\mathrm{div}\,\boldsymbol{u} = \boldsymbol{0}. \tag{3.68}$$

This is a first-order system of $\boldsymbol{H}$ which takes a similar form with the mass conservation equation (3.40) and has a decoupled principal part. So now the magnetohydrodynamics system is composed of a first-order symmetric hyperbolic system with a decoupled principal part, coupled with a symmetric parabolic system, and it is still a quasi-linear symmetric hyperbolic-parabolic coupled system, except that the number of hyperbolic equations increases, while the number of parabolic equations decreases.

If there is no dissipation process in the fluid, then $\overline{\mu} = \overline{\mu}' = \kappa = 0$ and $\sigma = +\infty$; this kind of fluid is called an *ideal magnetofluid*. Now we focus on this case; i.e., we discuss the *ideal magnetohydrodynamics system*.

Here, we first write the momentum equation (3.56) as

$$\begin{aligned}
&\rho \frac{\partial u_1}{\partial t} + \sum_{k=1}^{3} \rho u_k \frac{\partial u_1}{\partial x_k} + \frac{\partial p}{\partial x_1} \\
&+ \mu_0 \left(H_2 \left(\frac{\partial H_2}{\partial x_1} - \frac{\partial H_1}{\partial x_2} \right) - H_3 \left(\frac{\partial H_1}{\partial x_3} - \frac{\partial H_3}{\partial x_1} \right) \right) \\
&= \rho F_1,
\end{aligned} \tag{3.69}$$

$$\begin{aligned}
&\rho \frac{\partial u_2}{\partial t} + \sum_{k=1}^{3} \rho u_k \frac{\partial u_2}{\partial x_k} + \frac{\partial p}{\partial x_2} \\
&+ \mu_0 \left(H_3 \left(\frac{\partial H_3}{\partial x_2} - \frac{\partial H_2}{\partial x_3} \right) - H_1 \left(\frac{\partial H_2}{\partial x_1} - \frac{\partial H_1}{\partial x_2} \right) \right) \\
&= \rho F_2,
\end{aligned} \tag{3.70}$$

$$\begin{aligned}
&\rho \frac{\partial u_3}{\partial t} + \sum_{k=1}^{3} \rho u_k \frac{\partial u_3}{\partial x_k} + \frac{\partial p}{\partial x_3} \\
&+ \mu_0 \left(H_1 \left(\frac{\partial H_1}{\partial x_3} - \frac{\partial H_3}{\partial x_1} \right) - H_2 \left(\frac{\partial H_3}{\partial x_2} - \frac{\partial H_2}{\partial x_3} \right) \right) \\
&= \rho F_3;
\end{aligned} \tag{3.71}$$

then in the same way as in section 2.1 of Chapter 2, write the mass conservation equation (3.55) as

$$\frac{1}{\rho\widetilde{c}^2}\frac{\partial p}{\partial t} + \sum_{k=1}^{3}\frac{\partial u_k}{\partial x_k} + \sum_{k=1}^{3}\frac{u_k}{\rho\widetilde{c}^2}\frac{\partial p}{\partial x_k} = 0, \quad (3.72)$$

where $\widetilde{c}^2 = \frac{\partial p}{\partial \rho}$ is the square of the local speed of sound (see (2.32) in Chapter 2; here we use $\widetilde{c}$ to denote the local speed of sound to avoid confusion with light speed c). Equation (3.53) can be written as

$$\mu_0 \frac{\partial H_1}{\partial t} + \sum_{k=1}^{3} \mu_0 u_k \frac{\partial H_1}{\partial x_k}$$
$$- \mu_0 \left(H_2 \frac{\partial u_1}{\partial x_2} + H_3 \frac{\partial u_1}{\partial x_3} \right) + \mu_0 H_1 \left(\frac{\partial u_2}{\partial x_2} + \frac{\partial u_3}{\partial x_3} \right)$$
$$= 0, \quad (3.73)$$

$$\mu_0 \frac{\partial H_2}{\partial t} + \sum_{k=1}^{3} \mu_0 u_k \frac{\partial H_2}{\partial x_k}$$
$$- \mu_0 \left(H_1 \frac{\partial u_2}{\partial x_1} + H_3 \frac{\partial u_2}{\partial x_3} \right) + \mu_0 H_2 \left(\frac{\partial u_1}{\partial x_1} + \frac{\partial u_3}{\partial x_3} \right)$$
$$= 0, \quad (3.74)$$

$$\mu_0 \frac{\partial H_3}{\partial t} + \sum_{k=1}^{3} \mu_0 u_k \frac{\partial H_3}{\partial x_k}$$
$$- \mu_0 \left(H_1 \frac{\partial u_3}{\partial x_1} + H_2 \frac{\partial u_3}{\partial x_2} \right) + \mu_0 H_3 \left(\frac{\partial u_1}{\partial x_1} + \frac{\partial u_2}{\partial x_2} \right)$$
$$= 0, \quad (3.75)$$

and the energy equation (3.57) can be simply replaced by the following conservation equation of entropy:

$$\frac{\partial S}{\partial t} + \sum_{k=1}^{3} u_k \frac{\partial S}{\partial x_k} = 0. \quad (3.76)$$

Hence, if we take $(u_1, u_2, u_3, p, H_1, H_2, H_3, S)$ as unknown variables and rearrange the system of ideal magnetohydrodynamics according to the above order, then system (3.69)–(3.76) can be written in the vector form

$$A_0 \frac{\partial U}{\partial t} + A_1 \frac{\partial U}{\partial x_1} + A_2 \frac{\partial U}{\partial x_2} + A_3 \frac{\partial U}{\partial x_3} = C, \quad (3.77)$$

where

$$U = (u_1, u_2, u_3, p, H_1, H_2, H_3, S)^{\mathrm{T}},$$
$$A_0 = \mathrm{diag}\{\rho, \rho, \rho, \rho^{-1}\widetilde{c}^{-2}, \mu_0, \mu_0, \mu_0, 1\},$$

3.4 Mathematical Structure of Magnetohydrodynamics System

$$A_1 = \begin{pmatrix} \rho u_1 & 0 & 0 & 1 & 0 & \mu_0 H_2 & \mu_0 H_3 & 0 \\ 0 & \rho u_1 & 0 & 0 & 0 & -\mu_0 H_1 & 0 & 0 \\ 0 & 0 & \rho u_1 & 0 & 0 & 0 & -\mu_0 H_1 & 0 \\ 1 & 0 & 0 & \frac{u_1}{\rho c^2} & 0 & 0 & 0 & 0 \\ 0 & 0 & 0 & 0 & u_1 & 0 & 0 & 0 \\ \mu_0 H_2 & -\mu_0 H_1 & 0 & 0 & 0 & u_1 & 0 & 0 \\ \mu_0 H_3 & 0 & -\mu_0 H_1 & 0 & 0 & 0 & u_1 & 0 \\ 0 & 0 & 0 & 0 & 0 & 0 & 0 & u_1 \end{pmatrix},$$

$$A_2 = \begin{pmatrix} \rho u_2 & 0 & 0 & 0 & -\mu_0 H_2 & 0 & 0 & 0 \\ 0 & \rho u_2 & 0 & 1 & \mu_0 H_1 & 0 & \mu_0 H_3 & 0 \\ 0 & 0 & \rho u_2 & 0 & 0 & 0 & -\mu_0 H_2 & 0 \\ 0 & 1 & 0 & \frac{u_2}{\rho c^2} & 0 & 0 & 0 & 0 \\ -\mu_0 H_2 & \mu_0 H_1 & 0 & 0 & u_2 & 0 & 0 & 0 \\ 0 & 0 & 0 & 0 & 0 & u_2 & 0 & 0 \\ 0 & \mu_0 H_3 & -\mu_0 H_2 & 0 & 0 & 0 & u_2 & 0 \\ 0 & 0 & 0 & 0 & 0 & 0 & 0 & u_2 \end{pmatrix},$$

$$A_3 = \begin{pmatrix} \rho u_3 & 0 & 0 & 0 & -\mu_0 H_3 & 0 & 0 & 0 \\ 0 & \rho u_3 & 0 & 0 & 0 & -\mu_0 H_3 & 0 & 0 \\ 0 & 0 & \rho u_3 & 1 & \mu_0 H_1 & \mu_0 H_2 & 0 & 0 \\ 0 & 0 & 1 & \frac{u_3}{\rho c^2} & 0 & 0 & 0 & 0 \\ -\mu_0 H_3 & 0 & \mu_0 H_1 & 0 & u_3 & 0 & 0 & 0 \\ 0 & -\mu_0 H_3 & \mu_0 H_2 & 0 & 0 & u_3 & 0 & 0 \\ 0 & 0 & 0 & 0 & 0 & 0 & u_3 & 0 \\ 0 & 0 & 0 & 0 & 0 & 0 & 0 & u_3 \end{pmatrix},$$

and

$$C = (\rho F_1, \rho F_2, \rho F_3, 0, 0, 0, 0, 0)^{\mathrm{T}}.$$

It is easy to see that, in the region where no vacuum occurs ($\rho \neq 0$), A_0 is a positively definite symmetric matrix, and A_1, A_2, A_3 are all symmetric matrices; therefore, the ideal magnetohydrodynamics system (3.77) is a first-order quasi-linear symmetric hyperbolic system.

The above arguments are restricted to the range where the solution is continuously differentiable. But because of the existence of electromagnetic shock waves, we have to consider discontinuous solutions. For this, we have to use the system in the form of conservation laws.

For the ideal magnetofluid, system (3.33)–(3.37) can be reduced to the following form:

$$\frac{\partial \rho}{\partial t} + \text{div}(\rho \boldsymbol{u}) = 0, \tag{3.78}$$

$$\frac{\partial}{\partial t}(\rho \boldsymbol{u}) + \text{div}(\rho \boldsymbol{u} \otimes \boldsymbol{u} - p\boldsymbol{I}) - \mu_0 \text{rot}\, \boldsymbol{H} \times \boldsymbol{H} = \rho \boldsymbol{F}, \tag{3.79}$$

$$\frac{\partial}{\partial t}\left(\rho e + \frac{1}{2}\rho u^2 + \frac{1}{2}\mu_0 H^2\right) + \text{div}\left(\left(\rho e + \frac{1}{2}\rho u^2\right)\boldsymbol{u} - p\boldsymbol{u}\right)$$
$$-\text{div}(\mu_0(\boldsymbol{u}\times\boldsymbol{H})\times\boldsymbol{H}) = \rho \boldsymbol{F}\cdot\boldsymbol{u}, \tag{3.80}$$

$$\frac{\partial \boldsymbol{H}}{\partial t} - \text{rot}(\boldsymbol{u}\times\boldsymbol{H}) = \boldsymbol{0}, \tag{3.81}$$

$$\text{div}\, \boldsymbol{H} = 0. \tag{3.82}$$

As mentioned before, (3.82) can be reduced to the requirement of the initial conditions, and we need only focus on (3.78)–(3.81). $\mu_0\,\text{rot}\,\boldsymbol{H}\times\boldsymbol{H}$ in (3.79) does not have divergence form, so this equation does not possess a form of conservation laws. But from (3.16), it is easy to rewrite (3.79) in the following form of conservation laws:

$$\frac{\partial}{\partial t}(\rho u_i) + \sum_{k=1}^{3} \frac{\partial}{\partial x_k}(\rho u_i u_k - \mu_0 H_i H_k)$$
$$+ \frac{\partial}{\partial x_i}\left(p + \frac{1}{2}\mu_0 H^2\right) = \rho F_i \quad (i=1,2,3). \tag{3.83}$$

Hence, the ideal magnetohydrodynamics system composed of equations (3.78), (3.83), (3.80), and (3.81) is exactly a first-order quasi-linear system of partial differential equations in the form of conservation laws.

However, can the above system be reduced to a first-order symmetric hyperbolic system that is still in the form of conservation laws by means of introducing new unknown variables? This can be done with the help of the method used in section 2.1 of Chapter 2 for the system of ideal fluid dynamics. We refer the interested reader to reference [7].

3.5 System of One-Dimensional Magnetohydrodynamics

3.5.1 System of One-Dimensional Magnetohydrodynamics

When the quantities involved in the magnetohydrodynamics system depend only on time t and one spatial variable (e.g., x_1), this system is one-dimensional. Generally speaking, now we cannot expect, as in fluid dynamics, that the fluid flows only in the x_1 direction, namely, that the components of its velocity in the x_2 and x_3 directions are zero, since if so, the components H_2 and H_3 of the magnetic field would depend generically on x_2 and x_3.

For simplicity, denote x_1 by x. Then (3.39) is reduced to

$$\frac{\partial H_1}{\partial x} = 0; \tag{3.84}$$

3.5 System of One-Dimensional Magnetohydrodynamics

i.e., H_1 is independent of x. From this, it is easy to reduce the first equation in system (3.38) to

$$\frac{\partial H_1}{\partial t} = 0, \tag{3.85}$$

so H_1 is always a constant. Thus the other two equations in (3.38) can be written as

$$\frac{\partial H_2}{\partial t} + u_1 \frac{\partial H_2}{\partial x} + H_2 \frac{\partial u_1}{\partial x} - H_1 \frac{\partial u_2}{\partial x} = \frac{1}{\sigma \mu_0} \frac{\partial^2 H_2}{\partial x^2}, \tag{3.86}$$

$$\frac{\partial H_3}{\partial t} + u_1 \frac{\partial H_3}{\partial x} + H_3 \frac{\partial u_1}{\partial x} - H_1 \frac{\partial u_3}{\partial x} = \frac{1}{\sigma \mu_0} \frac{\partial^2 H_3}{\partial x^2}. \tag{3.87}$$

Besides, system (3.40)–(3.42) describing the motion of the fluid can be written as

$$\frac{\partial \rho}{\partial t} + u_1 \frac{\partial \rho}{\partial x} + \rho \frac{\partial u_1}{\partial x} = 0, \tag{3.88}$$

$$\frac{\partial u_1}{\partial t} + u_1 \frac{\partial u_1}{\partial x} + \frac{1}{\rho} \frac{\partial p}{\partial x} - \frac{1}{\rho} \frac{\partial}{\partial x} \left(\left(\frac{4}{3}\overline{\mu} + \overline{\mu}' \right) \frac{\partial u_1}{\partial x} \right)$$
$$+ \frac{\mu_0}{\rho} \left(H_2 \frac{\partial H_2}{\partial x} + H_3 \frac{\partial H_3}{\partial x} \right) = F_1, \tag{3.89}$$

$$\frac{\partial u_2}{\partial t} + u_1 \frac{\partial u_2}{\partial x} - \frac{1}{\rho} \frac{\partial}{\partial x} \left(\overline{\mu} \frac{\partial u_2}{\partial x} \right) - \frac{\mu_0}{\rho} H_1 \frac{\partial H_2}{\partial x} = F_2, \tag{3.90}$$

$$\frac{\partial u_3}{\partial t} + u_1 \frac{\partial u_3}{\partial x} - \frac{1}{\rho} \frac{\partial}{\partial x} \left(\overline{\mu} \frac{\partial u_3}{\partial x} \right) - \frac{\mu_0}{\rho} H_1 \frac{\partial H_3}{\partial x} = F_3, \tag{3.91}$$

$$\rho T \frac{\partial S}{\partial t} + \rho T u_1 \frac{\partial S}{\partial x} - \left(\frac{4}{3}\overline{\mu} + \overline{\mu}' \right) \left(\frac{\partial u_1}{\partial x} \right)^2 - \overline{\mu} \left(\frac{\partial u_2}{\partial x} \right)^2$$
$$- \overline{\mu} \left(\frac{\partial u_3}{\partial x} \right)^2 = \frac{\partial}{\partial x} \left(\kappa \frac{\partial T}{\partial x} \right). \tag{3.92}$$

In particular, for the ideal magnetofluid, namely, in the case of $\overline{\mu} = \overline{\mu}' = \kappa = 0$ with $\sigma = +\infty$, we get

$$\frac{\partial H_2}{\partial t} + u_1 \frac{\partial H_2}{\partial t} + H_2 \frac{\partial u_1}{\partial x} - H_1 \frac{\partial u_2}{\partial x} = 0, \tag{3.93}$$

$$\frac{\partial H_3}{\partial t} + u_1 \frac{\partial H_3}{\partial x} + H_3 \frac{\partial u_1}{\partial x} - H_1 \frac{\partial u_3}{\partial x} = 0, \tag{3.94}$$

$$\frac{\partial \rho}{\partial t} + u_1 \frac{\partial \rho}{\partial x} + \rho \frac{\partial u_1}{\partial x} = 0, \tag{3.95}$$

$$\frac{\partial u_1}{\partial t} + u_1 \frac{\partial u_1}{\partial x} + \frac{\widetilde{c}^2}{\rho} \frac{\partial \rho}{\partial x} + \frac{1}{\rho} \frac{\partial p}{\partial S} \frac{\partial S}{\partial x}$$
$$+ \frac{\mu_0}{\rho} \left(H_2 \frac{\partial H_2}{\partial x} + H_3 \frac{\partial H_3}{\partial x} \right) = F_1, \tag{3.96}$$

$$\frac{\partial u_2}{\partial t} + u_1 \frac{\partial u_2}{\partial x} - \frac{\mu_0}{\rho} H_1 \frac{\partial H_2}{\partial x} = F_2, \tag{3.97}$$

$$\frac{\partial u_3}{\partial t} + u_1 \frac{\partial u_3}{\partial x} - \frac{\mu_0}{\rho} H_1 \frac{\partial H_3}{\partial x} = F_3, \tag{3.98}$$

$$\frac{\partial S}{\partial t} + u_1 \frac{\partial S}{\partial x} = 0, \tag{3.99}$$

where $\tilde{c}^2 = \frac{\partial p}{\partial \rho}$.

Let

$$U = (H_2, H_3, \rho, u_1, u_2, u_3, S)^{\mathrm{T}}.$$

The above system can be written as

$$\frac{\partial U}{\partial t} + A(U) \frac{\partial U}{\partial x} = C, \tag{3.100}$$

where

$$A(U) = \begin{pmatrix} u_1 & 0 & 0 & H_2 & -H_1 & 0 & 0 \\ 0 & u_1 & 0 & H_3 & 0 & -H_1 & 0 \\ 0 & 0 & u_1 & \rho & 0 & 0 & 0 \\ \frac{\mu_0}{\rho} H_2 & \frac{\mu_0}{\rho} H_3 & \frac{\tilde{c}^2}{\rho} & u_1 & 0 & 0 & \frac{1}{\rho}\frac{\partial p}{\partial S} \\ -\frac{\mu_0}{\rho} H_1 & 0 & 0 & 0 & u_1 & 0 & 0 \\ 0 & -\frac{\mu_0}{\rho} H_1 & 0 & 0 & 0 & u_1 & 0 \\ 0 & 0 & 0 & 0 & 0 & 0 & u_1 \end{pmatrix} \tag{3.101}$$

and

$$C = (0,0,0,F_1,F_2,F_3,0)^{\mathrm{T}}.$$

Of course, we can rewrite system (3.100) as a symmetric hyperbolic system just like in section 3.4; however, we will consider its hyperbolicity from another point of view by focusing on the particularity of the one-dimensional case. First, we discuss the eigenvalues of the matrix $A(U)$ given by (3.101). From

$$\det(A(U) - \lambda I) = \begin{vmatrix} u_1-\lambda & 0 & 0 & H_2 & -H_1 & 0 & 0 \\ 0 & u_1-\lambda & 0 & H_3 & 0 & -H_1 & 0 \\ 0 & 0 & u_1-\lambda & \rho & 0 & 0 & 0 \\ \frac{\mu_0}{\rho} H_2 & \frac{\mu_0}{\rho} H_3 & \frac{\tilde{c}^2}{\rho} & u_1-\lambda & 0 & 0 & \frac{1}{\rho}\frac{\partial p}{\partial S} \\ -\frac{\mu_0}{\rho} H_1 & 0 & 0 & 0 & u_1-\lambda & 0 & 0 \\ 0 & -\frac{\mu_0}{\rho} H_1 & 0 & 0 & 0 & u_1-\lambda & 0 \\ 0 & 0 & 0 & 0 & 0 & 0 & u_1-\lambda \end{vmatrix} = 0,$$

3.5 System of One-Dimensional Magnetohydrodynamics

after calculating, we can obtain

$$(u_1 - \lambda)\left((u_1 - \lambda)^2 - \frac{\mu_0}{\rho}H_1^2\right)\left\{((u_1 - \lambda)^2 - \widetilde{c}^2)\right.$$
$$\left. \cdot \left((u_1 - \lambda)^2 - \frac{\mu_0}{\rho}H_1^2\right) - \frac{\mu_0}{\rho}(u_1 - \lambda)^2(H_2^2 + H_3^2)\right\} = 0.$$

So we have

$$u_1 - \lambda = 0 \qquad (3.102)$$

or

$$(u_1 - \lambda)^2 - \frac{\mu_0}{\rho}H_1^2 = 0 \qquad (3.103)$$

or

$$((u_1 - \lambda)^2 - \widetilde{c}^2)\left((u_1 - \lambda)^2 - \frac{\mu_0}{\rho}H_1^2\right)$$
$$- \frac{\mu_0}{\rho}(u_1 - \lambda)^2(H_2^2 + H_3^2) = 0.$$

The above formula can also be written as

$$(u_1 - \lambda)^4 - \left(\widetilde{c}^2 + \frac{\mu_0}{\rho}H^2\right)(u_1 - \lambda)^2 + \frac{\mu_0 \widetilde{c}^2}{\rho}H_1^2 = 0. \qquad (3.104)$$

From (3.102), (3.103), and (3.104), the eigenvalues of $A(U)$ are given by

$$\begin{aligned} &\lambda_1 = u_1 - C_f, \quad \lambda_2 = u_1 - C_a, \quad \lambda_3 = u_1 - C_s, \\ &\lambda_4 = u_1, \quad \lambda_5 = u_1 + C_s, \quad \lambda_6 = u_1 + C_a, \\ &\lambda_7 = u_1 + C_f, \end{aligned} \qquad (3.105)$$

where

$$C_a^2 = \frac{\mu_0}{\rho}H_1^2,$$
$$C_{f,s}^2 = \frac{1}{2}\left\{\widetilde{c}^2 + \frac{\mu_0}{\rho}H^2 \pm \sqrt{\left(\widetilde{c}^2 + \frac{\mu_0}{\rho}H^2\right)^2 - \frac{4\mu_0 \widetilde{c}^2}{\rho}H_1^2}\right\}, \qquad (3.106)$$

in which C_f is given by taking the "+" sign, while C_s is given by taking the "−" sign. Thus, we obtain seven real eigenvalues of $A(U)$. It is not hard to verify directly that

$$C_s^2 \leq C_a^2 \leq C_f^2. \qquad (3.107)$$

C_f, C_s, and C_a are called the *fast, slow,* and *Alfvén characteristic speed*, respectively.

If $H_1 \neq 0$ and $H_2^2 + H_3^2 \neq 0$, it can be verified that the strict sign of inequality holds in (3.107), i.e.,

$$0 < C_s^2 < C_a^2 < C_f^2 \qquad (3.108)$$

(see exercise 7). Then, the seven eigenvalues given above are not equal to each other, and so the first-order quasi-linear system (3.100) is strictly hyperbolic.

If $H_1 = 0$ or $H_2 = H_3 = 0$, then matrix $A(U)$ has multiple eigenvalues. Here we discuss only the case of $H_1 = 0$,

$$\lambda_1 = u_1 - \sqrt{\tilde{c}^2 + \frac{\mu_0}{\rho}H^2}, \quad \lambda_7 = u_1 + \sqrt{\tilde{c}^2 + \frac{\mu_0}{\rho}H^2},$$
$$\lambda_2 = \lambda_3 = \lambda_4 = \lambda_5 = \lambda_6 = u_1,$$

and $\lambda = u_1$ is an eigenvalue with multiplicity 5. Now we show that, for the abovementioned seven eigenvalues, their corresponding left eigenvectors still constitute a complete set. Here, the left eigenvector (row vector) ζ corresponding to the eigenvalue λ of $A(U)$ is defined by the following formula:

$$\zeta A = \lambda \zeta.$$

To prove the completeness of the left eigenvectors, we need only discuss the case of multiple eigenvalue $\lambda = u_1$. Suppose that

$$\zeta = (\zeta_1, \zeta_2, \ldots, \zeta_7)$$

is the left eigenvector corresponding to $\lambda = u_1$; then we have

$$(\zeta_1, \zeta_2, \ldots, \zeta_7) \begin{pmatrix} 0 & 0 & 0 & H_2 & 0 & 0 & 0 \\ 0 & 0 & 0 & H_3 & 0 & 0 & 0 \\ 0 & 0 & 0 & \rho & 0 & 0 & 0 \\ \frac{\mu_0}{\rho}H_2 & \frac{\mu_0}{\rho}H_3 & \frac{\tilde{c}^2}{\rho} & 0 & 0 & 0 & \frac{1}{\rho}\frac{\partial p}{\partial S} \\ 0 & 0 & 0 & 0 & 0 & 0 & 0 \\ 0 & 0 & 0 & 0 & 0 & 0 & 0 \\ 0 & 0 & 0 & 0 & 0 & 0 & 0 \end{pmatrix} = 0.$$

It yields

$$\zeta_4 = 0,$$
$$H_2 \zeta_1 + H_3 \zeta_2 + \rho \zeta_3 = 0.$$

So there exist the following five linearly independent left eigenvectors:

$$\left(1, 0, -\frac{1}{\rho}H_2, 0, 0, 0, 0\right),$$
$$\left(0, 1, -\frac{1}{\rho}H_3, 0, 0, 0, 0\right),$$
$$(0, 0, 0, 0, 1, 0, 0),$$
$$(0, 0, 0, 0, 0, 1, 0),$$
$$(0, 0, 0, 0, 0, 0, 1).$$

3.5 System of One-Dimensional Magnetohydrodynamics

Therefore, now matrix $A(U)$ still has seven linearly independent left eigenvectors. According to the definition of one-dimensional first-order quasi-linear hyperbolic system in section 2.1.4 of Chapter 2, system (3.100) is still hyperbolic. But due to the multiple eigenvalues, the system is no longer strictly hyperbolic.

In summary, we conclude that, when $H_1 \neq 0$ and $H_2^2 + H_3^2 \neq 0$, the one-dimensional ideal magnetohydrodynamics system is strictly hyperbolic, while, when $H_1 = 0$, although its coefficient matrix $A(U)$ possesses an eigenvalue with multiplicity 5, the left eigenvectors still constitute a complete set, and then the system is still hyperbolic.

3.5.2 System of One-Dimensional Magnetohydrodynamics in Lagrangian Representation

For the system of one-dimensional magnetohydrodynamics, although u_2, u_3 may not be zero, we can still start from the continuity equation (3.88) to introduce Lagrangian coordinates (t', m), just as what we did in section 2.5 of Chapter 2. Using the transform rules (see (2.233), (2.234) in Chapter 2)

$$\frac{\partial}{\partial t} = \frac{\partial}{\partial t'} - \rho u_1 \frac{\partial}{\partial m}, \quad \frac{\partial}{\partial x} = \rho \frac{\partial}{\partial m},$$

from (3.86)–(3.92) we can obtain the following Lagrangian representation of the one-dimensional magnetohydrodynamics system:

$$\frac{\partial \tau}{\partial t'} - \frac{\partial u_1}{\partial m} = 0, \tag{3.109}$$

$$\frac{\partial H_2}{\partial t'} + \rho H_2 \frac{\partial u_1}{\partial m} - \rho H_1 \frac{\partial u_2}{\partial m} = \frac{1}{\sigma \mu_0} \rho \frac{\partial}{\partial m}\left(\rho \frac{\partial H_2}{\partial m}\right), \tag{3.110}$$

$$\frac{\partial H_3}{\partial t'} + \rho H_3 \frac{\partial u_1}{\partial m} - \rho H_1 \frac{\partial u_3}{\partial m} = \frac{1}{\sigma \mu_0} \rho \frac{\partial}{\partial m}\left(\rho \frac{\partial H_3}{\partial m}\right), \tag{3.111}$$

$$\frac{\partial u_1}{\partial t'} + \frac{\partial p}{\partial m} - \frac{\partial}{\partial m}\left(\left(\frac{4}{3}\overline{\mu} + \overline{\mu}'\right)\rho \frac{\partial u_1}{\partial m}\right)$$
$$+ \mu_0 \left(H_2 \frac{\partial H_2}{\partial m} + H_3 \frac{\partial H_3}{\partial m}\right) = F_1, \tag{3.112}$$

$$\frac{\partial u_2}{\partial t'} - \frac{\partial}{\partial m}\left(\overline{\mu} \rho \frac{\partial u_2}{\partial m}\right) - \mu_0 H_1 \frac{\partial H_2}{\partial m} = F_2, \tag{3.113}$$

$$\frac{\partial u_3}{\partial t'} - \frac{\partial}{\partial m}\left(\overline{\mu} \rho \frac{\partial u_3}{\partial m}\right) - \mu_0 H_1 \frac{\partial H_3}{\partial m} = F_3, \tag{3.114}$$

$$T \frac{\partial S}{\partial t'} - \left(\frac{4}{3}\overline{\mu} + \overline{\mu}'\right)\rho \left(\frac{\partial u_1}{\partial m}\right)^2 - \overline{\mu}\rho \left(\frac{\partial u_2}{\partial m}\right)^2$$
$$- \overline{\mu}\rho \left(\frac{\partial u_3}{\partial m}\right)^2 = \frac{\partial}{\partial m}\left(\kappa \rho \frac{\partial T}{\partial m}\right). \tag{3.115}$$

In particular, for the ideal magnetofluid, the equations in Lagrangian representation are given by

$$\frac{\partial \tau}{\partial t'} - \frac{\partial u_1}{\partial m} = 0, \tag{3.116}$$

$$\frac{\partial H_2}{\partial t'} + \rho H_2 \frac{\partial u_1}{\partial m} - \rho H_1 \frac{\partial u_2}{\partial m} = 0, \tag{3.117}$$

$$\frac{\partial H_3}{\partial t'} + \rho H_3 \frac{\partial u_1}{\partial m} - \rho H_1 \frac{\partial u_3}{\partial m} = 0, \tag{3.118}$$

$$\frac{\partial u_1}{\partial t'} + \frac{\partial}{\partial m}\left(p + \frac{1}{2}\mu_0(H_2^2 + H_3^2)\right) = F_1, \tag{3.119}$$

$$\frac{\partial u_2}{\partial t'} - \mu_0 H_1 \frac{\partial H_2}{\partial m} = F_2, \tag{3.120}$$

$$\frac{\partial u_3}{\partial t'} - \mu_0 H_1 \frac{\partial H_3}{\partial m} = F_3, \tag{3.121}$$

$$\frac{\partial S}{\partial t'} = 0. \tag{3.122}$$

This system has a relatively simple form.

Exercises

1. Suppose that the magnetic field H has only one nonzero component. Prove that

$$(H \cdot \nabla)H = 0.$$

2. Suppose that for the steady-state (namely, $\frac{\partial u}{\partial t} = 0$), incompressible (assume $\rho \equiv 1$) ideal fluid, the only volume force acting on it is the gravity. Suppose furthermore that the magnetic field satisfies the condition $(H \cdot \nabla)H = 0$. We take x_3 as the perpendicular coordinate starting from the ground and pointing upward. Prove that along the streamline we have

$$\frac{u^2}{2} + p + \frac{1}{2}\mu_0 H^2 + gx_3 = C,$$

where g is the gravitational acceleration and C is a constant along the same streamline.

3. Suppose that ϕ and A are the scalar and vector potentials of the electromagnetic field, respectively (see section 1.6 in Chapter 1). Prove that if ϕ and A satisfy the condition

$$\phi + \frac{1}{\sigma\mu_0} \operatorname{div} A = 0,$$

then (3.33) can be written in the following form:

$$\frac{\partial A}{\partial t} = u \times \operatorname{rot} A + \frac{1}{\sigma\mu_0}\Delta A.$$

4. Prove that for the ideal magnetofluid, the energy conservation equation (3.80) can be written in the following form:

$$\frac{\partial}{\partial t}\left(\rho e + \frac{1}{2}\rho u^2 + \frac{1}{2}\mu_0 H^2\right) + \sum_{k=1}^{3} \frac{\partial}{\partial x_k}\left\{\rho u_k\left(e + \frac{1}{2}u^2 - \frac{p}{\rho}\right)\right.$$
$$\left. + \mu_0 u_k H^2 - \mu_0 H_k(\boldsymbol{u} \cdot \boldsymbol{H})\right\} = \rho \boldsymbol{F} \cdot \boldsymbol{u}.$$

5. Reduce the one-dimensional ideal magnetohydrodynamics system (3.93)–(3.99) into the form of a first-order quasi-linear symmetric hyperbolic system.

6. Discuss the type of one-dimensional ideal magnetohydrodynamics system (3.116)–(3.122) in Lagrangian representation.

7. Prove that when $H_1 \neq 0$ and $H_2^2 + H_3^2 \neq 0$, the fast, slow, and Alfvén characteristic speeds C_f, C_s, and C_a satisfy

$$0 < C_s^2 < C_a^2 < C_f^2.$$

Bibliography

[1] Cowling T. G. *Magnetohydrodynamics*, Second Edition. London: Hilger, 1976.

[2] Department of Mathematics, Fudan University. *Continuum Mechanics* (in Chinese). Shanghai: Shanghai Press of Science and Technology, 1960.

[3] Landau L. D., Lifshitz E. M. *Electrodynamics of Continuous Media*, First English Edition. Reading, MA: Addison-Wesley, 1960.

[4] Hu Wenrui. *Cosmic Magnetohydrodynamics* (in Chinese). Beijing: Science Press, 1987.

[5] Polovin R. V., Demutskii V. P. *Fundamentals of Magnetohydrodynamics*. New York: Consultants Bureau, 1990.

[6] Moreau R. *Magnetohydrodynamics*. Dordrecht: Kluwer Academic Publishers, 1990.

[7] Godunov S. K. *Lois de conservation et intégrales d'énergie des équations hyperboliques*. In Nonlinear Hyperbolic Problems. Eds. C. Carasso, P.-A. Raviat, and D. Serre, Lecture Notes in Mathematics 1270, New York: Springer–Verlag, 1987, 135–149.

Chapter 4
Reacting Fluid Dynamics

4.1 Introduction

In this chapter we will discuss combustion phenomena from a general point of view. More specifically, we will discuss the case of combustible fluids accompanied by the combustion phenomenon in the flow process. Nowadays, this is a hot research topic in the field of applied mathematics, including the applied partial differential equations and computational mathematics, and is of great significance in both theory and application. To reveal the mechanism and laws of combustion phenomena, many people are engaged in this aspect of research, and many international conferences are held about this. So far, this is still a burgeoning subject.

There are some very interesting laws in combustion phenomena. Consider a pipeline full of combustible gas and ignited on one end. Under normal circumstances, the flame spreads forward at a low speed of a few meters per second; this is called *deflagration*. However, in some cases, the slow burning will turn into a very fast process, and the flame will spread forward at a speed of 2000 or more meters per second; this is called *detonation*.

It is amazing that there exist two types of combustion propagation processes so different from each other, but they are both observed from experiments. Besides gases, similar phenomena can be observed for solid explosives as well.

David Leonard Chapman and Jacques Charles Emile Jouguet gave a simplified but convincing explanation for the above phenomena in 1899 and 1905, respectively, which laid the foundation for combustion theory. Their simplifying assumption is that the chemical reaction occurs and completes instantaneously; that is, a wavefront goes into the unburned gas and instantaneously turns it into burned gas. Considering the conservation laws of mass, momentum, and energy in these processes, and noticing the difference in the chemical properties between the burned and the unburned gases, i.e., they have different chemical energy (the energy of atoms inside the molecules), and the expressions of the total energy and the momentum balance conditions need to be amended appropriately. The above phenomena can be explained after complicated computations (see [1]).

However, the Chapman–Jouguet assumption is extremely idealized. It is usually considered to be satisfied for detonation, while it is difficult to be met by slow burning (deflagration). Generally speaking, the fluid micelle is usually a mixture of burned and unburned gases in the combustion process, and it cannot be assumed that there exists a clear interface between the burned and unburned parts. In order to further reveal the laws of combustion phenomena, we have to start from the general framework of fluid dynamics involving the chemical reaction process, and then combine the specific features to give an explanation. This task is still far from being accomplished. In this chapter, we will focus on the establishment of the system of partial differential equations corresponding to this general framework, which will lay the foundation for further research.

4.2 System of Reacting Fluid Dynamics

4.2.1 System of Viscous and Heat-Conductive Reacting Fluid Dynamics

Now we establish the corresponding system of partial differential equations for the reacting fluid with viscosity and heat conduction.

Different from ordinary fluids, due to the combustion (chemical reaction) the reacting fluid micelle is usually made up of a mixture of the burned and unburned parts, and we will establish equations for this kind of mixed fluid. A simple way is to consider the mixed gases as a whole and to introduce the corresponding density ρ, velocity $\boldsymbol{u}$ (the two parts are assumed to move with the same macroscopic speed), pressure p, etc. Meanwhile, we introduce a function $Z = Z(t,\boldsymbol{x})$ ($\boldsymbol{x} = (x_1, x_2, x_3)$) to denote the percentage of the unburned fluid in the fluid micelle so as to distinguish these two kinds of fluids and thus to describe the transformation of the fluid from unburned to burned. Obviously, we have

$$0 \leq Z \leq 1; \tag{4.1}$$

$Z = 1$ means completely unburned, while $Z = 0$ means completely burned.

Here we need to pay our attention to the following:

(1) In the combustion process, energy is released through a chemical reaction, and this energy is transformed from the chemical energy of the unburned gas, so we need to take the chemical energy into consideration as well, which is different from the usual situation when we consider only the internal energy (the kinetic and potential energies of molecules) per unit mass. Therefore, instead of the previous internal energy e, we now introduce the *total energy*

$$E = e + g, \tag{4.2}$$

where g represents the chemical energy per unit mass. Usually the combustion process is exothermal, so the chemical energy of the burned gas is less than that of the unburned gas. We should consider the total energy when discussing the conservation law of energy.

(2) The equation of state of the fluid is generally relevant to the proportion Z of the unburned gas, since the mixed fluid is different when Z is different. Hence, we should have

$$p = p(\rho, T, Z), \tag{4.3}$$

etc.; likewise, the total energy E is also relevant to Z:

$$E = E(\rho, T, Z), \tag{4.4}$$

4.2 System of Reacting Fluid Dynamics

etc. When $Z = 1$, it is reduced to the equation of state of the unburned fluid while, when $Z = 0$, it is reduced to the equation of state of the burned fluid. The specific expressions of these equations of state are to be established, which will be further explained below.

Now we establish the reacting fluid dynamics system.

The conservation laws of mass and momentum still take the original forms (for mixed fluid)

$$\frac{\partial \rho}{\partial t} + \text{div}(\rho \boldsymbol{u}) = 0, \tag{4.5}$$

$$\frac{\partial}{\partial t}(\rho \boldsymbol{u}) + \text{div}(\rho \boldsymbol{u} \otimes \boldsymbol{u} - \boldsymbol{P}) = \rho \boldsymbol{F} \tag{4.6}$$

(see (2.132) and (2.133) in Chapter 2), where $\boldsymbol{P} = \{p_{ij}\}$ is given as in (2.130) in Chapter 2. In the conservation law of energy, e will be replaced by E, that is,

$$\frac{\partial}{\partial t}\left(\rho E + \frac{1}{2}\rho u^2\right) + \text{div}\left(\left(\rho E + \frac{1}{2}\rho u^2\right)\boldsymbol{u} - \boldsymbol{P}\boldsymbol{u}\right)$$
$$= \text{div}(\kappa \, \text{grad}\, T) + \rho \boldsymbol{F} \cdot \boldsymbol{u} \tag{4.7}$$

(see (2.139) in Chapter 2).

In addition to the above equations, we still need an equation of mass balance for the unburned fluid to determine Z.

The mass increment of the unburned fluid during the time period $[t, t + dt]$ in any given domain Ω is

$$\int_{\Omega} \rho Z(t+dt, \boldsymbol{x}) dx - \int_{\Omega} \rho Z(t, \boldsymbol{x}) dx = \int_{\Omega} \frac{\partial}{\partial t}(\rho Z) dx dt, \tag{4.8}$$

where $dx = dx_1 dx_2 dx_3$. It should be equal to the mass I of the unburned fluid coming into Ω through the boundary S in $[t, t + dt]$, subtracting the decrement II of the unburned mass due to the combustion in this time period. From the definition of the mass density vector,

$$\text{I} = -\int_S \rho Z \boldsymbol{u} \cdot \boldsymbol{n} dS dt = -\int_{\Omega} \text{div}(\rho Z \boldsymbol{u}) dx dt. \tag{4.9}$$

On the other hand, the mass of the unburned fluid to be burned in time period $[t, t + dt]$ in any infinitesimal element dx should be in proportion to the time dt and the mass $\rho Z dx$ of the unburned fluid, whose proportional coefficient is denoted by $\bar{k}$, which reflects the reaction rate possibly relevant to ρ, p, and Z. Then

$$\text{II} = \int_{\Omega} \bar{k}(\rho, p, Z) \rho Z dx dt. \tag{4.10}$$

Thus, from (4.8)–(4.10) we obtain

$$\int_{\Omega} \frac{\partial}{\partial t}(\rho Z) dx dt$$
$$= -\int_{\Omega} \text{div}(\rho Z \boldsymbol{u}) dx dt - \int_{\Omega} \bar{k}(\rho, p, Z) \rho Z dx dt. \tag{4.11}$$

The arbitrariness of Ω yields the equation reflecting the mass balance of the unburned fluid

$$\frac{\partial}{\partial t}(\rho Z) + \mathrm{div}(\rho Z \boldsymbol{u}) = -\bar{k}(\rho, p, Z)\rho Z. \tag{4.12}$$

Hence, we obtain the following *system of reacting fluid dynamics*:

$$\frac{\partial \rho}{\partial t} + \mathrm{div}(\rho \boldsymbol{u}) = 0, \tag{4.13}$$

$$\frac{\partial}{\partial t}(\rho \boldsymbol{u}) + \mathrm{div}(\rho \boldsymbol{u} \otimes \boldsymbol{u} - \boldsymbol{P}) = \rho \boldsymbol{F}, \tag{4.14}$$

$$\frac{\partial}{\partial t}\left(\rho E + \frac{1}{2}\rho u^2\right) + \mathrm{div}\left(\left(\rho E + \frac{1}{2}\rho u^2\right)\boldsymbol{u} - \boldsymbol{P}\boldsymbol{u}\right)$$
$$= \mathrm{div}(\kappa \,\mathrm{grad}\, T) + \rho \boldsymbol{F} \cdot \boldsymbol{u}, \tag{4.15}$$

$$\frac{\partial}{\partial t}(\rho Z) + \mathrm{div}(\rho Z \boldsymbol{u}) = -\bar{k}(\rho, p, Z)\rho Z, \tag{4.16}$$

where the unknown variables can be taken as, say, ρ, $\boldsymbol{u}$, T, and Z. Together with the equations of state

$$p = p(\rho, T, Z), \tag{4.17}$$
$$E = E(\rho, T, Z), \tag{4.18}$$

(4.13)–(4.16) constitute a closed system of equations.

4.2.2 Reduction of the System of Reacting Fluid Dynamics

Just as in the case without the chemical reaction, by using the continuity equation (4.13), the momentum equation (4.14) can be rewritten into

$$\frac{du_i}{dt} + \frac{1}{\rho}\frac{\partial p}{\partial x_i} = \frac{1}{\rho}\sum_{j=1}^{3}\frac{\partial}{\partial x_j}\left(\mu\left(\frac{\partial u_i}{\partial x_j} + \frac{\partial u_j}{\partial x_i}\right)\right)$$
$$+ \frac{1}{\rho}\frac{\partial}{\partial x_i}\left(\left(\mu' - \frac{2}{3}\mu\right)\sum_{l=1}^{3}\frac{\partial u_l}{\partial x_l}\right) + F_i \quad (i=1,2,3), \tag{4.19}$$

where we denote, just as in Chapter 2,

$$\frac{d}{dt} = \frac{\partial}{\partial t} + \sum_{j=1}^{3} u_j \frac{\partial}{\partial x_j}.$$

Equation (4.16) can also be rewritten as follows, using the equation of continuity:

$$\frac{dZ}{dt} = -\bar{k}(\rho, p, Z)Z. \tag{4.20}$$

4.2 System of Reacting Fluid Dynamics

The energy equation (4.15) can be reduced as follows after using the continuity equation (4.13) and the momentum equation (4.14):

$$\rho \frac{dE}{dt} + p \operatorname{div} \boldsymbol{u} - \mu \sum_{i,j=1}^{3} \left(\frac{\partial u_i}{\partial x_j} + \frac{\partial u_j}{\partial x_i} \right) \frac{\partial u_j}{\partial x_i}$$
$$- \left(\mu' - \frac{2}{3}\mu \right) (\operatorname{div} \boldsymbol{u})^2 = \operatorname{div}(k \operatorname{grad} T). \tag{4.21}$$

Using again the continuity equation, the above formula can also be written as

$$\frac{dE}{dt} + p \frac{d\tau}{dt} - \frac{\mu}{\rho} \sum_{i,j=1}^{3} \left(\frac{\partial u_i}{\partial x_j} + \frac{\partial u_j}{\partial x_i} \right) \frac{\partial u_j}{\partial x_i}$$
$$- \frac{1}{\rho} \left(\mu' - \frac{2}{3}\mu \right) (\operatorname{div} \boldsymbol{u})^2 = \frac{1}{\rho} \operatorname{div}(\kappa \operatorname{grad} T). \tag{4.22}$$

From the thermodynamics (see Appendix B or (2.26) in Chapter 2), for fixed Z (i.e., for mixed gas with fixed proportion), the entropy $S = S(\rho, T, Z)$ satisfies

$$T d_Z S = de + p d\tau. \tag{4.23}$$

Regarding Z as a parameter in the above formula, $d_Z S$ represents the differential with respect to S when fixing Z. Noticing that the chemical energy g remains unchanged when Z is fixed, and thus $dE = de$, we see that (4.23) can be written into

$$T d_Z S = dE + p d\tau. \tag{4.24}$$

Consequently, regarding the change of Z and using (4.20), we have

$$\frac{dS}{dt} = \frac{d_Z S}{dt} + \frac{\partial S}{\partial Z} \frac{dZ}{dt} = \frac{d_Z S}{dt} - f(\rho, p, Z)Z, \tag{4.25}$$

where we denote

$$f(\rho, p, Z) = \frac{\partial S}{\partial Z} \overline{k}(\rho, p, Z). \tag{4.26}$$

Using (4.24)–(4.25), the energy equation (4.22) can also be written as

$$T \frac{dS}{dt} - \frac{\mu}{\rho} \sum_{i,j=1}^{3} \left(\frac{\partial u_i}{\partial x_j} + \frac{\partial u_j}{\partial x_i} \right) \frac{\partial u_j}{\partial x_i} - \frac{1}{\rho} \left(\mu' - \frac{2}{3}\mu \right) (\operatorname{div} \boldsymbol{u})^2$$
$$= \frac{1}{\rho} \operatorname{div}(\kappa \operatorname{grad} T) - \widetilde{f}(\rho, p, Z)Z, \tag{4.27}$$

where

$$\widetilde{f}(\rho, p, Z) = f(\rho, p, Z)T. \tag{4.28}$$

So the reacting fluid dynamics system can also be written as

$$\frac{d\rho}{dt} + \rho \operatorname{div} \boldsymbol{u} = 0, \tag{4.29}$$

$$\frac{dZ}{dt} = -\bar{k}(\rho, p, Z)Z, \tag{4.30}$$

$$\frac{du_i}{dt} + \frac{1}{\rho}\frac{\partial p}{\partial x_i} = \frac{1}{\rho}\sum_{j=1}^{3}\frac{\partial}{\partial x_j}\left(\mu\left(\frac{\partial u_i}{\partial x_j} + \frac{\partial u_j}{\partial x_i}\right)\right)$$
$$+ \frac{1}{\rho}\frac{\partial}{\partial x_i}\left(\left(\mu' - \frac{2}{3}\mu\right)\operatorname{div}\boldsymbol{u}\right) + F_i \quad (i=1,2,3), \tag{4.31}$$

$$T\frac{dS}{dt} - \frac{\mu}{\rho}\sum_{i,j=1}^{3}\left(\frac{\partial u_i}{\partial x_j} + \frac{\partial u_j}{\partial x_i}\right)\frac{\partial u_j}{\partial x_i} - \frac{1}{\rho}\left(\mu' - \frac{2}{3}\mu\right)(\operatorname{div}\boldsymbol{u})^2$$
$$= \frac{1}{\rho}\operatorname{div}(\kappa\operatorname{grad}T) - \widetilde{f}(\rho, p, Z)Z. \tag{4.32}$$

In particular, under the circumstance when the viscosity and heat conduction are neglected, the corresponding system can be written as

$$\frac{d\rho}{dt} + \rho \operatorname{div} \boldsymbol{u} = 0, \tag{4.33}$$

$$\frac{dZ}{dt} = -\bar{k}(\rho, p, Z)Z, \tag{4.34}$$

$$\frac{d\boldsymbol{u}}{dt} + \frac{1}{\rho}\operatorname{grad} p = \boldsymbol{F}, \tag{4.35}$$

$$\frac{dS}{dt} = -f(\rho, p, Z)Z, \tag{4.36}$$

where f is given as in (4.26).

4.2.3 Equations of State of Mixed Gas

We still consider the case of polytropic gases. For a single gas, we know that (see Appendix B)

$$p = (\gamma - 1)\exp\left(\frac{S - S_0}{c_V}\right)\rho^\gamma, \tag{4.37}$$

$$e = \exp\left(\frac{S - S_0}{c_V}\right)\rho^{\gamma - 1}, \tag{4.38}$$

where $\gamma > 1$. Then we have

$$p = (\gamma - 1)\rho e. \tag{4.39}$$

For mixed gases, there is $E = e + g$. Assume that the chemical energy of the burned gas is zero, and that the chemical energy of the unburned gas is g_0. For the mixed gas with

4.2 System of Reacting Fluid Dynamics

the value Z, we can assume that

$$E = e + Zg_0. \tag{4.40}$$

Thus from (4.39), the equation of state of this gas is

$$p = (\gamma - 1)\rho(E - Zg_0), \tag{4.41}$$

where $\gamma > 1$, while, $g_0 > 0$ is the chemical energy of the unburned gas per unit mass released after complete combustion, which is a constant.

Here we assume that $\gamma > 1$ is independent of Z; in particular, for the burned and unburned gases, we can take the same value of γ. This is adopted from the assumption of A. Chorin, P. Colella, A. Majda, V. Roytburd, etc., used in their studies on the combustion problem (see [2], [3]). For the case which is closer to the real situation, γ may depend on $Z : \gamma = \gamma(Z)$.

The temperature T can be determined by

$$p = R\rho T. \tag{4.42}$$

Noting (4.39)–(4.40), we have

$$T = \frac{(\gamma-1)e}{R} = \frac{(\gamma-1)}{R}(E - Zg_0). \tag{4.43}$$

Now we derive the expression of the entropy. When Z is fixed, we have

$$\begin{aligned}
dE + pd\tau &= de + pd\tau \\
&= de + \frac{(1-\gamma)}{\rho}ed\rho \\
&= \rho^{\gamma-1}d(\rho^{1-\gamma}e) \\
&= \frac{(\gamma-1)e}{R}d\left(\frac{R}{\gamma-1}\ln(\rho^{1-\gamma}e)\right) \\
&= Td\left(\frac{R}{\gamma-1}\ln(\rho^{1-\gamma}e)\right).
\end{aligned} \tag{4.44}$$

We used (4.43) in the above derivation. Comparing (4.24) with (4.44), we can see that the entropy S can be taken as

$$S = \frac{R}{\gamma-1}\ln(\rho^{1-\gamma}e) + S_0,$$

i.e.,

$$S = \frac{R}{\gamma-1}\ln(\rho^{1-\gamma}(E - Zg_0)) + S_0, \tag{4.45}$$

where S_0 is a constant.

In the above derivation process, Z is always assumed to be fixed, so the above formula still holds when $\gamma = \gamma(Z)$.

In the above state equations (4.41), (4.43), and (4.45), p, T, and S are all represented as functions of ρ, E, and Z. Therefore, the unknown variables for the reacting fluid dynamics system can be taken as ρ, E, Z, and $\boldsymbol{u}$.

Now we look at the expression of the reaction rate function $\bar{k}(\rho, p, Z)$. Whether the fluid is able to burn depends on whether the temperature is high enough. Assuming that T_c is the critical temperature, the fluid burns only when $T > T_c$. According to the suggestion of A. Chorin and others, we can simply assume that

$$\bar{k} = K H(T - T_c), \qquad (4.46)$$

where $K > 0$ is a constant, and

$$H(T) = \begin{cases} 1, & T > 0, \\ 0, & T \leq 0 \end{cases} \qquad (4.47)$$

is the Heaviside function. Thus from (4.26) and noting (4.45), we have

$$\begin{aligned} f(\rho, p, Z) &= \frac{\partial S}{\partial Z} \bar{k}(\rho, p, Z) \\ &= -\frac{R}{\gamma - 1} \frac{g_0}{(E - Z g_0)} \bar{k}(\rho, p, Z) \\ &= -\frac{K g_0}{T} H(T - T_c). \end{aligned} \qquad (4.48)$$

Just as $\bar{k}$, it is a function with discontinuity as well. From (4.28), we also have

$$\tilde{f}(\rho, p, Z) = -K g_0 H(T - T_c). \qquad (4.49)$$

4.2.4 Mathematical Structure of the System of Reacting Fluid Dynamics

For the case of viscosity and heat conduction, although p and S are functions of ρ, T, and Z instead of functions of ρ and T, the momentum and energy equations (4.31)–(4.32) can still be written as a symmetric parabolic system just as in Chapter 2, while the continuity equation (4.29) and equation (4.30) describing the mass balance of the unburned gas are both first-order equations with decoupled principal part. So this system is still a quasi-linear symmetric hyperbolic-parabolic coupled system. But from the above discussion, its right-hand side functions may have discontinuity.

Now we show that the above system of equations constitutes a first-order quasi-linear symmetric hyperbolic system when neglecting the viscosity and the heat conduction. For this, we take $\boldsymbol{u}$, p, S, and Z as unknown variables. Combining the mass equation (4.33), the entropy equation (4.36), and equation (4.34) for Z, we have

$$\frac{\partial p}{\partial \rho}\left(\frac{\partial \rho}{\partial t}+\sum_{k=1}^{3} u_k \frac{\partial \rho}{\partial x_k}+\rho \sum_{k=1}^{3} \frac{\partial u_k}{\partial x_k}\right)$$

$$+\frac{\partial p}{\partial S}\left(\frac{\partial S}{\partial t}+\sum_{k=1}^{3} u_k \frac{\partial S}{\partial x_k}+f(\rho,p,Z)Z\right)$$

$$+\frac{\partial p}{\partial Z}\left(\frac{\partial Z}{\partial t}+\sum_{k=1}^{3} u_k \frac{\partial Z}{\partial x_k}+\bar{k}(\rho,p,Z)Z\right)=0.$$

Accordingly, we have

$$\frac{\partial p}{\partial t}+\sum_{k=1}^{3} u_k \frac{\partial p}{\partial x_k}+\rho c^2 \sum_{k=1}^{3} \frac{\partial u_k}{\partial x_k}=-G(\rho,p,Z)Z, \qquad (4.50)$$

where $c^2 = \frac{\partial p}{\partial \rho}$, and

$$G(\rho,p,Z)=\frac{\partial p}{\partial S}f(\rho,p,Z)+\frac{\partial p}{\partial Z}\bar{k}(\rho,p,Z). \qquad (4.51)$$

The above formula can also be written as

$$\frac{1}{\rho c^2}\frac{\partial p}{\partial t}+\sum_{k=1}^{3}\frac{\partial u_k}{\partial x_k}+\sum_{k=1}^{3}\frac{u_k}{\rho c^2}\frac{\partial p}{\partial x_k}=\widetilde{G}(\rho,p,Z)Z, \qquad (4.52)$$

where

$$\widetilde{G}(\rho,p,Z)=G(\rho,p,Z)/\rho c^2. \qquad (4.53)$$

Equation (4.52) has a form similar to (2.32) in Chapter 2, while its right-hand side may have discontinuity. It can be used to replace the mass equation (4.33).

Rearranging the above system in the order of the momentum equation (4.35), the mass equation (4.52), the entropy equation (4.36), and equation (4.34) for Z, it is not hard to see that it is a first-order quasi-linear symmetric hyperbolic system whose right-hand side term may have discontinuity.

4.3 System of One-Dimensional Reacting Fluid Dynamics

4.3.1 System of One-Dimensional Reacting Fluid Dynamics

Now we consider the one-dimensional flow of the fluid in a pipeline. Suppose that the direction of the flow is $x = x_1$; then $u = u_1$, $u_2 = u_3 = 0$, while all the state variables are independent of x_2 and x_3 and are functions only of t and x.

Thus, system (4.13)–(4.16) can be reduced into

$$\frac{\partial \rho}{\partial t} + \frac{\partial}{\partial x}(\rho u) = 0, \tag{4.54}$$

$$\frac{\partial}{\partial t}(\rho u) + \frac{\partial}{\partial x}\left(\rho u^2 + p - \left(\frac{4}{3}\mu + \mu'\right)\frac{\partial u}{\partial x}\right) = \rho F, \tag{4.55}$$

$$\frac{\partial}{\partial t}\left(\rho E + \frac{1}{2}\rho u^2\right) + \frac{\partial}{\partial x}\left(\left(\rho E + \frac{1}{2}\rho u^2 + p\right)u\right)$$
$$- \left(\frac{4}{3}\mu + \mu'\right)u\frac{\partial u}{\partial x}\right) = \frac{\partial}{\partial x}\left(\kappa \frac{\partial T}{\partial x}\right) + \rho F u, \tag{4.56}$$

$$\frac{\partial}{\partial t}(\rho Z) + \frac{\partial}{\partial x}(\rho Z u) = -\bar{k}(\rho, p, Z)\rho Z. \tag{4.57}$$

Or we can rewrite it, from (4.29)–(4.32), as

$$\frac{\partial \rho}{\partial t} + \frac{\partial}{\partial x}(\rho u) = 0, \tag{4.58}$$

$$\frac{\partial u}{\partial t} + u\frac{\partial u}{\partial x} + \frac{1}{\rho}\frac{\partial p}{\partial x} - \frac{1}{\rho}\frac{\partial}{\partial x}\left(\left(\frac{4}{3}\mu + \mu'\right)\frac{\partial u}{\partial x}\right) = F, \tag{4.59}$$

$$T\frac{\partial S}{\partial t} + Tu\frac{\partial S}{\partial x} - \frac{1}{\rho}\left(\frac{4}{3}\mu + \mu'\right)\left(\frac{\partial u}{\partial x}\right)^2$$
$$= \frac{1}{\rho}\frac{\partial}{\partial x}\left(\kappa \frac{\partial T}{\partial x}\right) - \tilde{f}Z, \tag{4.60}$$

$$\frac{\partial Z}{\partial t} + u\frac{\partial Z}{\partial x} = -\bar{k}(\rho, p, Z)Z. \tag{4.61}$$

From (4.21), (4.60) can be replaced by

$$\rho \frac{\partial E}{\partial t} + \rho u \frac{\partial E}{\partial x} + p\frac{\partial u}{\partial x} - \left(\frac{4}{3}\mu + \mu'\right)\left(\frac{\partial u}{\partial x}\right)^2 = \frac{\partial}{\partial x}\left(k\frac{\partial T}{\partial x}\right). \tag{4.62}$$

Especially in the case of $\mu = \mu' = \kappa = 0$, we have

$$\frac{\partial \rho}{\partial t} + \frac{\partial}{\partial x}(\rho u) = 0, \tag{4.63}$$

$$\frac{\partial u}{\partial t} + u\frac{\partial u}{\partial x} + \frac{1}{\rho}\frac{\partial p}{\partial x} = F, \tag{4.64}$$

$$\frac{\partial S}{\partial t} + u\frac{\partial S}{\partial x} = -f(\rho, p, Z)Z, \tag{4.65}$$

$$\frac{\partial Z}{\partial t} + u\frac{\partial Z}{\partial x} = -\bar{k}(\rho, p, Z)Z, \tag{4.66}$$

where $p = p(\rho, S, Z)$. Just as normal fluids, we also have

$$c^2 = \frac{\partial p}{\partial \rho} > 0,$$

4.3 System of One-Dimensional Reacting Fluid Dynamics

and c is the local speed of sound. In fact, for polytropic gases, it is not hard to see, from (4.41) and (4.45), that

$$p = (\gamma - 1)\exp\left(\frac{S - S_0}{\frac{R}{\gamma - 1}}\right)\rho^\gamma. \tag{4.67}$$

It is of the same form as in the case of a single gas.

4.3.2 System of One-Dimensional Reacting Fluid Dynamics in Lagrangian Representation

We take the Lagrangian coordinates (t', m). Since

$$\frac{\partial}{\partial t} = \frac{\partial}{\partial t'} - \rho u \frac{\partial}{\partial m}, \quad \frac{\partial}{\partial x} = \rho \frac{\partial}{\partial m} \tag{4.68}$$

(see section 2.5 in Chapter 2), system (4.58)–(4.61) for the case with viscosity and heat conduction can be written as

$$\frac{\partial \tau}{\partial t'} - \frac{\partial u}{\partial m} = 0, \tag{4.69}$$

$$\frac{\partial u}{\partial t'} + \frac{\partial p}{\partial m} - \frac{\partial}{\partial m}\left(\left(\frac{4}{3}\mu + \mu'\right)\rho \frac{\partial u}{\partial m}\right) = F, \tag{4.70}$$

$$T\frac{\partial S}{\partial t'} - \left(\frac{4}{3}\mu + \mu'\right)\rho\left(\frac{\partial u}{\partial m}\right)^2 = \frac{\partial}{\partial m}\left(\kappa\rho \frac{\partial T}{\partial m}\right)$$
$$- \widetilde{f}(\rho, p, Z)Z, \tag{4.71}$$

$$\frac{\partial Z}{\partial t'} = -\overline{k}(\rho, p, Z)Z; \tag{4.72}$$

while system (4.63)–(4.66) for the case neglecting the viscosity and the heat conduction can be reduced to

$$\frac{\partial \tau}{\partial t'} - \frac{\partial u}{\partial m} = 0, \tag{4.73}$$

$$\frac{\partial u}{\partial t'} + \frac{\partial p}{\partial m} = F, \tag{4.74}$$

$$\frac{\partial S}{\partial t'} = -f(\rho, p, Z)Z, \tag{4.75}$$

$$\frac{\partial Z}{\partial t'} = -\overline{k}(\rho, p, Z)Z. \tag{4.76}$$

The latter is of a very simple form.

4.3.3 Mathematical Structure of the System of One-Dimensional Reacting Fluid Dynamics

Now we take only the equations in Lagrangian representation (4.73)–(4.76) for the case without viscosity and heat conduction as an example. Denoting

$$U = (\tau, u, S, Z)^T, \tag{4.77}$$

and representing (t', m) as (t, x), the above system can be written in the following form:

$$\frac{\partial U}{\partial t} + A \frac{\partial U}{\partial x} = C, \tag{4.78}$$

where

$$A = \begin{pmatrix} 0 & -1 & 0 & 0 \\ \frac{\partial p}{\partial \tau} & 0 & \frac{\partial p}{\partial S} & \frac{\partial p}{\partial Z} \\ 0 & 0 & 0 & 0 \\ 0 & 0 & 0 & 0 \end{pmatrix}, \tag{4.79}$$

$$C = (0, F, -fZ, -\bar{k}Z)^\mathrm{T},$$

and

$$\frac{\partial p}{\partial \tau} = -\frac{1}{\tau^2} \frac{\partial p}{\partial \rho} = -c^2 \rho^2 < 0. \tag{4.80}$$

Now we find the eigenvalues of A. It is easy to obtain, from (4.79), that

$$\det(A - \lambda I) = \lambda^2 \left(\lambda^2 + \frac{\partial p}{\partial \tau} \right),$$

so A has four real eigenvalues

$$\lambda_{1,2} = 0, \quad \lambda_{3,4} = \pm \sqrt{-\frac{\partial p}{\partial \tau}} = \pm c\rho, \tag{4.81}$$

and their corresponding left eigenvectors constitute a complete set. In fact, by denoting $(\zeta_1, \zeta_2, \zeta_3, \zeta_4)$ the left eigenvector corresponding to the double eigenvalue $\lambda_{1,2} = 0$, we then have

$$\begin{cases} \frac{\partial p}{\partial \tau} \zeta_2 = 0, \\ \zeta_1 = 0. \end{cases}$$

So there exist two linearly independent left eigenvectors $(0, 0, 1, 0)$ and $(0, 0, 0, 1)$ corresponding to this eigenvalue.

Hence, system (4.78) is a first-order quasi-linear hyperbolic system (but with multiple eigenvalues, so is not strictly hyperbolic).

For some research on this system, see, e.g., [4], [5].

Exercises

1. Prove that the momentum equation (4.14) and the mass balance equation (4.16) for the unburned fluid can be reduced to the forms of (4.19) and (4.20) with the help of the continuity equation.
2. Prove that the energy conservation equation (4.15) can be reduced to the form of (4.21) with the aid of the continuity equation and the momentum equation.

3. Prove that the system (4.63)–(4.66) of one-dimensional reacting fluid dynamics in Eulerian representation is also a first-order quasi-linear hyperbolic system.

4. Write out the system of conservation laws for the one-dimensional reacting fluid dynamics when neglecting viscosity and heat conduction, i.e., in the case with $\mu = \mu' = k = 0$. Then find the Rankine–Hugoniot conditions satisfied by the solution on the strong discontinuity (see section 2.4 in Chapter 2), and prove that function Z remains continuous across the strong discontinuity.

Bibliography

[1] Courant R., Friedrichs K. O. *Supersonic Flow and Shock Waves*. New York: Interscience Publishers, 1948.

[2] Chorin A. *Random choice methods with applications for reacting gas flow*. J. Comput. Phys., 25 (1977), 253–272.

[3] Colella P., Majda A., Roytburd V. *Theoretical and numerical structure for reacting shock waves*. SIAM J. Sci. Stat. Comput., 7 (1986), 1059–1080.

[4] Ben-Artzi M. *The generalized Riemann problem for reacting flows*. J. Comput. Phys., 81 (1989), 70–101.

[5] Ben-Artzi M., Birman A. *Computation of reactive duct flows in external fields*. J. Comput. Phys., 86 (1990), 225–255.

Chapter 5
Elastic Mechanics

5.1 Introduction

Elastic mechanics is a discipline aimed at studying the laws satisfied by internal force (stress) and deformation of an elastic body under loads. These laws are mainly reflected through discussions on the deformation and motion law of an elastic body. The so-called loads mainly refer to mechanical forces acting on an elastic body, or to temperature and various physical factors that can lead to deformation of an elastic body. In this chapter we consider only the case when an elastic body is acted on by mechanical forces, while the influence of the temperature load will be specially discussed in later chapters. What is an *elastic body*? An elastic body is an object which deforms elastically under loads; meanwhile, the deformation disappears immediately and the object returns to its original shape when the load is dropped. Metal, rock, glass, wood, quartz, etc. can all be regarded as an elastic body in a certain range of deformation. The fact that the deformation of the elastic body disappears immediately after dropping the load reflects in theory a certain function relation between deformation and stress. This relation is determined by properties of the materials constituting this object; it is usually called the *constitutive relation*.

The constitutive relation can be linear or nonlinear, which depends on properties of the material and size of deformation. When the stress is less than the elastic limit, the constitutive relation of considerably many materials can be approximately regarded as a linear relation, that is, subject to the generalized Hooke's law. Then this object is called the *linear elastic body*. Sometimes the constitutive relation of an elastic body is given by a nonlinear function; the nonlinearity resulting from this is called the *material nonlinearity*.

The deformation of elastic body can be depicted by the first-order partial derivatives of the displacement. In the case of small deformation (i.e., the deformation is very small compared to the geometric dimension of the object), the deformation (strain) can be given by a linear function of the first-order partial derivatives of the displacement, while, for large deformation, the deformation has to be expressed by a nonlinear function of the first-order partial derivatives of the displacement. The resulting nonlinearity is called the *geometric nonlinearity*.

The elasticity theory aimed at the discussion on the small deformation of an elastic body is called the *linear elasticity theory*. While the elasticity theory considering geometric nonlinearity or material nonlinearity or both is called the *nonlinear elasticity theory* or the *finite elasticity theory*. This chapter is focused on the mathematical model of general nonlinear elasticity, including both geometric nonlinearity and material nonlinearity. For the discussion on linear elasticity, see sections 5.5.1 and 5.6.1, or refer to Chapter 2 in [10].

We know that the quantity relevant to continuous media usually has two such representation methods as Eulerian and Lagrangian (see Chapter 2). In elastic mechanics, we often use Lagrangian representation. Here, we first give a brief introduction to this description method.

Assume that we have designated a rectangular coordinate system (e_1, e_2, e_3) in the space located by the object. Suppose that the elastic body occupies a region $\Omega \subset \mathbb{R}^3$ (at a certain moment, say, $t = 0$) in this space before deformation. This region is usually called the *reference configuration*. The object particles in this region can be expressed by the coordinate vector

$$x = (x_1, x_2, x_3)$$

in the above coordinate system. Assuming that the elastic body deforms after this moment, its law of motion can be expressed by

$$y = y(t, x), \qquad (5.1)$$

where t represents the time and $y = (y_1, y_2, y_3)$ represents the position vector at time t of the particle which is located at point x at $t = 0$. Obviously,

$$x = y(0, x). \qquad (5.2)$$

Denoting by Ω_t the region occupied by the elastic body at time t, for any given $t \geq 0$, (5.1) is an $\Omega \to \Omega_t$ one-to-one and onto mapping, namely, a bijection. The arguments assume that

$$J = \det\left(\frac{\partial y_i}{\partial x_j}\right) > 0. \qquad (5.3)$$

The physical meaning is obvious. It means that in the process of deformation, it is impossible for any part of the elastic body with nonzero volume to be compressed to that with zero volume or expanded to that with infinitely large volume.

$x = (x_1, x_2, x_3)$ are usually called *material coordinates* or *Lagrangian coordinates* because different coordinate values correspond to different particles, while $y = (y_1, y_2, y_3)$ are called *spatial coordinates* or *Eulerian coordinates*, different coordinate values of which correspond to different spatial points. As a consequence, corresponding to these two different coordinates are two different descriptions for all physical quantities (scalar, vector, tensor) relevant to continuous media.

One description method is particle tracking. For example, a scalar θ can be expressed as a function of time t and material coordinate x:

$$\theta = \phi(t, x). \qquad (5.4)$$

This description method is called *material representation* or *Lagrangian representation*. It can also be expressed as a function of time t and space coordinate y when focusing on its change with spatial points:

$$\theta = \psi(t, y). \qquad (5.5)$$

5.2 Description of Deformation; Strain Tensor

This description method is called *spatial representation* or *Eulerian representation*. From (5.1), the relation between the above two description methods is given by

$$\psi(t, y(t, x)) = \phi(t, x). \tag{5.6}$$

For vector and tensor, the material and spatial description methods can also be given in a similar way.

For a physical quantity, its partial derivative with respect to t when x remains unchanged (i.e., the particle is fixed) is called the *material derivative* of that quantity, denoted by $\frac{d}{dt}$ just as in fluid mechanics (see (2.17) in Chapter 2). If a physical quantity (e.g., a scalar θ) is given in spatial representation (5.5), it is obvious that

$$\frac{d\theta}{dt} = \frac{\partial \psi}{\partial t} + (v \cdot \nabla_y)\psi, \tag{5.7}$$

where

$$v = \left(\frac{\partial y_1}{\partial t}, \frac{\partial y_2}{\partial t}, \frac{\partial y_3}{\partial t} \right)$$

is the velocity vector, and

$$\nabla_y = \left(\frac{\partial}{\partial y_1}, \frac{\partial}{\partial y_2}, \frac{\partial}{\partial y_3} \right)$$

is the gradient operator with respect to y.

Hereafter, for simplicity, the same physical quantity given by different description methods will not be distinguished any longer by ϕ and ψ, as in (5.4) and (5.5), as long as no confusion arises.

5.2 Description of Deformation; Strain Tensor

5.2.1 Deformation Gradient Tensor

The deformation of an object at a certain time t can be depicted by the deformation situation near each particle x in this object at this time. In order to describe the deformation at particle x, we need only figure out the behavior of each infinitesimal line element $dx = (dx_1, dx_2, dx_3)^T$ at x.

Suppose that the infinitesimal line element dx at x and $t = 0$ changes to the infinitesimal line element $dy = (dy_1, dy_2, dy_3)^T$ at $y = y(t, x)$ at time t; then

$$dy_i = \sum_{j=1}^{3} \frac{\partial y_i}{\partial x_j}(t, x) dx_j \qquad (i = 1, 2, 3), \tag{5.8}$$

which can also be written as

$$dy = F dx, \tag{5.9}$$

where

$$F = \nabla_x y = \left(\frac{\partial y_i}{\partial x_j}\right) \qquad (5.10)$$

is a second-order tensor, called the *deformation gradient tensor*. This tensor lets us know the deformation of the elastic body at time t.

5.2.2 Cauchy–Green Strain Tensor

In order to further explain the physical meaning of the deformation gradient tensor, we use the following linear algebra lemma to make the polar decomposition of the tensor.

Lemma 5.1. *Suppose that* $\det F \neq 0$; *then there exist an orthogonal matrix* R *and two positively definite symmetric matrices* U *and* V *such that*

$$F = RU = VR. \qquad (5.11)$$

Formula (5.11) *is called the* polar decomposition *of* F.

The proof of this lemma is not difficult and can be found in many textbooks on linear algebra. Here we leave it as an exercise for the reader.

Denote $U = (u_{ij})$, $R = (r_{ij})$. Using (5.11), (5.8) can be rewritten as

$$dy_i = \sum_{j,k=1}^{3} r_{ik} u_{kj} dx_j \qquad (i = 1, 2, 3)$$

or

$$dy_i = \sum_{k=1}^{3} r_{ik} dz_k \qquad (i = 1, 2, 3), \qquad (5.12)$$

where

$$dz_k = \sum_{j=1}^{3} u_{kj} dx_j \qquad (k = 1, 2, 3). \qquad (5.13)$$

This shows that the deformation from the infinitesimal line element dx to dy can be decomposed into the combination of the following two deformations: the deformation from dx to dz and the deformation from dz to dy. The former deformation is determined by tensor U, while the latter by tensor R. Since U is a positively definite symmetric matrix, there exist three mutually orthogonal principal directions and three positive principal values. Without loss of generality, we may assume that U is a diagonal matrix, that is,

$$u_{ii} > 0; u_{ij} = 0, \quad i \neq j,$$
$$(i, j = 1, 2, 3).$$

5.2 Description of Deformation; Strain Tensor

Then (5.13) can be written as

$$dz_i = u_{ii} dx_i, \quad (i = 1, 2, 3).$$

Therefore, this part of the deformation behaves as elongation or compression (depending on whether $u_{ii} > 1$ or < 1) along those three mutually orthogonal directions. Since $\boldsymbol{R}$ is an orthogonal matrix, the change from $d\boldsymbol{z}$ to $d\boldsymbol{y}$ expressed by (5.12) is just a rigid rotation.

We can give a similar explanation for the second formula $\boldsymbol{F} = \boldsymbol{V}\boldsymbol{R}$ in the polar decomposition (5.11).

Although the polar decomposition (5.11) allows us to exclude the rigid body rotation and detach the part expressing the local deformation from the deformation gradient $\boldsymbol{F}$, the direct calculation of $\boldsymbol{U}$ and $\boldsymbol{V}$ is very inconvenient. Noting that a positive definite symmetric matrix is uniquely determined by its square, we introduce

$$\boldsymbol{C} = \boldsymbol{F}^{\mathrm{T}} \boldsymbol{F} = \boldsymbol{U}^2, \tag{5.14}$$

$$\boldsymbol{B} = \boldsymbol{F} \boldsymbol{F}^{\mathrm{T}} = \boldsymbol{V}^2, \tag{5.15}$$

where $\boldsymbol{F}^{\mathrm{T}}$ is the transpose of $\boldsymbol{F}$, and $\boldsymbol{C}$ is called the *right Cauchy–Green strain tensor*, while $\boldsymbol{B}$ is called the *left Cauchy–Green strain tensor*.

Remark 5.1. In the case of steady state, the elastic body is in equilibrium, the deformation has nothing to do with time t, and all of the relevant variables are the functions of only $\boldsymbol{x}$. At this moment, if the Cauchy–Green strain tensor is given, to determine that the elastic deformation $\boldsymbol{y} = \boldsymbol{y}(\boldsymbol{x}) = (y_1(\boldsymbol{x}), y_2(\boldsymbol{x}), y_3(\boldsymbol{x}))$ is then reduced to solving the system of partial differential equations

$$\boldsymbol{F}^{\mathrm{T}} \boldsymbol{F} = J^{2/n} \boldsymbol{G}(\boldsymbol{x})$$

or

$$\boldsymbol{F} \boldsymbol{F}^{\mathrm{T}} = J^{2/n} \boldsymbol{G}(\boldsymbol{x}),$$

where $n = 3$, $\boldsymbol{F}$ and J are determined by (5.10) and (5.3), respectively, and $\boldsymbol{G}(\boldsymbol{x})$ is a given positively definite symmetric matrix with determinant 1. This system of partial differential equations has three unknown functions y_1, y_2, and y_3, while six equations are involved from the symmetry of matrix $\boldsymbol{G}(\boldsymbol{x})$, and thus it is an overdetermined (i.e., the number of equations is greater than the number of unknown functions) nonlinear system of partial differential equations called *three-dimensional Beltrami equations*. Similarly, we can define the *n-dimensional Beltrami equations*.

5.2.3 Displacement Gradient Tensor and Infinitesimal Strain Tensor

We call

$$\boldsymbol{u} = \boldsymbol{y} - \boldsymbol{x} \tag{5.16}$$

the *displacement vector*. Using this displacement vector, $\boldsymbol{F}$ can be written as

$$\boldsymbol{F} = \boldsymbol{I} + \nabla_x \boldsymbol{u}, \tag{5.17}$$

where I is the second-order unit tensor, and

$$\nabla_x \boldsymbol{u} = \left(\frac{\partial u_i}{\partial x_j}\right) \tag{5.18}$$

is called the *displacement gradient tensor*. Hereafter, for the sake of simplicity, all ∇_x's will be simplified to ∇, as long as no confusion arises.

Now we use $\nabla \boldsymbol{u}$ to represent the right Cauchy–Green strain tensor $\boldsymbol{C}$. From (5.14) and (5.17), we have

$$\boldsymbol{C} = \boldsymbol{I} + \nabla \boldsymbol{u} + (\nabla \boldsymbol{u})^{\mathrm{T}} + (\nabla \boldsymbol{u})^{\mathrm{T}}(\nabla \boldsymbol{u}). \tag{5.19}$$

As mentioned before, tensor $\boldsymbol{C}$ can be used to describe the compression and elongation along each direction near the point under consideration in the process of deformation, while $\boldsymbol{C} = \boldsymbol{I}$ implies that there is no genuine deformation relative to the reference configuration at this point. Therefore, $\boldsymbol{C} - \boldsymbol{I}$ can be regarded as a criterion measuring the deformation of the object relative to the reference configuration. From (5.19),

$$\boldsymbol{C} - \boldsymbol{I} = \nabla \boldsymbol{u} + (\nabla \boldsymbol{u})^{\mathrm{T}} + (\nabla \boldsymbol{u})^{\mathrm{T}}(\nabla \boldsymbol{u}). \tag{5.20}$$

If the deformation is very small, namely,

$$\left|\frac{\partial u_i}{\partial x_j}\right| \ll 1 \quad (i,j = 1,2,3), \tag{5.21}$$

then ignoring higher order terms in (5.20), i.e., taking the linearization, we obtain

$$\boldsymbol{C} - \boldsymbol{I} = 2\boldsymbol{E}, \tag{5.22}$$

where the second-order symmetric tensor

$$\boldsymbol{E} = \frac{1}{2}(\nabla \boldsymbol{u} + (\nabla \boldsymbol{u})^{\mathrm{T}}) \tag{5.23}$$

is called the *infinitesimal strain tensor* or *Cauchy strain tensor*, whose components are

$$e_{ij} = \frac{1}{2}\left(\frac{\partial u_i}{\partial x_j} + \frac{\partial u_j}{\partial x_i}\right). \tag{5.24}$$

It should be noted here that, in the theory of infinitesimal deformation, due to assumption (5.21), and noting (5.16), we have

$$\frac{\partial u_i}{\partial x_j} = \sum_{k=1}^{3} \frac{\partial u_i}{\partial y_k} \frac{\partial y_k}{\partial x_j}$$

$$= \frac{\partial u_i}{\partial y_j} + \sum_{k=1}^{3} \frac{\partial u_i}{\partial y_k} \frac{\partial u_k}{\partial x_j} \approx \frac{\partial u_i}{\partial y_j}.$$

So the components of the infinitesimal strain tensor $\boldsymbol{E}$ can usually be taken as

$$e_{ij} = \frac{1}{2}\left(\frac{\partial u_i}{\partial y_j} + \frac{\partial u_j}{\partial y_i}\right). \tag{5.25}$$

5.2 Description of Deformation; Strain Tensor

Now we explain the geometric meaning of tensor $\boldsymbol{E} = (e_{ij})$ under infinitesimal deformation. Suppose that there are two infinitesimal line elements $d\boldsymbol{x}^1$ and $d\boldsymbol{x}^2$ at time $t = 0$:

$$d\boldsymbol{x}^1 = (dl_1, 0, 0)^T,$$
$$d\boldsymbol{x}^2 = (0, dl_2, 0)^T.$$

They deform into $d\boldsymbol{y}^1$ and $d\boldsymbol{y}^2$, respectively, and their lengths turn into $d\widetilde{l}_1$ and $d\widetilde{l}_2$ accordingly. Using (5.9) and (5.17), we have

$$d\boldsymbol{y}^1 = \left(1 + \frac{\partial u_1}{\partial x_1}, \frac{\partial u_2}{\partial x_1}, \frac{\partial u_3}{\partial x_1}\right)^T dl_1, \tag{5.26}$$

$$d\boldsymbol{y}^2 = \left(\frac{\partial u_1}{\partial x_2}, 1 + \frac{\partial u_2}{\partial x_2}, \frac{\partial u_3}{\partial x_2}\right)^T dl_2. \tag{5.27}$$

From (5.26) and ignoring higher order infinitesimals, we obtain

$$(d\widetilde{l}_1)^2 = \left(1 + 2\frac{\partial u_1}{\partial x_1}\right)(dl_1)^2.$$

Therefore, using again the small deformation assumption (5.21), we have

$$d\widetilde{l}_1 = \left(1 + \frac{\partial u_1}{\partial x_1}\right) dl_1,$$

i.e.,

$$\frac{d\widetilde{l}_1 - dl_1}{dl_1} = e_{11}. \tag{5.28}$$

Then, e_{11} represents the relative elongation of the infinitesimal line element which was along $\boldsymbol{e}_1$ originally. We give similar explanations for e_{22} and e_{33}.

Now we consider the change of the angle between two infinitesimal line elements under infinitesimal deformation. Let θ be the angle between $d\boldsymbol{y}^1$ and $d\boldsymbol{y}^2$. From (5.26) and (5.27) while ignoring higher order infinitesimals, we have

$$d\boldsymbol{y}^1 \cdot d\boldsymbol{y}^2 = \left(\frac{\partial u_1}{\partial x_2} + \frac{\partial u_2}{\partial x_1}\right) dl_1 dl_2.$$

So

$$d\widetilde{l}_1 d\widetilde{l}_2 \cos\theta = 2e_{12} dl_1 dl_2.$$

Using (5.28) and ignoring higher order infinitesimals, from the above formula we can obtain that

$$e_{12} = \frac{1}{2}\cos\theta.$$

Let $\gamma = \frac{\pi}{2} - \theta$. It is the decrement after deformation of the angle between $d\boldsymbol{x}^1$ and $d\boldsymbol{x}^2$. For small deformation, $\cos\theta = \sin\gamma \approx \gamma$, the above formula can also be written as

$$e_{12} = \frac{1}{2}\gamma. \tag{5.29}$$

Then, e_{12} represents half of the decrement of the angle between the two infinitesimal line elements which were along the $\boldsymbol{e}_1$ and $\boldsymbol{e}_2$ directions, respectively, before the infinitesimal deformation. Similar explanations can be given to e_{23} and e_{31}.

Finally, under infinitesimal deformation, we further have

$$J = \det \boldsymbol{F} = \det(\boldsymbol{I} + \nabla \boldsymbol{u}) \approx 1 + \operatorname{tr} \boldsymbol{E}, \tag{5.30}$$

where $\operatorname{tr} \boldsymbol{E} = e_{11} + e_{22} + e_{33}$ is the trace of $\boldsymbol{E}$. Equation (5.30) shows that $\operatorname{tr} \boldsymbol{E}$ represents the relative increment of the volume element in the process of infinitesimal deformation.

5.3 Conservation Laws; Stress Tensor

5.3.1 Conservation Law of Mass

Take any given subdomain G_t in the region Ω_t occupied by the elastic body after deformation. Suppose it corresponds to $G_0 \subset \Omega$ in the reference configuration (namely, before deformation). Denote by ρ_0 and ρ the mass densities of the elastic body before and after deformation, respectively; then the conservation law of mass can be represented as

$$\int_{G_t} \rho \, dy = \int_{G_0} \rho_0 \, dx, \tag{5.31}$$

where $dx = dx_1 dx_2 dx_3$ and so on. Changing the integral variable $\boldsymbol{y}$ on the left-hand side of the above formula to $\boldsymbol{x}$, and noticing that $J = \det \boldsymbol{F}$ is the Jacobian determinant of this transformation, we have

$$\int_{G_t} \rho \, dy = \int_{G_0} \rho J \, dx. \tag{5.32}$$

Thus from (5.31) it yields

$$\int_{G_0} (\rho J - \rho_0) dx = 0, \quad \forall G_0 \subset \Omega. \tag{5.33}$$

Noting the arbitrariness of G_0, we obtain from the above formula that

$$\rho J = \rho_0. \tag{5.34}$$

This is exactly the local form of the conservation law of mass.

Since ρ_0 depends only on $\boldsymbol{x}$, while independent of t, (5.34) can also be written as

$$\frac{d}{dt}(\rho J) = 0, \tag{5.35}$$

5.3 Conservation Laws; Stress Tensor

where $\frac{d}{dt}$ is the material derivative. It is not hard to verify, from direct calculation (see exercise 2), that

$$\frac{dJ}{dt} = J \operatorname{div}_y \boldsymbol{v}, \tag{5.36}$$

where div_y represents the divergence with respect to variable $\boldsymbol{y}$. Then (5.35) can be rewritten as

$$\frac{d\rho}{dt} + \rho \operatorname{div}_y \boldsymbol{v} = 0 \tag{5.37}$$

or

$$\frac{\partial \rho}{\partial t} + (\boldsymbol{v} \cdot \nabla_y)\rho + \rho \operatorname{div}_y \boldsymbol{v} = 0, \tag{5.38}$$

where we have used (5.7). Equation (5.38) can also be expressed into the following representation:

$$\frac{\partial \rho}{\partial t} + \operatorname{div}_y(\rho \boldsymbol{v}) = 0. \tag{5.39}$$

This is exactly the local form (differential form) of the conservation law of mass under spatial description—the *continuity equation*. This equation was already obtained in the discussion on fluid dynamics in Chapter 2 (see (2.10) in Chapter 2). The reader can also compare it with the form of conservation equation of charges obtained in Chapter 1 (see (1.20) in Chapter 1).

5.3.2 Stress

We discussed only the deformation of the elastic body from the kinematic point of view, while physical factors, i.e., the loads acting on the elastic body, which result in these deformations are not taken into consideration. In order to further discuss other conservation laws, e.g., conservation law of momentum, we have to specify these physical factors resulting in the deformation and to analyze deeply the change of forces caused by these factors inside the elastic body.

The loads on the elastic body can be classified into two types. One is external force, and the other is some physical factor such as temperature. As stated before, we first put aside the influence of temperature and discuss only the load in the form of external force. External forces received by the elastic body include *volume force* $\boldsymbol{b}$—the external force per unit mass, such as gravitational force and so on, and *surface force* $\boldsymbol{\tau}$—external force per unit surface area.

The elastic body will deform under loads, and the deformation changes the relative positions among molecules inside the body and forms an additional field of internal force inside the body. When this field of internal force is strong enough to balance the external force, the elastic body stops deforming and reaches equilibrium. In order to precisely describe and analyze this field of internal force, Cauchy introduced the notion of *stress*.

Let M be a point in the elastic body Ω_t after deformation. Take a surface S through M to divide the elastic body into two parts I and II (see Figure 5.1). Now consider the action

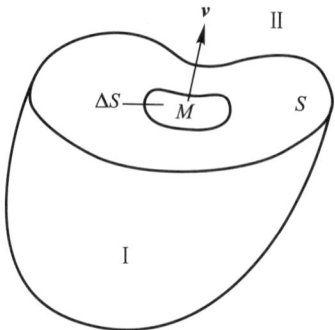

Figure 5.1.

of part II on part I. Choose the unit normal vector ν of surface S at point M, directing to part II. Take area element ΔS in a neighborhood of point M on surface S, and let Δf be the force acting on part I received from part II through ΔS. Define

$$\sigma = \lim_{\Delta S \to 0} \frac{\Delta f}{\Delta S} \tag{5.40}$$

as the *stress vector* at point M in the direction ν (note that the direction of σ is generally not that of ν).

To rationalize the above definition, we make the following assumption.

Cauchy Stress Principle. *At any given point M (its coordinate is assumed to be y) in the elastic body, for all surface S with the same normal ν, the stress vector at point M in the direction ν defined as above at a given time is the same, i.e., independent of the choice of S. Thus, the stress vector can be written as*

$$\sigma = \sigma(t, y, \nu). \tag{5.41}$$

This principle shows that stress vector σ has nothing to do with surface S, while depending only on the normal direction of S at the point under consideration. From Newton's third law, we immediately have

$$\sigma(t, y, -\nu) = -\sigma(t, y, \nu). \tag{5.42}$$

5.3.3 Integral Form of Conservation Laws of Momentum

For any given subdomain G_t in Ω_t, we consider the change of the momentum inside it. The momentum in G_t is

$$\int_{G_t} \rho \nu \, dy.$$

5.3 Conservation Laws; Stress Tensor

Assume that S_t is the boundary of G_t; then the forces on G_t include the volume force $\boldsymbol{b}$ on it and the stress $\boldsymbol{\sigma}$ on S_t received from the elastic body outside of G_t, which are in total

$$\int_{S_t} \boldsymbol{\sigma}(t, \boldsymbol{y}, \boldsymbol{v}) \mathrm{d}S_t + \int_{G_t} \rho \boldsymbol{b}(t, \boldsymbol{y}) \mathrm{d}y.$$

From Newton's second law, we have

$$\frac{\mathrm{d}}{\mathrm{d}t} \int_{G_t} \rho \boldsymbol{v} \mathrm{d}y = \int_{S_t} \boldsymbol{\sigma} \mathrm{d}S_t + \int_{G_t} \rho \boldsymbol{b} \mathrm{d}y, \tag{5.43}$$

where $\frac{\mathrm{d}}{\mathrm{d}t}$ is the material derivative. Since G_t changes with t, the derivative $\frac{\mathrm{d}}{\mathrm{d}t}$ on the left-hand side of the above formula cannot be simply moved into the integral, but it should be handled with the aid of the following lemma.

Lemma 5.2. *For any continuously differentiable function ϕ, we have*

$$\frac{\mathrm{d}}{\mathrm{d}t} \int_{G_t} \rho \phi \mathrm{d}y = \int_{G_t} \rho \frac{\mathrm{d}\phi}{\mathrm{d}t} \mathrm{d}y. \tag{5.44}$$

Proof. Just as how we obtained (5.32), we have

$$\int_{G_t} \rho \phi \mathrm{d}y = \int_{G_0} \rho J \phi \mathrm{d}x.$$

From the mass conservation equation (3.4), the above formula can be written as

$$\int_{G_t} \rho \phi \mathrm{d}y = \int_{G_0} \rho_0 \phi \mathrm{d}x.$$

So

$$\frac{\mathrm{d}}{\mathrm{d}t} \int_{G_t} \rho \phi \mathrm{d}y = \int_{G_0} \rho_0 \frac{\mathrm{d}\phi}{\mathrm{d}t} \mathrm{d}x.$$

Then (5.44) is obtained from (5.34) while changing the integral variable on the right-hand side of the above formula back to $\boldsymbol{y}$. The proof is finished. □

Using Lemma 5.2, (5.43) can be written as

$$\int_{G_t} \rho \frac{\mathrm{d}\boldsymbol{v}}{\mathrm{d}t} \mathrm{d}y = \int_{S_t} \boldsymbol{\sigma} \mathrm{d}S_t + \int_{G_t} \rho \boldsymbol{b} \mathrm{d}y. \tag{5.45}$$

This is exactly the integral form of the conservation law of momentum.

In order to use Green's formula to turn the surface integral of the first term on the right-hand side of the above formula into a volume integral so as to obtain the local (differential) form of conservation law of momentum, we need to further investigate the properties of $\boldsymbol{\sigma}$.

5.3.4 Integral Form of Conservation Laws of Momentum Moment

Let G_t and S_t be as mentioned above. The momentum moment of G_t with respect to the origin is

$$\int_{G_t} (\mathbf{y} \times \rho \mathbf{v}) \mathrm{d}y.$$

The total moment of force received by G_t should be composed of the moment of volume force acting on it and the moment of stress acting on S_t, that is,

$$\int_{S_t} \mathbf{y} \times \boldsymbol{\sigma} \mathrm{d}S_t + \int_{G_t} (\mathbf{y} \times \rho \mathbf{b}) \mathrm{d}y.$$

From the conservation law of momentum moment, we have

$$\frac{\mathrm{d}}{\mathrm{d}t} \int_{G_t} \rho(\mathbf{y} \times \mathbf{v}) \mathrm{d}y = \int_{S_t} \mathbf{y} \times \boldsymbol{\sigma} \mathrm{d}S_t + \int_{G_t} \rho(\mathbf{y} \times \mathbf{b}) \mathrm{d}y,$$

where $\frac{\mathrm{d}}{\mathrm{d}t}$ is still the material derivative. Applying Lemma 5.2, and noting $\frac{\mathrm{d}\mathbf{y}}{\mathrm{d}t} = \mathbf{v}$ and $\mathbf{v} \times \mathbf{v} = \mathbf{0}$, we have

$$\frac{\mathrm{d}}{\mathrm{d}t} \int_{G_t} \rho(\mathbf{y} \times \mathbf{v}) \mathrm{d}y = \int_{G_t} \rho \left(\mathbf{y} \times \frac{\mathrm{d}\mathbf{v}}{\mathrm{d}t} \right) \mathrm{d}y. \tag{5.46}$$

Thus we have

$$\int_{G_t} \rho \left(\mathbf{y} \times \frac{\mathrm{d}\mathbf{v}}{\mathrm{d}t} \right) \mathrm{d}y = \int_{S_t} (\mathbf{y} \times \boldsymbol{\sigma}) \mathrm{d}S_t + \int_{G_t} \rho(\mathbf{y} \times \mathbf{b}) \mathrm{d}y. \tag{5.47}$$

This is exactly the integral form of the conservation law of momentum moment.

5.3.5 Cauchy Stress Tensor

The Cauchy stress principle tells us that the stress vector $\boldsymbol{\sigma}$ at each point in the elastic body depends only on the direction $\mathbf{v}$ under consideration and has nothing to do with a particularly chosen surface S, which enables us to define the stress vector. But there are infinitely many directions $\mathbf{v}$ at each point in the space, so it is very inconvenient to represent the stress like this. We expect to depict the stress in the form of only one variable at each point in the elastic body. This variable is the Cauchy stress tensor.

Theorem 5.1. *There exists a second-order tensor $\mathbf{T}(\mathbf{y})$ such that*

$$\boldsymbol{\sigma}(\mathbf{y}, \mathbf{v}) = \mathbf{T}(\mathbf{y})\mathbf{v}. \tag{5.48}$$

Here, for simplicity we do not indicate the dependence on t of relative quantities.

Proof. Without loss of generality, we assume $\mathbf{y} = \mathbf{0}$. Construct a tetrahedron ΔG with the origin as its vortex such that three of its surfaces are lying on the coordinate planes

5.3 Conservation Laws; Stress Tensor

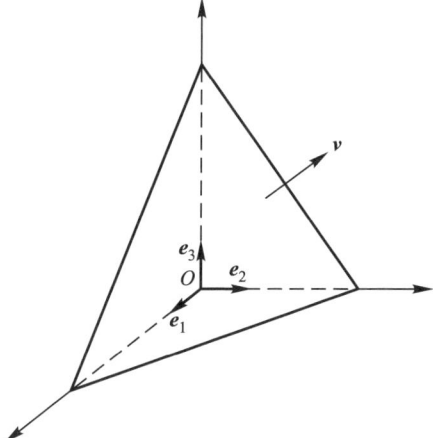

Figure 5.2.

passing through the origin, while the normal vector of the fourth surface ΔS is exactly v (see Figure 5.2). Denote by ΔS_i ($i = 1, 2, 3$) those three surfaces on the coordinate planes whose unit outward normal vectors are $-e_i$ ($i = 1, 2, 3$), respectively. In what follows, these surfaces and their areas will be denoted by the same notation. Obviously we have

$$\Delta S_i = v_i \Delta S \qquad (i = 1, 2, 3), \tag{5.49}$$

where v_i ($i = 1, 2, 3$) are the components of v.

Applying the momentum conservation equation (5.45) to ΔG, we have

$$\int_{\Delta G} \rho \frac{dv}{dt} dy = \int_{\Delta S} \sigma(y, v) dS$$
$$+ \sum_{i=1}^{3} \int_{\Delta S_i} \sigma(y, -e_i) dS + \int_{\Delta G} \rho b dy.$$

Dividing both sides of the above formula by ΔS, we obtain

$$\frac{1}{\Delta S} \int_{\Delta G} \rho \frac{dv}{dt} dy = \frac{1}{\Delta S} \int_{\Delta S} \sigma(y, v) dS$$
$$+ \frac{1}{\Delta S} \sum_{i=1}^{3} \int_{\Delta S_i} \sigma(y, -e_i) dS + \frac{1}{\Delta S} \int_{\Delta G} \rho b dy. \tag{5.50}$$

Let the distance h from ΔS to the origin tend to zero when keeping the normal vector of ΔS always at v. Since the volume of ΔG is an infinitesimal of order 3, it is easy to see, under the assumption that the integrand is bounded, that, as $h \to 0$,

$$\frac{1}{\Delta S} \int_{\Delta G} \rho \frac{dv}{dt} dy, \quad \frac{1}{\Delta S} \int_{\Delta G} \rho b dy \to \mathbf{0}. \tag{5.51}$$

Noting (5.49), as $h \to 0$ we have

$$\frac{1}{\Delta S}\int_{\Delta S}\sigma(y,v)\mathrm{d}S \to \sigma(0,v), \tag{5.52}$$

$$\frac{1}{\Delta S}\int_{\Delta S_i}\sigma(y,-e_i)\mathrm{d}S \to v_i\sigma(0,-e_i)$$
$$(i=1,2,3). \tag{5.53}$$

Thus, noting (5.42) and sending $h \to 0$, it follows from (5.50) that

$$\sigma(0,v) = \sum_{i=1}^{3} v_i\sigma(0,e_i).$$

In general, we have

$$\sigma(y,v) = \sum_{i=1}^{3} v_i\sigma(y,e_i). \tag{5.54}$$

Suppose that

$$\sigma(y,e_i) = \sum_{j=1}^{3} t_{ji}(y)e_j \quad (i=1,2,3);$$

plugging this into (5.54), we then obtain

$$\sigma(y,v) = \sum_{i,j=1}^{3} t_{ij}(y)v_j e_i.$$

Thus, taking $T(y) = (t_{ij}(y))$, the above formula leads to (5.48). From the tensor recognition theorem (see Appendix A), T is a second-order tensor. □

$T(y) = (t_{ij}(y))$ is called the *Cauchy stress tensor*. Its components t_{11}, t_{22}, t_{33} are called the *normal stresses*, and $t_{12}, \ldots, t_{23}$ are called the *shear stresses*.

5.3.6 Differential Form of Conservation Laws of Momentum in Spatial Description; Symmetry of Cauchy Stress Tensor

By using Theorem 5.1, the conservation equation (5.45) of momentum can be written as

$$\int_{G_t} \rho \frac{\mathrm{d}v_i}{\mathrm{d}t}\mathrm{d}y = \sum_{j=1}^{3}\int_{S_t} t_{ij}v_j\mathrm{d}S_t + \int_{G_t}\rho b_i\mathrm{d}y \quad (i=1,2,3). \tag{5.55}$$

From Green's formula

$$\int_{S_t} t_{ij}v_j\mathrm{d}S_t = \int_{G_t}\frac{\partial t_{ij}}{\partial y_j}\mathrm{d}y,$$

5.3 Conservation Laws; Stress Tensor

(5.55) can be reduced to

$$\int_{G_t} \left(\rho \frac{dv_i}{dt} - \sum_{j=1}^{3} \frac{\partial t_{ij}}{\partial y_j} - \rho b_i \right) dy = 0 \quad (i = 1, 2, 3).$$

Since the above formula holds for any given $G_t \subset \Omega_t$, we have

$$\rho \frac{dv_i}{dt} - \sum_{j=1}^{3} \frac{\partial t_{ij}}{\partial y_j} - \rho b_i = 0 \quad (i = 1, 2, 3) \tag{5.56}$$

or the vector form

$$\rho \frac{d\boldsymbol{v}}{dt} - \mathrm{div}_y \boldsymbol{T} - \rho \boldsymbol{b} = \boldsymbol{0}, \tag{5.57}$$

where

$$\mathrm{div}_y \boldsymbol{T} = \left(\sum_{j=1}^{3} \frac{\partial t_{ij}}{\partial y_j} \right)$$

is a first-order tensor, namely, a vector.

Equation (5.57) is the differential form of the conservation laws of momentum in spatial description. It has the same form as the equations we obtained for fluid dynamics in Chapter 2. As a matter of fact, so far we have not used the assumption that the object under consideration is an elastic body.

Now let us explore the differential form of the conservation laws of momentum moment. The component form of (5.47) is

$$\sum_{j,k=1}^{3} \int_{G_t} \rho \varepsilon_{ijk} y_j \frac{dv_k}{dt} dy = \sum_{j,k=1}^{3} \int_{S_t} \varepsilon_{ijk} y_j \sigma_k dS_t$$
$$+ \sum_{j,k=1}^{3} \int_{G_t} \rho \varepsilon_{ijk} y_j b_k dy \quad (i = 1, 2, 3), \tag{5.58}$$

where σ_k and b_k are components of $\boldsymbol{\sigma}$ and $\boldsymbol{b}$, respectively, and

$$\varepsilon_{ijk} = \begin{cases} 1 & \text{if } (i,j,k) \text{ is an even permutation of } (1,2,3), \\ -1 & \text{if } (i,j,k) \text{ is an odd permutation of } (1,2,3), \\ 0 & \text{if at least two of } (i,j,k) \text{ are equal.} \end{cases}$$

By using Theorem 5.1 and Green's formula, we have

$$\int_{S_t} \varepsilon_{ijk} y_j \sigma_k dS_t = \sum_{l=1}^{3} \int_{S_t} \varepsilon_{ijk} y_j t_{kl} v_l dS_t$$
$$= \sum_{l=1}^{3} \int_{G_t} \varepsilon_{ijk} \frac{\partial}{\partial y_l} (y_j t_{kl}) dy.$$

Substituting the above formula into (5.58) and noticing the arbitrariness of G_t, we obtain the differential form of the conservation laws of momentum moment as

$$\rho \sum_{j,k=1}^{3} \varepsilon_{ijk} y_j \frac{dv_k}{dt} - \sum_{j,k,l=1}^{3} \varepsilon_{ijk} \frac{\partial}{\partial y_l}(y_j t_{kl})$$

$$-\rho \sum_{j,k=1}^{3} \varepsilon_{ijk} y_j b_k = 0 \quad (i = 1,2,3). \tag{5.59}$$

Now we simplify the above formula by using the conservation equation (5.56) of momentum. Obviously,

$$\sum_{j,k,l=1}^{3} \varepsilon_{ijk} \frac{\partial}{\partial y_l}(y_j t_{kl}) = \sum_{j,k,l=1}^{3} \varepsilon_{ijk} \left(\delta_{jl} t_{kl} + y_j \frac{\partial t_{kl}}{\partial y_l} \right)$$

$$= \sum_{j,k=1}^{3} \varepsilon_{ijk} t_{kj} + \sum_{j,k,l=1}^{3} \varepsilon_{ijk} y_j \frac{\partial t_{kl}}{\partial y_l},$$

where δ_{jl} is the Kronecker symbol:

$$\delta_{jl} = \begin{cases} 1 & \text{if } j = l, \\ 0 & \text{if } j \neq l. \end{cases}$$

Substituting the above formula into (5.59), it follows immediately that

$$\sum_{j,k=1}^{3} \varepsilon_{ijk} y_j \left(\rho \frac{dv_k}{dt} - \sum_{l=1}^{3} \frac{\partial t_{kl}}{\partial y_l} - \rho b_k \right) - \sum_{j,k=1}^{3} \varepsilon_{ijk} t_{kj} = 0 \quad (i = 1,2,3).$$

Noting the conservation equation (5.56) of momentum, the above formula can be reduced to

$$\sum_{j,k=1}^{3} \varepsilon_{ijk} t_{kj} = 0 \quad (i = 1,2,3).$$

Taking $i = 1, 2, 3$, respectively, we get

$$t_{23} = t_{32}, \quad t_{31} = t_{13}, \quad t_{12} = t_{21}.$$

In summary, we have the following theorem.

Theorem 5.2. *The Cauchy stress tensor $T = (t_{ij})$ is a second-order symmetric tensor, i.e.,*

$$t_{ij} = t_{ji} \quad (i, j = 1,2,3). \tag{5.60}$$

From the above discussion we can see that, under the precondition that the conservation equation (5.56) of momentum is satisfied, **the conservation laws of momentum moment are equivalent to the symmetry of the Cauchy stress tensor**.

5.3.7 Piola Stress Tensor; Differential Form of Conservation Laws of Momentum in Material Description

We pointed out at the beginning of this chapter that, in elastic mechanics, the more conveniently used method of description is the material description instead of the spatial description. Now we first consider what difficulties we are going to face in order to obtain the differential form of the conservation laws of momentum in material description. To obtain this equation, it is necessary to turn the integral in (5.55) into an integral in the domain G_0 and its boundary S_0 corresponding to the reference configuration Ω. For the volume integral therein, using the change of variables and (5.34), we have

$$\int_{G_t} \rho \frac{d\boldsymbol{v}}{dt} d y = \int_{G_0} \rho_0 \frac{d\boldsymbol{v}}{dt} d x, \tag{5.61}$$

$$\int_{G_t} \rho \boldsymbol{b} d y = \int_{G_0} \rho_0 \boldsymbol{b} d x. \tag{5.62}$$

What is difficult to handle is how to reduce the surface integral

$$\int_{S_t} \boldsymbol{T}(\boldsymbol{y}) \boldsymbol{v} d S_t$$

into a surface integral on the boundary S_0 of G_0 with the following form:

$$\int_{S_0} \boxed{?} \, \boldsymbol{n} d S_0,$$

where $\boldsymbol{n}$ is the unit outward normal vector on S_0. Only in this way is it possible to use Green's formula to turn the above integral into a volume integral in G_0 and then to obtain the differential equation in material description. For this purpose, we need a transform formula about the surface area element.

Lemma 5.3. *Suppose that the surface infinitesimal area element* dS_0 *at* $\boldsymbol{x}$ *in* Ω *(its unit normal vector is* $\boldsymbol{n}$*) corresponds to the infinitesimal area element* dS_t *(its unit normal vector is* $\boldsymbol{v}$*) in* Ω_t *under the deformation*

$$\boldsymbol{y} = \boldsymbol{y}(t, \boldsymbol{x});$$

then

$$\boldsymbol{v} d S_t = J \boldsymbol{F}^{-T} \boldsymbol{n} d S_0, \tag{5.63}$$

where $\boldsymbol{F}^{-T}$ *represents the transpose of* $\boldsymbol{F}^{-1}$*, and* J *is defined by* (5.30).

Proof. Suppose that two line elements $d\boldsymbol{x}^1$ and $d\boldsymbol{x}^2$ at $\boldsymbol{x}$ in Ω form a parallelogram area element dS_0 with normal vector $\boldsymbol{n}$; correspondingly, under deformation, the two line elements $d\boldsymbol{y}^1$ and $d\boldsymbol{y}^2$ at $\boldsymbol{y}$ in Ω_t form the surface element dS_t with normal vector $\boldsymbol{v}$. Then we have

$$\boldsymbol{n} d S_0 = d\boldsymbol{x}^1 \times d\boldsymbol{x}^2, \tag{5.64}$$

$$\boldsymbol{v} d S_t = d\boldsymbol{y}^1 \times d\boldsymbol{y}^2. \tag{5.65}$$

Equation (5.65) can be written into the following component form:

$$v_i \mathrm{d} S_t = \sum_{j,k=1}^{3} \varepsilon_{ijk} \mathrm{d} y_j^1 \mathrm{d} y_k^2$$

$$= \sum_{j,k,q,r=1}^{3} \varepsilon_{ijk} \frac{\partial y_j}{\partial x_q} \frac{\partial y_k}{\partial x_r} \mathrm{d} x_q^1 \mathrm{d} x_r^2,$$

where y_j^1 represents the component of $\boldsymbol{y}^1$, etc., and ε_{ijk} is defined as in (5.58). Multiplying both sides of the above formula by $\frac{\partial y_i}{\partial x_p}$, and summing over i from 1 to 3, we get

$$\sum_{i=1}^{3} v_i \frac{\partial y_i}{\partial x_p} \mathrm{d} S_t = \sum_{i,j,k,q,r=1}^{3} \varepsilon_{ijk} \frac{\partial y_i}{\partial x_p} \frac{\partial y_j}{\partial x_q} \frac{\partial y_k}{\partial x_r} \mathrm{d} x_q^1 \mathrm{d} x_r^2. \tag{5.66}$$

From the definition of determinant, we have

$$J = \sum_{p,q,r=1}^{3} \varepsilon_{pqr} \frac{\partial y_1}{\partial x_p} \frac{\partial y_2}{\partial x_q} \frac{\partial y_3}{\partial x_r},$$

and then it is not hard to verify that

$$\sum_{i,j,k=1}^{3} \varepsilon_{ijk} \frac{\partial y_i}{\partial x_p} \frac{\partial y_j}{\partial x_q} \frac{\partial y_k}{\partial x_r} = \varepsilon_{pqr} J. \tag{5.67}$$

Thus, noting (5.64), (5.66) can be written as

$$\sum_{i=1}^{3} v_i \frac{\partial y_i}{\partial x_p} \mathrm{d} S_t = \sum_{q,r=1}^{3} \varepsilon_{pqr} J \mathrm{d} x_q^1 \mathrm{d} x_r^2$$

$$= J n_p \mathrm{d} S_0, \tag{5.68}$$

where n_p is the component of $\boldsymbol{n}$. Noticing the definition (5.10) of the deformation gradient tensor $\boldsymbol{F}$, we can also rewrite (5.68) as

$$\sum_{i=1}^{3} f_{ip} v_i \mathrm{d} S_t = J n_p \mathrm{d} S_0,$$

where f_{ip} is the component of $\boldsymbol{F}$. The above formula holds for $p = 1, 2, 3$, and so

$$\boldsymbol{F}^{\mathrm{T}} \boldsymbol{v} \mathrm{d} S_t = J \boldsymbol{n} \mathrm{d} S_0.$$

Thus (5.63) is proved. The proof of Lemma 5.3 is finished. □

Denoting

$$\boldsymbol{P} = J \boldsymbol{T} \boldsymbol{F}^{-\mathrm{T}}, \tag{5.69}$$

5.3 Conservation Laws; Stress Tensor

by Lemma 5.3, we have

$$\int_{S_t} \boldsymbol{Tv} \mathrm{d}S_t = \int_{S_0} \boldsymbol{Pn} \mathrm{d}S_0. \tag{5.70}$$

$\boldsymbol{P}$ given by (5.69) is called the *Piola stress tensor*; correspondingly, $\boldsymbol{Pn}$ is called the *Piola stress vector*.

It is necessary to note that the Piola stress tensor describes the stress of the deformed elastic body although it is defined on the reference configuration Ω. For the same particle (at point $\boldsymbol{y}$ after deformation), the direction of Piola stress vector $\boldsymbol{Pn}$ is consistent with that of Cauchy stress vector $\boldsymbol{Tv}$; just the former is measured by the undeformed unit area, while the latter is measured by the deformed unit area (see Figure 5.3).

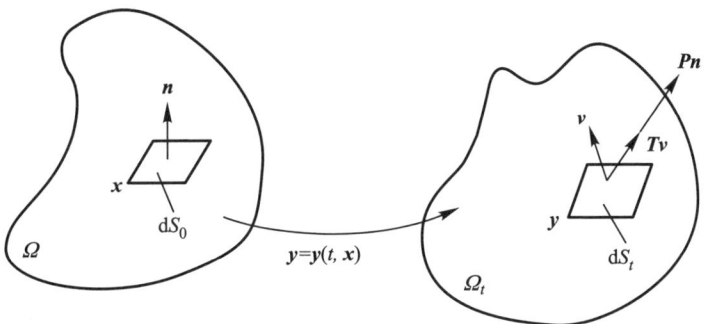

Figure 5.3.

It can be seen from (5.69) that the Piola stress tensor is generally not symmetric. Denote

$$\boldsymbol{\Sigma} = \boldsymbol{F}^{-1} \boldsymbol{P}. \tag{5.71}$$

Obviously, $\boldsymbol{\Sigma}$ is a symmetric tensor, called the *second Piola stress tensor*.

After introducing the Piola stress tensor, it is not difficult to get the differential form of the conservation laws of momentum under material description.

Using (5.61), (5.62), and (5.70), and changing the variables of integration, (5.55) is reduced to

$$\int_{G_0} \rho_0 \frac{\mathrm{d}\boldsymbol{v}}{\mathrm{d}t} \mathrm{d}x - \int_{S_0} \boldsymbol{Pn} \mathrm{d}S_0 - \int_{G_0} \rho_0 \boldsymbol{b} \mathrm{d}x = \boldsymbol{0}. \tag{5.72}$$

From Green's formula, we have

$$\int_{S_0} \boldsymbol{Pn} \mathrm{d}S_0 = \int_{G_0} \mathrm{div}_x \, \boldsymbol{P} \mathrm{d}x.$$

From (5.72), and noting the arbitrariness of G_0, we immediately get

$$\rho_0 \frac{\partial \boldsymbol{v}}{\partial t} = \mathrm{div}_x \, \boldsymbol{P} + \rho_0 \boldsymbol{b} \tag{5.73}$$

or, written in component form,

$$\rho_0 \frac{\partial v_i}{\partial t} = \sum_{j=1}^{3} \frac{\partial p_{ij}}{\partial x_j} + \rho_0 b_i \qquad (i = 1, 2, 3), \tag{5.74}$$

where p_{ij} are the components of P, while $\frac{d}{dt}$ as quantities under material description are consistent with $\frac{\partial}{\partial t}$. Equation (5.73) is exactly the differential form of the conservation laws of momentum under material description.

Under material description, the conservation laws of momentum moment are equivalent to symmetry of the second Piola stress tensor (see exercise 3).

5.4 Constitutive Equation: Relationship Between Stress and Deformation

5.4.1 General Form of Constitutive Relation

What we discussed so far does not involve in what material the object of research is composed of and what character of reaction to loads this material possesses, so it is applicable to any continuous media. In order to close the conservation equation (5.73) of momentum (if in spatial description, then the momentum equation (5.57) and the continuity equation (5.39)), we have to establish the relation between stress and deformation reflecting the instinct character of the material composing of the object under consideration. This relation is called the *constitutive relation*.

As stated before, the deformation of an object can be described by the deformation gradient tensor $F = \left(\frac{\partial y_i}{\partial x_j}\right)$. The fact that an object is an elastic body is reflected as follows: The stress of the object at a certain time point and any given particle point is uniquely determined by the value of the deformation gradient at this time and particle point; that is to say, the stress should be a function of the deformation gradient. Therefore, the constitutive equation of the elastic body should take the following form:

$$T(y) = \hat{T}(x, F(x)), \tag{5.75}$$

where the explicit dependence on t of corresponding quantities is not shown.

Definition 5.1. *The material whose Cauchy stress tensor T satisfies assumption* (5.75) *is called the* elastic material *or the* Cauchy elastic material, *where $\hat{T}$ is called the* response function. *If the response function $\hat{T}$ does not depend on x explicitly, we say that the elastic body is* homogeneous; *otherwise it is* inhomogeneous.

It is worth mentioning that whether an elastic body is homogeneous or not is not determined only by this elastic body itself, but it also depends on the selection of reference configuration as well. That is to say, an elastic body is homogeneous under a certain reference configuration, while it is not necessarily so under another "deformed" reference configuration.

Although most of the following discussion also applies to inhomogeneous elastic bodies, for simplicity we do not take into account the explicit dependence on x of the

5.4 Constitutive Equation: Relationship Between Stress and Deformation

response function; i.e., we suppose that

$$T(y) = \hat{T}(F(x)). \tag{5.76}$$

Equation (5.76) shows only that the Cauchy stress tensor depends on the deformation gradient. In order to make the response function $\hat{T}$ of the elastic body reflect the true relation between deformation and stress, it has to meet the following *objectivity hypothesis*.

We know that the elastic body is not deformed at all under rigid motion, and consequently the distribution of stress does not change. That is to say, under rigid motion, the stress vector at one point in the elastic body with respect to a certain direction should turn into the stress vector at the corresponding point with respect to the corresponding direction. Suppose that the elastic body turns from the reference configuration Ω into Ω_t and then into Ω_t^* under a rigid motion; i.e., y turns into

$$y^* = a + Qy, \tag{5.77}$$

where a is a constant vector, and Q is an orthogonal matrix.

Now we look at the stress vector $T(y)v$ at a point y in Ω_t with respect to the direction v. Under the rigid motion (5.77), y turns into a point y^* in Ω_t^*, the direction v at y turns into the direction

$$v^* = Qv \tag{5.78}$$

at y^*, and the stress vector $T(y)v$ turns into the stress vector $T^*(y^*)v^*$ at point y^* with respect to the direction v^* (see Figure 5.4), where T^* is the stress tensor corresponding to the configuration Ω_t^*. Then

$$T^*(y^*)v^* = QT(y)v. \tag{5.79}$$

Noting (5.78), the above formula can be written as

$$T^*(y^*)Qv = QT(y)v.$$

Since the above formula is satisfied for any direction v, we then have $T^*(y^*)Q = QT(y)$, i.e.,

$$T^*(y^*) = QT(y)Q^{\mathrm{T}}. \tag{5.80}$$

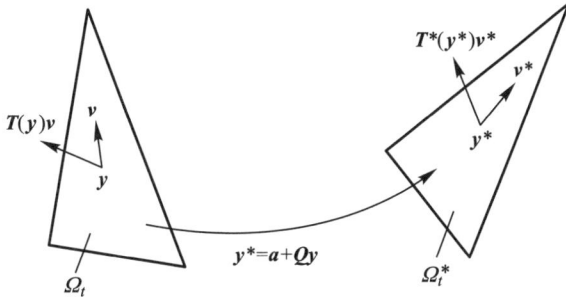

Figure 5.4.

It is easy to see from (5.77) that

$$F^* \stackrel{\mathrm{d}}{=} \left(\frac{\partial y_i^*}{\partial x_j}\right) = QF.$$

Thus, using the response function $\hat{T}$ given by (5.76), (5.80) can be rewritten as

$$\hat{T}(QF) = Q\hat{T}(F)Q^{\mathrm{T}}. \tag{5.81}$$

Definition 5.2. *If the response function $\hat{T}$ in the constitutive relation satisfies (5.81), then we say that it satisfies the* objectivity hypothesis.

A relation of form (5.76) cannot be the constitutive equation of a kind of elastic material unless its response function $\hat{T}$ satisfies (5.81), namely, the objectivity hypothesis is satisfied.

Now we discuss what kind of form the response function in the constitutive relation should take in the case that the objectivity hypothesis is satisfied.

From Lemma 5.1, the deformation gradient tensor F can be represented as $F = RU$, where R is an orthogonal matrix and U is a symmetric positively definite matrix. Taking $Q = R^{\mathrm{T}}$ in (5.81), we have

$$\hat{T}(F) = R\hat{T}(U)R^{\mathrm{T}}. \tag{5.82}$$

By using the right Cauchy–Green strain tensor $C = U^2$, the above formula can be rewritten as

$$\hat{T}(F) = R\widetilde{T}(C)R^{\mathrm{T}}, \tag{5.83}$$
$$\hat{T}(F) = F\overline{T}(C)F^{\mathrm{T}}, \tag{5.84}$$

where

$$\widetilde{T}(C) = \hat{T}(C^{1/2}), \tag{5.85}$$
$$\overline{T}(C) = C^{-1/2}\hat{T}(C^{1/2})C^{-1/2}. \tag{5.86}$$

Therefore, the constitutive equation (5.76) satisfying the objectivity hypothesis should take the form

$$T = R\widetilde{T}(C)R^{\mathrm{T}} \tag{5.87}$$

or

$$T = F\overline{T}(C)F^{\mathrm{T}}. \tag{5.88}$$

Now let us look at the Piola stress tensor P. From the definition (5.69) of P and using (5.88), we have

$$P = JF\overline{T}(C). \tag{5.89}$$

Noting that $\det C = J^2$, the above formula can also be written as

$$P = F\overline{P}(C), \tag{5.90}$$

5.4 Constitutive Equation: Relationship Between Stress and Deformation

where

$$\overline{P}(C) = \sqrt{\det C} \, \overline{T}(C). \tag{5.91}$$

For some elastic materials, their constitutive equations can also be determined by means of the stored-energy function.

Definition 5.3. *If there exists a scalar function* $W = \hat{W}(F)$ *of F such that*

$$p_{ij} = \frac{\partial \hat{W}(F)}{\partial f_{ij}} \quad (i, j = 1, 2, 3), \tag{5.92}$$

where p_{ij} are components of the Piola stress tensor, then this material is called hyperelastic. *Function $W = \hat{W}(F)$ is called the* stored-energy function *or the* strain energy function.

Obviously, hyperelastic materials must be elastic.

For hyperelastic materials, the objectivity hypothesis is given by the formula

$$\hat{W}(QF) = \hat{W}(F), \tag{5.93}$$

where Q is any given orthogonal matrix. It is not hard to verify that if the stored-energy function $\hat{W}$ satisfies (5.93), then the Cauchy stress tensor T determined by it satisfies (5.81) (see exercise 5).

5.4.2 Constitutive Equation of Isotropic Materials

Many of the materials we have contact with are isotropic, such as rubber, metal, etc. But there are exceptions; for instance, wood is a typical anisotropic material. Now we first give the precise definition of isotropic material and then inspect how the feature of isotropy is reflected in the properties of the response function describing the constitutive relation of the elastic body.

From physical intuition, when an elastic body is isotropic at one point, it means that no matter what the direction of the deformation is at this point, the stress response arising therefrom should be essentially the same.

Let us observe the situation at a particle point $x^0 \in \Omega$ in the elastic body. First, we look at the following uniform deformation:

$$y(x) = x^0 + F(x - x^0), \tag{5.94}$$

where F is a constant tensor. Under this deformation, particle x^0 remains unchanged, and the Cauchy stress tensor at x^0 is $\hat{T}(F)$.

Second, we consider the deformation as follows. We first take a rotation transformation of Ω based at point x^0, assuming its transformation matrix to be Q (orthogonal matrix), and then we make a uniform deformation on it, keeping x^0 unchanged, with deformation gradient F; that is, we turn point x into (see Figure 5.5)

$$y^*(x) = x^0 + FQ(x - x^0).$$

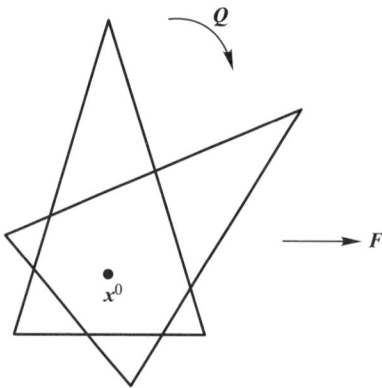

Figure 5.5.

The deformation gradient tensor of this deformation is FQ. Under this deformation, the Cauchy stress tensor at particle x^0 is $\hat{T}(FQ)$.

In both of the above cases, we make the same deformation at x^0 of the elastic body along different directions. If the material has the same stress response along each direction, then in both cases, the stress tensors at x^0 should be the same, i.e.,

$$\hat{T}(FQ) = \hat{T}(F). \tag{5.95}$$

Definition 5.4. *If the response function $\hat{T}$ in the constitutive equation (5.76) of the elastic material satisfies (5.95) for any given orthogonal matrix Q, then we call this material isotropic.*

Now we discuss the form of the constitutive equation for the isotropic material.
From Lemma 5.1, F can be expressed as

$$F = VR,$$

where R is an orthogonal matrix. Thus, from the definition (5.95) of the isotropic material, we have

$$\hat{T}(F) = \hat{T}(VR) = \hat{T}(V). \tag{5.96}$$

Noting the definition (5.85) of $\widetilde{T}$ and (5.15), we also have

$$\hat{T}(F) = \hat{T}(V) = \widetilde{T}(B).$$

Therefore, for the isotropic material, the Cauchy stress tensor can be expressed as a function of V or B; i.e., the constitutive relation takes the following form:

$$T = \hat{T}(V) \tag{5.97}$$

or

$$T = \widetilde{T}(B). \tag{5.98}$$

5.4 Constitutive Equation: Relationship Between Stress and Deformation

Through a more detailed discussion we can prove that (5.98) can actually be represented more specifically; see the following theorem.

Theorem 5.3. *For the isotropic material, its constitutive relation takes the following form:*

$$T = \beta_0(I_B)I + \beta_1(I_B)B + \beta_2(I_B)B^2, \tag{5.99}$$

where β_0, β_1, and β_2 are the scalar functions of the three principal invariants I_B of the symmetric tensor B.

Proof. We refer the reader to [4] for the proof of this theorem. □

Now we discuss the Piola stress tensor. Since P is not a symmetric tensor, it is impossible to give an expression similar to (5.99). But we still have a similar result for the second Piola stress tensor Σ of the isotropic material, i.e., it can be expressed as a quadratic matrix polynomial of the right Cauchy–Green strain tensor with coefficients depending only on the three principal invariants of C. See the following theorem for details.

Theorem 5.4. *For the isotropic material, its constitutive equation can be represented in the following form:*

$$\Sigma = \gamma_0(I_C)I + \gamma_1(I_C)C + \gamma_2(I_C)C^2, \tag{5.100}$$

where γ_0, γ_1, and γ_2 are the scalar functions of the three principal invariants I_C of the symmetric tensor C.

Proof. From (5.69) and (5.71), we have

$$\Sigma = JF^{-1}TF^{-T}. \tag{5.101}$$

Applying Theorem 5.3 to T, and noting (5.14)–(5.15), the above formula leads to

$$\Sigma = J\left(\beta_0(I_B)C^{-1} + \beta_1(I_B)I + \beta_2(I_B)C\right). \tag{5.102}$$

From the polar decomposition of F (see Lemma 5.1), it is easy to get $B = RCR^T$, where R is an orthogonal matrix. So,

$$I_B = I_C. \tag{5.103}$$

Besides, from the Cayley–Hamilton theorem, it holds that

$$C^3 - I_1(C)C^2 + I_2(C)C - I_3(C)I = 0, \tag{5.104}$$

where $I_i(C)$ ($i = 1, 2, 3$) represents the ith principal invariant of C. Therefore,

$$C^{-1} = I_3(C)^{-1}(I_2(C)I - I_1(C)C + C^2). \tag{5.105}$$

Moreover, it is clear that

$$J = I_3(C)^{1/2}. \tag{5.106}$$

Plugging (5.103), (5.105), and (5.106) into (5.102), we get (5.100). □

Theorem 5.5. *Suppose that the reference configuration is the natural state, i.e., $\hat{T}(I) = 0$; then for the deformation near this reference configuration, the constitutive equation of the isotropic material takes the following form:*

$$\Sigma = \lambda(\operatorname{tr} \widetilde{E})I + 2\mu \widetilde{E} + o(|\widetilde{E}|), \tag{5.107}$$

where constants λ and μ are called **Lamé constants**, $\widetilde{E} = \frac{1}{2}(C - I)$, *and $o(|\widetilde{E}|)$ represents the higher order terms of $|\widetilde{E}|$.*

Proof. It is easy to verify directly that

$$\operatorname{tr} C = 3 + 2 \operatorname{tr} \widetilde{E},$$
$$\operatorname{tr} C^2 = 3 + 4 \operatorname{tr} \widetilde{E} + o(|\widetilde{E}|),$$
$$\operatorname{tr} C^3 = 3 + 6 \operatorname{tr} \widetilde{E} + o(|\widetilde{E}|).$$

Now we use the above formulas to compute the three principal invariants of C. First, it is easy to see that

$$I_1(C) = \operatorname{tr} C = 3 + 2 \operatorname{tr} \widetilde{E},$$
$$I_2(C) = \frac{1}{2}((\operatorname{tr} C)^2 - \operatorname{tr} C^2) = 3 + 4 \operatorname{tr} \widetilde{E} + o(|\widetilde{E}|).$$

To compute $I_3(C)$, taking traces on both sides of (5.104), and using both of the above formulas, we get

$$I_3(C) = \frac{1}{6}(\operatorname{tr} C)^3 - \frac{1}{2}\operatorname{tr} C \operatorname{tr} C^2 + \frac{1}{3}\operatorname{tr} C^3$$
$$= 1 + 2\operatorname{tr} \widetilde{E} + o(|\widetilde{E}|).$$

For any scalar function γ of I_C, from Taylor's expansion, it is easy to know that

$$\gamma(I_C) = \gamma(I_C)\Big|_{C=I} + \frac{\partial}{\partial I_1}\gamma(I_C)\Big|_{C=I} \cdot 2\operatorname{tr} \widetilde{E}$$
$$+ \frac{\partial}{\partial I_2}\gamma(I_C)\Big|_{C=I} \cdot 4\operatorname{tr} \widetilde{E} + \frac{\partial}{\partial I_3}\gamma(I_C)\Big|_{C=I} \cdot 2\operatorname{tr} \widetilde{E}$$
$$+ o(|\widetilde{E}|). \tag{5.108}$$

Taking $C = I$ in (5.100), we have

$$\gamma_0(I_C) + \gamma_1(I_C) + \gamma_2(I_C) = 0 \quad \text{when} \quad C = I. \tag{5.109}$$

Representing $\gamma_0(I_C)$, $\gamma_1(I_C)$, and $\gamma_2(I_C)$ in the form of (5.108), respectively, substituting them into (5.100), and noting (5.109), we get (5.107). □

If we neglect the higher order term of $\widetilde{E}$ in the constitutive relation (5.107), then we get

$$\Sigma = \lambda(\operatorname{tr} \widetilde{E})I + 2\mu \widetilde{E}. \tag{5.110}$$

5.4 Constitutive Equation: Relationship Between Stress and Deformation

It is not hard to see that this constitutive relation still satisfies the objectivity hypothesis, and the material represented by it is isotropic. The material with the constitutive relation of the form (5.110) is called the *St. Venant–Kirchhoff material*. This is one of the simplest nonlinear elastic materials.

For the hyperelastic material, the isotropy is defined by

$$\hat{W}(FQ) = \hat{W}(F), \tag{5.111}$$

where Q is any given orthogonal matrix. It is not difficult to prove that if the stored-energy function of the hyperelastic material satisfies (5.111), then the Cauchy stress tensor given by it satisfies (5.95) (see exercise 6).

5.4.3 Examples of Stored-Energy Function

For the St. Venant–Kirchhoff material given in the last section, it can be verified that its stored-energy function is given by the following formula:

$$W = \frac{\lambda}{2}(\operatorname{tr}\widetilde{E})^2 + \mu \operatorname{tr}\widetilde{E}^2. \tag{5.112}$$

In order to further provide some useful stored-energy functions, we first explore the form of the stored-energy function for the isotropic material. Suppose that this function is given by

$$W = \hat{W}(F). \tag{5.113}$$

Taking $Q = R^T$ in (5.93) and noting (5.11) and (5.14), we have

$$\hat{W}(F) = \hat{W}(R^T F) = \hat{W}(U) = \widetilde{W}(C);$$

i.e., the stored-energy function depends only on C. Here, similarly to (5.85), define $\widetilde{W}(C) = \hat{W}(C^{\frac{1}{2}})$. Along with (5.111) and the above formula, and noting (5.14), we get

$$\hat{W}(F) = \hat{W}(FQ) = \widetilde{W}((FQ)^T(FQ)) = \widetilde{W}(Q^T C Q);$$

i.e., the stored-energy function can be written as

$$W = \widetilde{W}(Q^T C Q), \tag{5.114}$$

where Q is any given orthogonal matrix. Since C is a positively definite symmetric matrix, there always exists an orthogonal matrix Q such that $Q^T C Q = \operatorname{diag}(\lambda_1, \lambda_2, \lambda_3)$, where λ_1, λ_2, and λ_3 are the principal values of C. Then the stored-energy function can also be written as

$$W = \widetilde{W}(\operatorname{diag}(\lambda_1, \lambda_2, \lambda_3)). \tag{5.115}$$

This implies that, in fact, W depends only on the principal values of C, and thus W depends only on the principal values μ_1, μ_2, and μ_3 of U. Obviously, $\mu_i = \lambda_i^{\frac{1}{2}}$ ($i = 1, 2, 3$).

For the incompressible case, namely, when the deformation satisfies the constraint condition

$$J = 1, \tag{5.116}$$

Ogden put forth the stored-energy function in the following form (*Ogden material*):

$$W = \sum_{i=1}^{M} a_i(\mu_1^{\alpha_i} + \mu_2^{\alpha_i} + \mu_3^{\alpha_i} - 3)$$
$$+ \sum_{i=1}^{N} b_i((\mu_2\mu_3)^{\beta_i} + (\mu_3\mu_1)^{\beta_i} + (\mu_1\mu_2)^{\beta_i} - 3), \quad (5.117)$$

where constants $a_i \geq 0$, $b_i \geq 0$, $\alpha_i \geq 0$, $\beta_i \geq 0$. The -3 appearing in each term of the above formula is inessential; it is there just to make $W = 0$ as $\mu_1 = \mu_2 = \mu_3 = 1$. Experiments show that the stored-energy function of a certain vulcanized rubber can be given by (5.117), where

$$M = 2, \quad N = 1, \quad \alpha_1 = 5, \quad \alpha_2 = 1.3, \quad \beta_1 = 2;$$
$$a_1 = 24, \quad a_2 = 4.8 \times 10^4, \quad b_1 = 5 \times 10^2 (\text{kg/m}^2).$$

As special examples of (5.117), we have *Neo–Hookean material*

$$W = a_1(\mu_1^2 + \mu_2^2 + \mu_3^2 - 3) \quad (5.118)$$

and *Mooney–Rivlin material*

$$W = a_1(\mu_1^2 + \mu_2^2 + \mu_3^2 - 3)$$
$$+ b_1((\mu_2\mu_3)^2 + (\mu_3\mu_1)^2 + (\mu_1\mu_2)^2 - 3). \quad (5.119)$$

For the compressible case, the deformation is not constrained by (5.116); Ogden proposed adding a term $\Gamma(\mu_1\mu_2\mu_3) = \Gamma(\det \boldsymbol{F})$ on the right-hand side of (5.117), where Γ should satisfy the following condition:

$$\text{if } \xi \to +0, \quad \Gamma(\xi) \to +\infty. \quad (5.120)$$

This implies that the volume of the material under consideration can be compressed to zero only by infinitely large energy. See Figure 5.6 for the graph of function Γ. Thus, the stored-energy function of the *compressible Ogden material* is given by

$$W = \sum_{i=1}^{M} a_i(\mu_1^{\alpha_i} + \mu_2^{\alpha_i} + \mu_3^{\alpha_i} - 3)$$
$$+ \sum_{i=1}^{N} b_i((\mu_2\mu_3)^{\beta_i} + (\mu_3\mu_1)^{\beta_i} + (\mu_1\mu_2)^{\beta_i} - 3) + \Gamma(\mu_1\mu_2\mu_3). \quad (5.121)$$

Ciarlet and Geymonat gave the stored-energy function,

$$W = a(\mu_1^2 + \mu_2^2 + \mu_3^2) + b((\mu_2\mu_3)^2 + (\mu_3\mu_1)^2 + (\mu_1\mu_2)^2)$$
$$+ \Gamma(\mu_1\mu_2\mu_3), \quad (5.122)$$

for a kind of material, where $a > 0$, $b > 0$, $\Gamma(\xi) = c\xi^2 - d\ln\xi + e$ and $c > 0$, $d > 0$, $e \in \mathbb{R}$ (see [4], [5]).

5.4 Constitutive Equation: Relationship Between Stress and Deformation

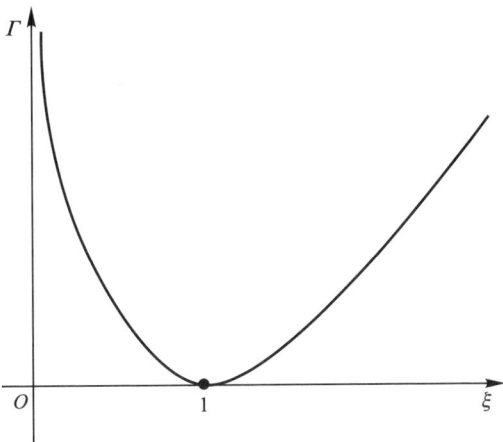

Figure 5.6.

All of the above stored-energy functions are expressed in terms of the principal values of the tensor U (or C). In order to represent the stored-energy function by the deformation gradient tensor F, we need only notice that

$$\begin{cases} \mu_1^2 + \mu_2^2 + \mu_3^2 = \mathrm{tr}(F^T F), \\ (\mu_2 \mu_3)^2 + (\mu_3 \mu_1)^2 + (\mu_1 \mu_2)^2 = \mathrm{tr}(\mathrm{cof}(F^T F)), \\ \mu_1 \mu_2 \mu_3 = \det F, \end{cases} \quad (5.123)$$

where $\mathrm{cof}\, A$ represents the cofactor matrix of A, i.e., the matrix generated by the algebraic cofactor corresponding to the entries of A. The first and the third formulas in (5.123) are obvious. For the proof of the second formula, the reader is referred to exercise 9. In particular, when $\alpha_i = \beta_i = 2$, the right-hand side of (5.121) can be easily expressed in terms of F by using the formulas in (5.123).

5.4.4 Linear Elasticity: Generalized Hook's Law

Suppose that the reference configuration is the natural state, i.e., $\hat{T}(I) = 0$. For a small deformation near the reference configuration, it can be approximated by its linearization.

Denote

$$\varepsilon = |\nabla u| = \left(\sum_{i,j=1}^{3} \left| \frac{\partial u_i}{\partial x_j} \right|^2 \right)^{1/2}. \quad (5.124)$$

For the small deformation, we always assume that

$$\varepsilon \ll 1. \quad (5.125)$$

Expanding $\overline{P}(C)$ in (5.90) near $C = I$, and noting

$$\overline{P}(I) = 0, \quad (5.126)$$

we have
$$\overline{P}(C) = \frac{1}{2}A(C - I) + O(\varepsilon^2), \tag{5.127}$$

where $A = (a_{ijkl})$ is a fourth-order tensor whose components are
$$a_{ijkl} = 2\left.\frac{\partial \overline{p}_{ij}}{\partial c_{kl}}\right|_{C=I}, \tag{5.128}$$

where $\overline{p}_{ij}$ and c_{kl} are components of $\overline{P}$ and C, respectively, and $A(C - I)$ is achieved by first taking the product of tensors A and $C - I$ and then contracting (see Appendix A); this is a second-order tensor with components
$$\sum_{k,l=1}^{3} a_{ijkl}(c_{kl} - \delta_{kl}).$$

Noting (5.22), we have
$$\frac{1}{2}(C - I) = E + O(\varepsilon^2). \tag{5.129}$$

Plugging (5.127) into (5.90), it is easy to get
$$P = (I + \nabla u)\left(\frac{1}{2}A(C - I) + O(\varepsilon^2)\right)$$
$$= AE + O(\varepsilon^2), \tag{5.130}$$

where AE can be explained in the same sense as $A(C - I)$. Neglecting $O(\varepsilon^2)$ in the above formula, namely, taking the linearized approximation, we get
$$P = AE \tag{5.131}$$

or, written in the component form,
$$p_{ij} = \sum_{k,l=1}^{3} a_{ijkl} e_{kl}, \tag{5.132}$$

where
$$e_{ij} = \frac{1}{2}\left(\frac{\partial u_i}{\partial x_j} + \frac{\partial u_j}{\partial x_i}\right)$$

are components of the infinitesimal strain tensor (see (5.24)).

Equation (5.131) or (5.132) is the linearized approximation of the constitutive relation in the case of small deformation near the natural state. It is the basis of the classical linear elasticity, called the *generalized Hooke's law*. It is necessary to mention that this generalized Hooke's law is not a special example of the aforementioned general theory of

5.4 Constitutive Equation: Relationship Between Stress and Deformation

elasticity, but is only an approximation in the case of small deformation, since it does not satisfy the objectivity hypothesis.

Since C is a symmetric tensor, from the definition (5.128) of A we have

$$a_{ijkl} = a_{ijlk}.$$

If the material is assumed to be hyperelastic, then it is not hard to verify that (see exercise 4)

$$a_{ijkl} = a_{klij}.$$

Thus, the fourth-order tensor A possesses the following symmetry:

$$a_{ijkl} = a_{klij} = a_{ijlk} = a_{jikl}. \tag{5.133}$$

Now we discuss the case of isotropic materials. Noting $P = F\Sigma$ (see (5.71)), from Theorem 5.5 it is easy to prove that

$$p_{ij} = \lambda(e_{11} + e_{22} + e_{33})\delta_{ij} + 2\mu e_{ij} \tag{5.134}$$

to get

$$a_{ijkl} = \lambda \delta_{ij} \delta_{kl} + \mu(\delta_{ik}\delta_{jl} + \delta_{il}\delta_{jk}). \tag{5.135}$$

This result can also be derived from the general form of the fourth-order isotropic tensor (see Appendix A) by using the isotropy of the fourth-order tensor A. Equation (5.134) expresses the stress in terms of strain, usually called the *stress–strain relation*. We can solve out e_{ij} from (5.134) to obtain the following *strain–stress relation*:

$$e_{ij} = \frac{1}{2\mu} p_{ij} - \frac{\lambda}{2\mu(3\lambda + 2\mu)}(p_{11} + p_{22} + p_{33})\delta_{ij}. \tag{5.136}$$

In order to explain further the physical meaning of Lamé constants, we first review the simplest form of Hooke's law. Let a uniform cylindrical component with length l and cross section S act on a tension deformation with elongation Δl under the axial pulling force f. Hooke's law implies that the relative elongation $\varepsilon_a = \Delta l/l$, when it is small, is proportional to the axial stress $\sigma = f/S$, i.e.,

$$\sigma = E\varepsilon_a, \tag{5.137}$$

where E is determined only by the material, called the *Young modulus*. In addition, experiments show that under certain conditions, the diameter d of the cross section of the component decreases with the increment of the tension load. Denote by Δd the increment of d (negative value) and $\varepsilon_d = \Delta d/d$; then $-\varepsilon_d$ is proportional to ε_a, i.e.,

$$-\varepsilon_d = \nu \varepsilon_a, \tag{5.138}$$

where ν is called the *Poisson's ratio*.

If we take the direction of the above-mentioned axial pulling force to be e_1, then the stress tensor of this stress state with uniaxial stretch is

$$P = \begin{pmatrix} \sigma & 0 & 0 \\ 0 & 0 & 0 \\ 0 & 0 & 0 \end{pmatrix},$$

and thus from (5.136) we have

$$e_{11} = \frac{(\lambda+\mu)\sigma}{\mu(3\lambda+2\mu)}, \quad e_{22} = e_{33} = -\frac{\lambda\sigma}{2\mu(3\lambda+2\mu)}, \quad e_{ij} = 0, \, i \neq j.$$

From the meaning of Young modulus and Poisson's ratio, we know that

$$\frac{\sigma}{e_{11}} = E, \quad -\frac{e_{22}}{e_{11}} = -\frac{e_{33}}{e_{11}} = \nu.$$

Therefore,

$$E = \frac{\mu(3\lambda+2\mu)}{\lambda+\mu}, \quad \nu = \frac{\lambda}{2(\lambda+\mu)} \tag{5.139}$$

or

$$\lambda = \frac{E\nu}{(1+\nu)(1-2\nu)}, \quad \mu = \frac{E}{2(1+\nu)}. \tag{5.140}$$

Hence, by using Young modulus E and Poisson's ratio ν, the strain–stress relation (5.136) can be written into the following form:

$$\begin{cases} e_{11} = \frac{1}{E}(p_{11} - \nu(p_{22}+p_{33})), \\ e_{22} = \frac{1}{E}(p_{22} - \nu(p_{33}+p_{11})), \\ e_{33} = \frac{1}{E}(p_{33} - \nu(p_{11}+p_{22})), \\ e_{ij} = \frac{1}{2\mu} p_{ij}, \quad i \neq j. \end{cases} \tag{5.141}$$

This is the strain–stress relation most often used in engineering. Its physical meaning is obvious. For instance, the first formula in (5.141) shows that the normal strain along the direction e_1 is the elongation produced by the normal stress p_{11} along the direction e_1 minus the contraction produced by the normal stresses along the directions e_2 and e_3.

Now we consider the shear strain e_{ij} and the shear stress p_{ij} ($i \neq j$). Take e_{12} and p_{12}, for example; substituting (5.29) into (5.141), we have

$$\mu = p_{12}/\gamma, \tag{5.142}$$

where γ is the decrement of the angle between the line elements along the directions e_1 and e_2 under deformation. Equation (5.142) implies that Lamé constant μ represents the ratio of the shear stress to the change of the angle caused by it, called the *shear modulus*.

In the case of linear elasticity, from (5.30) we have

$$J = 1 + \operatorname{tr} \boldsymbol{E};$$

then the rate of increase of the volume caused by the deformation is

$$J - 1 = \operatorname{tr} \boldsymbol{E}.$$

We call the ratio κ of the average normal stress $\frac{1}{3}\operatorname{tr} \boldsymbol{P}$ to the rate of increase of the volume caused by the deformation the *volume elastic modulus*. From (5.136) we obtain

$$\kappa = \lambda + \frac{2}{3}\mu. \tag{5.143}$$

From the above discussion we can see that Lamé constants λ and μ should be restricted such that the Young modulus E, the shear modulus μ, the Poisson's ratio ν, and the volume elastic modulus κ are all positive.

5.5 System of Elastodynamics and Its Mathematical Structure

5.5.1 System of Linear Elastodynamics

In this subsection, we will discuss the system of linear elastodynamics and its well-posed problems. At this time, the constitutive relation is given by (5.132). Plugging it into the conservation equation (5.74) of momentum, and noting that $\boldsymbol{v} = \frac{\partial \boldsymbol{u}}{\partial t}$ from (5.16), we obtain the following system:

$$\rho_0 \frac{\partial^2 u_i}{\partial t^2} = \frac{1}{2} \sum_{j,k,l=1}^{3} a_{ijkl} \left(\frac{\partial^2 u_k}{\partial x_j \partial x_l} + \frac{\partial^2 u_l}{\partial x_j \partial x_k} \right) + \rho_0 b_i$$
$$(i = 1, 2, 3). \tag{5.144}$$

Since $a_{ijkl} = a_{ijlk}$ (see (5.133)), we have

$$\sum_{j,k,l=1}^{3} a_{ijkl} \frac{\partial^2 u_l}{\partial x_j \partial x_k} = \sum_{j,k,l=1}^{3} a_{ijkl} \frac{\partial^2 u_k}{\partial x_j \partial x_l}.$$

Thus, (5.144) can be simplified into the following form:

$$\rho_0 \frac{\partial^2 u_i}{\partial t^2} = \sum_{j,k,l=1}^{3} a_{ijkl} \frac{\partial^2 u_k}{\partial x_j \partial x_l} + \rho_0 b_i \quad (i = 1, 2, 3). \tag{5.145}$$

Equation (5.145) is a system of second-order linear partial differential equations with unknown functions $\boldsymbol{u} = (u_1, u_2, u_3)$. To study the properties of this system and its well-posed problems, we have to be sure about its type. To this end, we have to make further physically reasonable assumptions on the linear elastic tensor $\boldsymbol{A} = (a_{ijkl})$.

In the case of free vibration of the elastic body, i.e., the external force $\boldsymbol{b} = \boldsymbol{0}$, we consider the following form of plane wave solutions of system (5.145):

$$\boldsymbol{u} = \boldsymbol{\xi} e^{\sqrt{-1}(\boldsymbol{\eta} \cdot \boldsymbol{x} + \lambda t)}, \tag{5.146}$$

where $\boldsymbol{\xi} = (\xi_1, \xi_2, \xi_3)$ and $\boldsymbol{\eta} = (\eta_1, \eta_2, \eta_3)$ are both real vectors. For the free vibration of pure elastic materials, since there is neither gain of energy nor loss of energy, namely, dissipation effect, we should expect that the amplitude of the plane wave (5.146) neither increases nor decreases in the process of propagation. Thus, the λ on the right-hand side of (5.146) should remain real, and then we have

$$\rho_0 \lambda^2 > 0. \tag{5.147}$$

Substituting the plane wave (5.146) into system (5.145) in the case of $\boldsymbol{b} = \boldsymbol{0}$, we obtain

$$\rho_0 \lambda^2 \xi_i = \sum_{j,k,l=1}^{3} a_{ijkl} \xi_k \eta_j \eta_l \quad (i=1,2,3).$$

Multiplying ξ_i on both sides of the above formula, and summing over i from 1 to 3, we have

$$\sum_{i,j,k,l=1}^{3} a_{ijkl} \xi_i \xi_k \eta_j \eta_l = \rho_0 \lambda^2 |\boldsymbol{\xi}|^2. \tag{5.148}$$

This implies that the left-hand side of (5.148) is always positive for any given nonzero vectors $\boldsymbol{\xi}$ and $\boldsymbol{\eta} \in \mathbb{R}^3$. According to this, we give the following definition.

Definition 5.5. *If there exists a constant $\alpha > 0$ such that*

$$\sum_{i,j,k,l=1}^{3} a_{ijkl} \xi_i \xi_k \eta_j \eta_l \geq \alpha |\boldsymbol{\xi}|^2 |\boldsymbol{\eta}|^2, \quad \forall \boldsymbol{\xi}, \boldsymbol{\eta} \in \mathbb{R}^3, \tag{5.149}$$

then we say that the fourth-order tensor $\boldsymbol{A} = (a_{ijkl})$ satisfies the strong ellipticity condition.

Below it is always assumed that the linear elastic tensor $\boldsymbol{A} = (a_{ijkl})$ satisfies the strong ellipticity condition. For isotropic materials, it can be proved that the strong ellipticity condition (5.149) is equivalent to

$$\mu > 0, \quad \lambda + 2\mu > 0 \tag{5.150}$$

for Lamé constants λ and μ. The proof of this conclusion is left as an exercise to the reader.

If the tensor $\boldsymbol{A} = (a_{ijkl})$ satisfies the strong ellipticity condition, then system (5.145) is called a *second-order hyperbolic system* (further explanation about the hyperbolicity of this system is left behind in section 5.5.3 in the discussion on the system of nonlinear elastodynamics).

Similarly to the wave equations, we can put forth two kinds of well-posed problems to the second-order hyperbolic system (5.145): the Cauchy problem and the mixed initial-boundary value problem.

The so-called Cauchy problem is to find a solution $\boldsymbol{u}(t,\boldsymbol{x})$ of system (5.145) in $t > 0$, $\boldsymbol{x} \in \mathbb{R}^3$ such that it satisfies the initial condition at $t = 0$:

$$u_i(0,\boldsymbol{x}) = u_i^0(\boldsymbol{x}), \quad \frac{\partial u_i}{\partial t}(0,\boldsymbol{x}) = u_i^1(\boldsymbol{x}) \quad (i=1,2,3), \tag{5.151}$$

where $u_i^0(\boldsymbol{x})$ and $u_i^1(\boldsymbol{x})$ ($i=1,2,3$) are suitably smooth known functions.

The mixed initial-boundary value problem should be posed as follows: For a domain Ω in $\mathbb{R}^3$, find a solution $\boldsymbol{u}(t,\boldsymbol{x})$ of system (5.145) in $t > 0$, $\boldsymbol{x} \in \Omega$ such that it satisfies

5.5 System of Elastodynamics and Its Mathematical Structure

both the initial condition of the form (5.151) at $t = 0$ and one of the following boundary conditions on the boundary $\partial\Omega$ of Ω:

$$u_i|_{\partial\Omega} = h_i(t,\boldsymbol{x}) \quad (i=1,2,3) \tag{5.152}$$

or

$$\left.\sum_{j,k,l=1}^{3} a_{ijkl}\frac{\partial u_k}{\partial x_l}n_j\right|_{\partial\Omega} = \sigma_i(t,\boldsymbol{x}) \quad (i=1,2,3), \tag{5.153}$$

where $T > 0$ is any given positive number, and h_i and σ_i ($i = 1, 2, 3$) are suitably smooth known functions satisfying certain compatibility conditions on $\{t = 0\} \times \partial\Omega$. The boundary condition (5.152) corresponds to giving the displacement $\boldsymbol{u}$ on the boundary $\partial\Omega$ of the elastic body, while the boundary condition (5.153) corresponds to giving the stress vector $\boldsymbol{Pn}$ on the boundary $\partial\Omega$. In application, we often encounter the case that these two kinds of conditions are mixed together: the condition of the form (5.152) is given on one part Γ_1 of $\partial\Omega$, while the condition of the form (5.153) is given on the other part Γ_2 of $\partial\Omega$.

The system (5.145) of linear elastodynamics is a second-order linear hyperbolic system, whose well-posed problems can be handled by using standard methods; for instance, it can be discussed by means of the theory of semigroup of linear operators (see [9]).

In particular, for the case of isotropic materials, many well-posed problems of system (5.145) can be simplified into the corresponding problems of wave equations. In fact, from (5.135) we have

$$a_{ijkl} = \lambda \delta_{ij}\delta_{kl} + \mu(\delta_{ik}\delta_{jl} + \delta_{il}\delta_{jk}),$$

and then system (5.145) can be reduced to

$$\frac{\partial^2 \boldsymbol{u}}{\partial t^2} = \mu \Delta \boldsymbol{u} + (\lambda + \mu)\operatorname{grad}\operatorname{div}\boldsymbol{u}. \tag{5.154}$$

Here, for simplicity we assume that $\rho_0 = 1$, $\boldsymbol{b} = \boldsymbol{0}$.

Now we consider the Cauchy problem of system (5.154). From Lemma 1.3 in Chapter 1, any vector field $\boldsymbol{u}$ can be decomposed into the superposition of a longitudinal field and a transverse field:

$$\boldsymbol{u} = \boldsymbol{v} + \boldsymbol{w}, \tag{5.155}$$

where $\boldsymbol{v}$ and $\boldsymbol{w}$ are the longitudinal field and transverse field, respectively, i.e., they satisfy the conditions

$$\operatorname{rot}\boldsymbol{v} = \boldsymbol{0}, \quad \operatorname{div}\boldsymbol{w} = 0. \tag{5.156}$$

This decomposition corresponds to decomposing the elastic wave into an irrotational expansion part and an isometric part. By Lemma 1.2 in Chapter 1, the longitudinal field $\boldsymbol{v}$ can be expressed by the gradient of a scalar field ϕ:

$$\boldsymbol{v} = \operatorname{grad}\phi. \tag{5.157}$$

Thus, for the longitudinal field v, system (5.154) is reduced to

$$\frac{\partial^2}{\partial t^2}\operatorname{grad}\phi = \mu\Delta\operatorname{grad}\phi + (\lambda+\mu)\operatorname{grad}\Delta\phi,$$

i.e.,

$$\frac{\partial^2 v}{\partial t^2} = a_1^2 \Delta v, \tag{5.158}$$

where

$$a_1^2 = \lambda + 2\mu, \tag{5.159}$$

while, for the transverse field w, system (5.154) is reduced to

$$\frac{\partial^2 w}{\partial t^2} = a_2^2 \Delta w, \tag{5.160}$$

where

$$a_2^2 = \mu. \tag{5.161}$$

Then the Cauchy problem of system (5.154) of the isotropic linear elastodynamics can be reduced to the corresponding problems of wave equations (5.158) and (5.160).

We assume that the initial condition is given by (5.151). The initial data u^0 and u^1 can be decomposed as the sum of longitudinal field and transverse field:

$$u^0 = u_L^0 + u_T^0, \tag{5.162}$$
$$u^1 = u_L^1 + u_T^1, \tag{5.163}$$

where u_L^0 and u_L^1 are longitudinal fields, and u_T^0 and u_T^1 are transverse fields. Suppose that v is the solution of the wave equation (5.158) under the initial condition

$$v\Big|_{t=0} = u_L^0, \qquad \frac{\partial v}{\partial t}\Big|_{t=0} = u_L^1, \tag{5.164}$$

and w is the solution of the wave equation (5.160) under the initial condition

$$w\Big|_{t=0} = u_T^0, \qquad \frac{\partial w}{\partial t}\Big|_{t=0} = u_T^1. \tag{5.165}$$

It is not hard to verify that v and w are still the longitudinal field and transverse field, respectively. So their superposition, i.e., (5.155), gives the solution to the Cauchy problem (5.154) and (5.151).

In the actual solving process, decomposing the initial data as the superposition of a longitudinal field and a transverse field is not always convenient. Utilizing (5.155), we can see that system (5.154) can also be reduced to the following fourth-order equation (left to

5.5 System of Elastodynamics and Its Mathematical Structure

the reader as an exercise):

$$\left(\frac{\partial^2}{\partial t^2} - a_1^2 \Delta\right)\left(\frac{\partial^2}{\partial t^2} - a_2^2 \Delta\right) u = 0. \tag{5.166}$$

Using the formulas for solving the Cauchy problem of homogeneous and inhomogeneous wave equations (see, for instance, [1], [2])), the expression of the solution to the Cauchy problem (5.154) and (5.151) can be given accordingly.

It is well known that, for the wave equation (5.158), the domain of dependence of the solution at time t and spatial position x is the sphere of radius $a_1 t$ centered at x (see [1], for instance); while, for the wave equation (5.160), the corresponding domain of dependence is the sphere of radius $a_2 t$ centered at x. However, it cannot be directly concluded that the corresponding domain of dependence of the solution to system (5.154) is made up of those two spheres centered at x with radii $a_1 t$ and $a_2 t$, respectively. This is because, in the decompositions (5.162) and (5.163), the values of the longitudinal field and the transverse field on the right-hand side at a certain point are not determined only by the local behavior near this point of the vector fields on the left-hand side; they depend on the vector fields decomposed in the whole space. But using the expression of the solution to the Cauchy problem of system (5.154) derived from system (5.166), we can prove that the domain of dependence of the solution to system (5.154) at time t and position x is the spherical shell domain surrounded by two spheres centered at x with radii $a_1 t$ and $a_2 t$, respectively (see [2], [8]).

Now we show that, in addition to the strong ellipticity condition (5.149), there is a stronger condition satisfied by the linear elastic tensor $A = (a_{ijkl})$. To derive this condition, we consider the conversion of the total mechanical energy in the elastic body. Taking any given $G_0 \subset \Omega$, in a certain period of time, the total work done by both the external force acting on G_0 and the stress on the boundary S_0 of G_0 is reflected partly as the increment of kinetic energy in G_0 and partly as the deformation energy (also called the strain energy) to be stored in G_0. Denoting by W the deformation energy per unit volume, the total mechanical energy in G_0 is

$$\frac{1}{2}\int_{G_0} \rho_0 \left|\frac{\partial u}{\partial t}\right|^2 dx + \int_{G_0} W dx.$$

The work done by the force acting on G_0 in unit time is (see (5.72))

$$\int_{G_0} \rho_0 b \cdot \frac{\partial u}{\partial t} dx + \int_{S_0} (Pn) \cdot \frac{\partial u}{\partial t} dS_0$$

$$= \int_{G_0} \rho_0 b \cdot \frac{\partial u}{\partial t} dx + \sum_{i,j=1}^{3} \int_{G_0} \frac{\partial}{\partial x_j}\left(p_{ij} \frac{\partial u_i}{\partial t}\right) dx.$$

In obtaining the right-hand side of the above formula, we used Green's formula. From the above, we have

$$\frac{d}{dt}\left(\frac{1}{2}\int_{G_0} \rho_0 \left|\frac{\partial u}{\partial t}\right|^2 dx + \int_{G_0} W dx\right)$$

$$= \sum_{i,j=1}^{3} \int_{G_0} \frac{\partial}{\partial x_j}\left(p_{ij} \frac{\partial u_i}{\partial t}\right) dx + \int_{G_0} \rho_0 b \cdot \frac{\partial u}{\partial t} dx. \tag{5.167}$$

Using the conservation equation (5.74) of momentum, and noting the symmetry of the stress tensor $\boldsymbol{P} = (p_{ij})$ (this can be seen from (5.132) and (5.133)), it is easy to obtain

$$\frac{d}{dt}\int_{G_0} W dx = \sum_{i,j=1}^{3} \int_{G_0} p_{ij} \frac{\partial^2 u_i}{\partial t \partial x_j} dx$$

$$= \sum_{i,j=1}^{3} \int_{G_0} p_{ij} \frac{\partial e_{ij}}{\partial t} dx.$$

Along with the generalized Hooke's law (5.132), the above formula can then be rewritten as

$$\frac{d}{dt}\int_{G_0} W dx = \frac{d}{dt} \sum_{i,j,k,l=1}^{3} \frac{1}{2} \int_{G_0} a_{ijkl} e_{ij} e_{kl} dx. \tag{5.168}$$

Taking the strain energy density function of the elastic body in the natural state (assumed to be at $t = 0$) to be $W = 0$, integrating the above formula from 0 to t, and noting the arbitrariness of $G_0 \subset \Omega$, we then have

$$W = \frac{1}{2} \sum_{i,j,k,l=1}^{3} a_{ijkl} e_{ij} e_{kl}. \tag{5.169}$$

This is the expression of the stored-energy function (namely, the strain energy density function) in the linear case. Physically, as long as the strain tensor $\boldsymbol{E} = (e_{ij})$ is not zero, W is always positive. Therefore, another reasonable assumption for the linear elastic tensor $\boldsymbol{A} = (a_{ijkl})$ is the following stability condition.

Definition 5.6. *If there exists a constant $\widetilde{\alpha} > 0$ such that*

$$\sum_{i,j,k,l=1}^{3} a_{ijkl} e_{ij} e_{kl} \geq \widetilde{\alpha} |\boldsymbol{E}|^2 \tag{5.170}$$

holds for any given symmetric matrix $\boldsymbol{E} = (e_{ij})$, where $|\boldsymbol{E}|^2 = \sum_{i,j=1}^{3} e_{ij}^2$, then we say that $\boldsymbol{A} = (a_{ijkl})$ satisfies the stability condition.

What is the relation between the strong ellipticity condition (5.149) and the stability condition (5.170)? Now we prove that the stability condition can infer the strong ellipticity condition. In fact, taking

$$e_{ij} = \frac{1}{2}(\xi_i \eta_j + \xi_j \eta_i) \qquad (i,j = 1,2,3),$$

and noting the symmetry (5.133) of $\boldsymbol{A}$, from the stability condition (5.170) we have

$$\sum_{i,j,k,l=1}^{3} a_{ijkl} \xi_i \xi_k \eta_j \eta_l = \sum_{i,j,k,l=1}^{3} a_{ijkl} e_{ij} e_{kl}$$

$$\geq \frac{\widetilde{\alpha}}{4} \sum_{i,j=1}^{3} (\xi_i \eta_j + \xi_j \eta_i)^2 \geq \frac{\widetilde{\alpha}}{2} |\boldsymbol{\xi}|^2 |\boldsymbol{\eta}|^2.$$

5.5 System of Elastodynamics and Its Mathematical Structure

This is exactly the strong ellipticity condition. However, the inverse is not true, which can be seen from the following discussion on the isotropic materials. For the isotropic materials, it can be proved that (see exercise 8) the stability condition (5.170) is equivalent to the requirement that the Lamé constants λ and μ satisfy

$$\mu > 0, \qquad \kappa = \lambda + \frac{2}{3}\mu > 0 \tag{5.171}$$

(comparing with (5.150)!).

In the following discussion on the steady-state problems, the stability condition will play an important role.

5.5.2 System of Nonlinear Elastodynamics

Suppose that the constitutive relation (5.76) of the elastic body is known; then from definition (5.69) of the Piola stress tensor, we have

$$P(x) = \hat{P}(F(x)), \tag{5.172}$$

where $\hat{P}(F) = \det F \cdot \hat{T}(F) F^{-T}$. Plugging (5.172) into the conservation equation (5.74) of momentum, we obtain the following fundamental system of nonlinear elastodynamics:

$$\rho_0 \frac{\partial^2 y_i}{\partial t^2} = \sum_{j,k,l=1}^{3} a_{ijkl}(\nabla y) \frac{\partial^2 y_k}{\partial x_j \partial x_l} + \rho_0 b_i \qquad (i = 1, 2, 3), \tag{5.173}$$

where

$$a_{ijkl}(F) = \frac{\partial p_{ij}}{\partial f_{kl}}. \tag{5.174}$$

The fourth-order tensor $A = (a_{ijkl}(F))$ is called the *elastic tensor*. Obviously, we also can write (5.173) as a system of the displacement vector $u = y - x$.

If the material is hyperelastic, then from (5.92) and (5.174) we have

$$a_{ijkl}(F) = \frac{\partial^2 \hat{W}}{\partial f_{ij} \partial f_{kl}}, \tag{5.175}$$

where $\hat{W}$ is the stored-energy function. At this time the elastic tensor A is obviously symmetric:

$$a_{ijkl}(F) = a_{klij}(F). \tag{5.176}$$

In the discussions about the systems of elastodynamics and elastostatics (i.e., steady-state case; see the next section), one of the fundamental questions is how to give physically reasonable assumptions on the response function, or the elastic tensor, such that these systems can be mathematically and effectively analyzed.

We first assume that A satisfies the following strong ellipticity condition (whose reasonableness will be dealt with below).

Definition 5.7. *If the elastic tensor* $A = (a_{ijkl}(F))$ *satisfies the following requirements:*

$$\sum_{i,j,k,l=1}^{3} a_{ijkl}(F)\xi_i \xi_k \eta_j \eta_l > 0,$$

$$\forall F \in \mathbb{R}^{3\times 3}, \quad \det F > 0, \quad \xi, \eta \in \mathbb{R}^3 \setminus \{0\}, \tag{5.177}$$

then we say that it satisfies the **strong ellipticity condition**, *where* $\mathbb{R}^{3\times 3}$ *represents the entire set of* 3×3 *real matrices.*

If the elastic tensor A satisfies the strong ellipticity condition, we say that system (5.173) is hyperbolic. Thus, (5.173) is then a *second-order quasi-linear hyperbolic system*. The formulation of its well-posed problems is similar to the above-mentioned linear case; i.e., we can propose the Cauchy (initial value) problem and the mixed initial-boundary value problem. The typical boundary conditions in the mixed initial-boundary value problem can be given similarly to the linear case; they can be

$$y|_{\partial\Omega} = h(t, x) \tag{5.178}$$

or

$$Pn|_{\partial\Omega} = \sigma(t, x) \tag{5.179}$$

or the mixed case

$$y|_{\Gamma_1} = h(t, x), \quad Pn|_{\Gamma_2} = \sigma(t, x), \tag{5.180}$$

where $\Gamma_1 \bigcup \Gamma_2 = \partial\Omega$. Here h and σ are known as suitably smooth vector functions of t and x, while sometimes σ may depend on the unknown function y and even its tangential partial derivatives (see Example 5.1 in section 5.6.2).

The study of the well-posed problems of the system of nonlinear elastodynamics is still not satisfactory. So far the main results are basically focused on the local existence of solutions, i.e., the existence of solutions near the initial time. However, for global solutions, i.e., solutions on a large scale of time t (whether they are classical solutions or weak solutions), the problem of existence is still quite open, except for a few special situations.

5.5.3 System of First-Order Nonlinear Elastodynamics of Conservation Laws

Sometimes the form of conservation laws brings us a lot of convenience when discussing the system of elastodynamics, and it is even more so when discussing those important problems such as shock waves.

We can reduce (5.173) to a first-order system of conservation laws as follows. Note that (5.10), (5.173) can be written in the following equivalent form:

$$\frac{\partial f_{kl}}{\partial t} - \frac{\partial v_k}{\partial x_l} = 0 \quad (k, l = 1, 2, 3), \tag{5.181}$$

$$\rho_0 \frac{\partial v_i}{\partial t} - \sum_{j=1}^{3} \frac{\partial}{\partial x_j} p_{ij}(F) - \rho_0 b_i = 0 \quad (i = 1, 2, 3). \tag{5.182}$$

5.5 System of Elastodynamics and Its Mathematical Structure

Setting $U = (f_{11}, f_{12}, \ldots, f_{33}, v_1, v_2, v_3)^T$ and taking $\rho_0 = 1$, $\boldsymbol{b} = \boldsymbol{0}$ for simplicity, system (5.181)–(5.182) can be written as the following first-order quasi-linear system of conservation laws:

$$\frac{\partial U}{\partial t} + \sum_{j=1}^{3} \frac{\partial}{\partial x_j} V_j(U) = 0, \tag{5.183}$$

where $V_1 = -(v_1, 0, 0, v_2, 0, 0, v_3, 0, 0, p_{11}, p_{21}, p_{31})^T$, $V_2 = -(0, v_1, 0, 0, v_2, 0, 0, v_3, 0, p_{12}, p_{22}, p_{32})^T$, $V_3 = -(0, 0, v_1, 0, 0, v_2, 0, 0, v_3, p_{13}, p_{23}, p_{33})^T$.

Now we discuss the type of system (5.183). For this, we write it in the following form:

$$\frac{\partial U}{\partial t} + \sum_{j=1}^{3} B_j(U) \frac{\partial U}{\partial x_j} = 0, \tag{5.184}$$

where

$$B_j(U) = \nabla_U V_j, \quad (j = 1, 2, 3); \tag{5.185}$$

here ∇_U represents taking the gradient with respect to U, so B_j is a 12×12 matrix.

Definition 5.8. *If for U in the domain under consideration and any given $\boldsymbol{\eta} \in \mathbb{R}^3 \setminus \{0\}$, the matrix*

$$\Lambda(U, \boldsymbol{\eta}) = \sum_{j=1}^{3} \eta_j B_j(U) \tag{5.186}$$

has $n(=12)$ real eigenvalues

$$\lambda_1(U, \boldsymbol{\eta}), \ldots, \lambda_n(U, \boldsymbol{\eta}),$$

and there are n linearly independent eigenvectors corresponding to these eigenvalues, then we say that system (5.184) *is a* hyperbolic system.

Now we prove that if the elastic tensor $A = (a_{ijkl})$ satisfies the strong ellipticity condition (5.177), then system (5.184) is hyperbolic in the sense of Definition 5.8. To this end, we first examine the eigenvalues of $\Lambda(U, \boldsymbol{\eta})$. From the definitions of B_j and V_j, it is not hard to see that

$$B_j(U) = \begin{pmatrix} 0 & -E_j \\ -A_j & 0 \end{pmatrix} \quad (j = 1, 2, 3), \tag{5.187}$$

where

$$A_j = \begin{pmatrix} a_{1j11} & a_{1j12} & \cdots & a_{1j33} \\ a_{2j11} & a_{2j12} & \cdots & a_{2j33} \\ a_{3j11} & a_{3j12} & \cdots & a_{3j33} \end{pmatrix} \quad (j = 1, 2, 3) \tag{5.188}$$

are 3 × 9 matrices, and

$$E_1 = \begin{pmatrix} 1 & 0 & 0 \\ 0 & 0 & 0 \\ 0 & 0 & 0 \\ 0 & 1 & 0 \\ 0 & 0 & 0 \\ 0 & 0 & 0 \\ 0 & 0 & 1 \\ 0 & 0 & 0 \\ 0 & 0 & 0 \end{pmatrix}, \quad E_2 = \begin{pmatrix} 0 & 0 & 0 \\ 1 & 0 & 0 \\ 0 & 0 & 0 \\ 0 & 0 & 0 \\ 0 & 1 & 0 \\ 0 & 0 & 0 \\ 0 & 0 & 0 \\ 0 & 0 & 1 \\ 0 & 0 & 0 \end{pmatrix}, \quad E_3 = \begin{pmatrix} 0 & 0 & 0 \\ 0 & 0 & 0 \\ 1 & 0 & 0 \\ 0 & 0 & 0 \\ 0 & 0 & 0 \\ 0 & 1 & 0 \\ 0 & 0 & 0 \\ 0 & 0 & 0 \\ 0 & 0 & 1 \end{pmatrix}. \tag{5.189}$$

So

$$\Lambda(U,\eta) = - \begin{pmatrix} 0 & \sum_{j=1}^{3} \eta_j E_j \\ \sum_{j=1}^{3} \eta_j A_j & 0 \end{pmatrix}. \tag{5.190}$$

Using the operation rules of determinants, it is not hard to verify that

$$\det(\Lambda(U,\eta) - \lambda I) = \lambda^6 \det\left(\lambda^2 \delta_{ik} - \sum_{j,l=1}^{3} a_{ijkl}\eta_j \eta_l\right), \tag{5.191}$$

where I is the unit matrix, and δ_{ik} is the Kronecker symbol. This implies that $\Lambda(U,\eta)$ has a zero eigenvalue with multiplicity 6. Besides, from (5.176) and the strong ellipticity condition (5.177), we know that the 3×3 matrix $\left(\sum_{j,k=1}^{3} a_{ijkl}\eta_j\eta_l\right)$ is symmetric and positively definite, and thus has three positive eigenvalues. Therefore, the other six eigenvalues of $\Lambda(U,\eta)$ are square roots of eigenvalues of $\left(\sum_{j,l=1}^{3} a_{ijkl}\eta_j\eta_l\right)$, respectively.

Now we look at the eigenvectors of $\Lambda(U,\eta)$.

Lemma 5.4. *For any given $\eta \in \mathbb{R}^3 \setminus \{0\}$, there are six linearly independent eigenvectors corresponding to the zero eigenvalues of matrix $\Lambda(U,\eta)$.*

Proof. Suppose that $(\omega_1, \omega_2, \ldots, \omega_{12})^T$ is the eigenvector of $\Lambda(U,\eta)$ corresponding to the zero eigenvalue. From the first nine rows of $\Lambda(U,\eta)$, we immediately have

$$\eta_i \omega_{10} = \eta_i \omega_{11} = \eta_i \omega_{12} = 0 \quad (i = 1,2,3).$$

Since η_1, η_2, η_3 are not zero at the same time, the above formula yields

$$\omega_{10} = \omega_{11} = \omega_{12} = 0.$$

Thus, $(\omega_1, \omega_2, \ldots, \omega_{12})$ has at most nine degrees of freedom. From expression (5.190) of $\Lambda(U,\eta)$ we can see that, to prove the conclusion of the lemma, we need only show that the last three rows of $\Lambda(U,\eta)$ are linearly independent. Suppose that there exist ξ_1, ξ_2, ξ_3 such that the sum of the 10th, 11th, and 12th rows of $\Lambda(U,\eta)$ multiplied by ξ_1, ξ_2, ξ_3, respectively, is equal to the zero vector, i.e.,

$$\sum_{i,j=1}^{3} a_{ij11}\xi_i \eta_j = \sum_{i,j=1}^{3} a_{ij12}\xi_i \eta_j = \cdots = \sum_{i,j=1}^{3} a_{ij33}\xi_i \eta_j = 0.$$

5.5 System of Elastodynamics and Its Mathematical Structure

Multiplying the first formula from the left by $\xi_1\eta_1$, the second by $\xi_1\eta_2,\ldots$, the ninth by $\xi_3\eta_3$, and summing them up, we have

$$\sum_{i,j,k,l=1}^{3} a_{ijkl}\xi_i\xi_k\eta_j\eta_l = 0.$$

Using the strong ellipticity condition (5.177), from the above formula we immediately obtain $\xi_1 = \xi_2 = \xi_3 = 0$. The proof of the lemma is finished. $\square$

Lemma 5.5. *Suppose that the material is hyperelastic; then for any given $\eta \in \mathbb{R}^3\setminus\{0\}$, matrix $\Lambda(U,\eta)$ possesses six linearly independent eigenvectors corresponding to nonzero eigenvalues.*

Proof. Let $(\omega_1,\omega_2,\ldots,\omega_{12})^T$ be an eigenvector of matrix $\Lambda(U,\eta)$ corresponding to the nonzero eigenvalue λ. From the first six rows of $\Lambda(U,\eta)$, we obtain

$$\begin{cases} \lambda\omega_1 + \eta_1\omega_{10} = \lambda\omega_2 + \eta_2\omega_{10} = \lambda\omega_3 + \eta_3\omega_{10} = 0, \\ \lambda\omega_4 + \eta_1\omega_{11} = \lambda\omega_5 + \eta_2\omega_{11} = \lambda\omega_6 + \eta_3\omega_{11} = 0, \\ \lambda\omega_7 + \eta_1\omega_{12} = \lambda\omega_8 + \eta_2\omega_{12} = \lambda\omega_9 + \eta_3\omega_{12} = 0. \end{cases} \quad (5.192)$$

Noting $\lambda \neq 0$, denote

$$\alpha_i = -\eta_i/\lambda \quad (i=1,2,3), \quad (5.193)$$

$$\xi_1 = \omega_{10}, \quad \xi_2 = \omega_{11}, \quad \xi_3 = \omega_{12}; \quad (5.194)$$

then from (5.192) we have

$$\begin{cases} \omega_1 = \alpha_1\xi_1, & \omega_2 = \alpha_2\xi_1, & \omega_3 = \alpha_3\xi_1, \\ \omega_4 = \alpha_1\xi_2, & \omega_5 = \alpha_2\xi_2, & \omega_6 = \alpha_3\xi_2, \\ \omega_7 = \alpha_1\xi_3, & \omega_8 = \alpha_2\xi_3, & \omega_9 = \alpha_3\xi_3. \end{cases} \quad (5.195)$$

Noticing (5.193)–(5.195), and corresponding to the 10th, 11th, and 12th rows of $\Lambda(U,\eta)$, the relations satisfied by the eigenvector should be

$$\sum_{j,k,l=1}^{3} a_{ijkl}\eta_j\eta_l\xi_k - \lambda^2\xi_i = 0 \quad (i=1,2,3). \quad (5.196)$$

This shows that vector $\boldsymbol{\xi} = (\xi_1,\xi_2,\xi_3)^T$ is an eigenvector of matrix $\left(\sum_{j,l=1}^{3} a_{ijkl}\eta_j\eta_l\right)$ corresponding to the eigenvalue λ^2. On the other hand, if vector $\boldsymbol{\xi} = (\xi_1,\xi_2,\xi_3)^T$ is an eigenvector of matrix $\left(\sum_{j,l=1}^{3} a_{ijkl}\eta_j\eta_l\right)$ corresponding to the eigenvalues $\lambda^2 > 0$, then vector $\omega = (\omega_1,\omega_2,\ldots,\omega_{12})^T$ given by (5.194), (5.195) is an eigenvector of matrix $\Lambda(U,\eta)$ corresponding to the eigenvalue λ. Then, for each eigenvector (corresponding to the eigenvalue $\lambda^2 > 0$) of matrix $\left(\sum_{j,l=1}^{3} a_{ijkl}\eta_j\eta_l\right)$, we obtain two eigenvectors (corresponding to eigenvalues λ and $-\lambda$, respectively) of matrix $\Lambda(U,\eta)$. As stated before, when the material is hyperelastic (i.e., (5.176) is satisfied at this time), matrix $\left(\sum_{j,l=1}^{3} a_{ijkl}\eta_j\eta_l\right)$ is symmetric and positive definite, and thus has three eigenvectors orthogonal to each other. Thus, we obtain six eigenvectors of $\Lambda(U, \eta)$ corresponding to nonzero eigenvalues. It is not hard to verify that they are linearly independent. The proof of the lemma is finished. $\square$

From the above discussion, we have the following theorem.

Theorem 5.6. *Assume that the material is hyperelastic, and the elastic tensor $A = (a_{ijkl})$ satisfies the strong ellipticity condition; then system (5.181)–(5.182) of first-order quasi-linear partial differential equations is hyperbolic in the sense of Definition 5.8.*

When discussing the purely elastic deformation, we do not take the heat effect into account, and we assume that all of the deformation is carried out under the isothermal condition. But if we want to study the discontinuous solutions of system (5.181)–(5.182), we cannot neglect the heat effect any longer because the stress and the strain will change suddenly near the jump of the solution. Since the heat effect is inevitably accompanied by the dissipation of the mechanical energy, the process is no longer invertible, and the physical entropy of the elastic body has to increase. So when studying discontinuous solutions of the first-order system of conservation laws (5.181), (5.182), we have to supply an inequality reflecting the increment of entropy, called the *entropy inequality*. In general, the entropy inequality takes the following form:

$$\frac{\partial}{\partial t}\eta(U) + \sum_{j=1}^{3} \frac{\partial}{\partial x_j} q_j(U) \leq 0, \tag{5.197}$$

where η is called the *entropy*, and $q = (q_1, q_2, q_3)$ is called the *entropy flux* (note that the entropy and entropy flux defined here are different up to a minus sign from the genuine physical entropy and entropy flux).

Now we consider how to determine the entropy and entropy flux. For any given domain $G_0 \subset \Omega$, the total mechanical energy in G_0 is

$$\int_{G_0} \left(\frac{1}{2} |v|^2 + \hat{W}(F) \right) dx,$$

where $\hat{W}(F)$ is the stored-energy function. The work done to G_0 by the stress on the boundary S_0 of G_0 in unit time is

$$\int_{S_0} (Pn) \cdot v \, dS_0 = \int_{G_0} \sum_{i,j=1}^{3} \frac{\partial}{\partial x_j}(p_{ij}v_i) dx.$$

Here we used Green's formula. Regarding the dissipation effect of the mechanical energy, we should have

$$\frac{d}{dt} \int_{G_0} \left(\frac{1}{2} |v|^2 + \hat{W}(F) \right) dx - \int_{G_0} \sum_{i,j=1}^{3} \frac{\partial}{\partial x_j}(p_{ij}v_i) dx \leq 0.$$

Since the above formula holds for any given $G_0 \subset \Omega$, we then have

$$\frac{\partial}{\partial t}\left(\frac{1}{2}|v|^2 + \hat{W}(F) \right) - \sum_{i,j=1}^{3} \frac{\partial}{\partial x_j}(p_{ij}v_i) \leq 0. \tag{5.198}$$

5.5 System of Elastodynamics and Its Mathematical Structure

Therefore, a natural way of choosing η and $\boldsymbol{q}$ is

$$\eta = \frac{1}{2}|\boldsymbol{v}|^2 + \hat{W}(\boldsymbol{F}), \tag{5.199}$$

$$q_j = -\sum_{i=1}^{3} p_{ij} v_i \qquad (j=1,2,3). \tag{5.200}$$

However, since $\hat{W}(\boldsymbol{F})$ is not a convex function of $\boldsymbol{F}$ in general (see section 5.6 in this chapter), the entropy given by (5.199) is generally not convex.

5.5.4 Reducing the System of Elastodynamics to First-Order Symmetric Hyperbolic System

An important and interesting problem is whether or not the system (5.173) of nonlinear elastodynamics can be reduced to a first-order quasi-linear symmetric hyperbolic system. Now we discuss a special case.

Suppose that an isotropic elastic body is deformed near the reference configuration which is the natural state. From Theorem 5.5 and (5.174), it is not difficult to obtain (see (5.135))

$$a_{ijkl}(\boldsymbol{I}) = \lambda \delta_{ij}\delta_{kl} + \mu(\delta_{ik}\delta_{jl} + \delta_{il}\delta_{jk}). \tag{5.201}$$

We know that the assumption that $(a_{ijkl}(\boldsymbol{I}))$ satisfies the strong ellipticity condition is equivalent to (5.150), and we further assume that

$$\lambda + 2\mu > \mu > 0. \tag{5.202}$$

Since the stored-energy function $\hat{W}(\boldsymbol{F})$ is not in general a convex function of $\boldsymbol{F}$ (see section 5.6 in this chapter), the elastic tensor $\boldsymbol{A} = (a_{ijkl})$ is generally not positively definite. However, some kind of positive definiteness is required in the following reduction. For this purpose, just as done in [8], instead of (a_{ijkl}), we introduce the tensor

$$\widetilde{a}_{ijkl}(\boldsymbol{F}) = a_{ijkl}(\boldsymbol{F}) + \mu(\delta_{ij}\delta_{kl} - \delta_{il}\delta_{jk}). \tag{5.203}$$

Tensor $(\widetilde{a}_{ijkl})$ still possesses the following symmetry:

$$\widetilde{a}_{ijkl}(\boldsymbol{F}) = \widetilde{a}_{klij}(\boldsymbol{F}), \tag{5.204}$$

and since

$$\sum_{j,k,l=1}^{3} (\delta_{ij}\delta_{kl} - \delta_{il}\delta_{jk})\frac{\partial f_{kl}}{\partial x_j}$$

$$= \sum_{k=1}^{3} \frac{\partial^2 y_k}{\partial x_i \partial x_k} - \sum_{k=1}^{3} \frac{\partial^2 y_k}{\partial x_k \partial x_i} = 0, \tag{5.205}$$

it is easy to see that system (5.182) can be written as (noting that we already assume $\rho_0 = 1$, $\boldsymbol{b} = \boldsymbol{0}$)

$$\frac{\partial v_i}{\partial t} - \sum_{j,k,l=1}^{3} \widetilde{a}_{ijkl}(\boldsymbol{F}) \frac{\partial}{\partial x_j} f_{kl} = 0 \quad (i = 1, 2, 3). \tag{5.206}$$

In addition, when $|\nabla \boldsymbol{u}|$ is sufficiently small, tensor $(\widetilde{a}_{ijkl}(\boldsymbol{F}))$ is positively definite; i.e., for any given $\boldsymbol{\zeta} = (\zeta_{ij}) \in \mathbb{R}^{3\times 3} \setminus \{\boldsymbol{0}\}$, it holds that

$$\sum_{i,j,k,l=1}^{3} \widetilde{a}_{ijkl}(\boldsymbol{F}) \zeta_{ij} \zeta_{kl} > 0. \tag{5.207}$$

In fact, from (5.201)–(5.203) we have

$$\sum_{i,j,k,l=1}^{3} \widetilde{a}_{ijkl}(\boldsymbol{I}) \zeta_{ij} \zeta_{kl} = (\lambda + \mu)(\operatorname{tr} \boldsymbol{\zeta})^2 + \mu |\boldsymbol{\zeta}|^2 > 0, \tag{5.208}$$

where $\operatorname{tr} \boldsymbol{\zeta} = \sum_{i=1}^{3} \zeta_{ii}$ and $|\boldsymbol{\zeta}|^2 = \sum_{i,j=1}^{3} |\zeta_{ij}|^2$. Equation (5.207) follows immediately from this when $|\nabla \boldsymbol{u}|$ is sufficiently small.

Regarding tensor $(\widetilde{a}_{ijkl}(\boldsymbol{F}))$ as a symmetric and positively definite 9×9 matrix $\widetilde{A}$, and multiplying system (5.181) by $\widetilde{A}$ from the left, we obtain the following system:

$$\sum_{p,q=1}^{3} \widetilde{a}_{klpq}(\boldsymbol{F}) \frac{\partial f_{pq}}{\partial t} - \sum_{p,q=1}^{3} \widetilde{a}_{klpq}(\boldsymbol{F}) \frac{\partial v_p}{\partial x_q} = 0$$
$$(k, l = 1, 2, 3). \tag{5.209}$$

From the positive definiteness of $\widetilde{A}$ we know that system (5.209) is equivalent to system (5.181).

Combining systems (5.209) and (5.206), we can write the form

$$K_0 \frac{\partial U}{\partial t} + \sum_{j=1}^{3} K_j(U) \frac{\partial U}{\partial x_j} = 0, \tag{5.210}$$

where

$$K_0 = \begin{pmatrix} \widetilde{A} & 0 \\ 0 & I_{3\times 3} \end{pmatrix}, \quad K_j = \begin{pmatrix} 0 & -\widetilde{A}_j^T \\ -\widetilde{A}_j & 0 \end{pmatrix}$$
$$(j = 1, 2, 3), \tag{5.211}$$

$I_{3\times 3}$ is the 3×3 unit matrix, and $\widetilde{A}_j$ is a 3×9 matrix which is obtained by replacing the entries of A_j given by (5.188) with the corresponding entries of $\widetilde{A}$. Obviously, K_0 is a positive definite symmetric matrix, and K_j ($j = 1, 2, 3$) are symmetric matrices. Therefore, (5.210) is a quasi-linear symmetric hyperbolic system. Thus we get that under assumption (5.202), the system of nonlinear elastodynamics for the isotropic material deformed

5.5 System of Elastodynamics and Its Mathematical Structure

near the natural state can always be reduced to a first-order symmetric hyperbolic system. The same conclusion can certainly be established for the system of linear elastodynamics.

Although we have reduced system (5.181)–(5.182) to the first-order quasi-linear symmetric hyperbolic system of (5.209) and (5.206), unfortunately, this new system does not possess the form of conservation laws. Besides, even if it can be rewritten in the form of conservation laws, it is not expected to have the same Rankine–Hugoniot conditions (for the higher dimensional first-order quasi-linear hyperbolic system of conservation laws, similar Rankine–Hugoniot conditions are satisfied on the strong jump surface of the solution similarly to that in section 2.4 of Chapter 2) on the strong jump as the original system (5.181), which is because system (5.209) is obtained by multiplying system (5.181) from the left by the 9×9 matrix $\widetilde{A} = (\widetilde{a}_{ijkl}(F))$. That is to say, the obtained system is equivalent only to the original system in the range of classical solutions even if it can be written in the form of conservation laws. But for the situation discussed in this section, under the condition that the original conservation laws are essentially unchanged, we can still reduce the system to a first-order quasi-linear symmetric hyperbolic system of conservation laws by introducing new unknown functions.

Instead of (5.199), take

$$\eta = \hat{W}(F) + \frac{\mu}{2} \sum_{i,j,k,l=1}^{3} (\delta_{ij}\delta_{kl} - \delta_{il}\delta_{jk}) f_{ij} f_{kl} + \frac{1}{2}|v|^2, \tag{5.212}$$

where $\hat{W}(F)$ is the stored-energy function of the material. It is easy to see that, after having made the above modification to the entropy function given by (5.199), the matrix $\frac{\partial^2 \eta}{\partial U^2}$ formed by the second-order partial derivatives of η given by (5.212) with respect to the components of $U = (f_{11}, \ldots, f_{33}, v_1, v_2, v_3)^{\mathrm{T}}$ is exactly the positively definite matrix K_0 in (5.211):

$$\frac{\partial^2 \eta}{\partial U^2} = K_0 > 0. \tag{5.213}$$

This implies that the entropy η given by (5.212) is convex.

Now we show that η given by (5.212) satisfies the following additional conservation law:

$$\frac{\partial \eta}{\partial t} + \sum_{j=1}^{3} \frac{\partial q_j}{\partial x_j} = 0, \tag{5.214}$$

where

$$q_j = -\sum_{i=1}^{3} p_{ij} v_i - \mu \sum_{i,k,l=1}^{3} (\delta_{ij}\delta_{kl} - \delta_{il}\delta_{jk}) v_i f_{kl} \quad (j = 1,2,3). \tag{5.215}$$

In fact, from the above discussion we can see that, in the range of classical solutions (at this time there is no dissipation of the mechanical energy), if η and q are given by (5.199) and (5.200), respectively, then the conservation law (5.214) is satisfied. Thus, to verify that the conservation law (5.214) is satisfied when η and q are given by (5.212) and (5.215), respectively, we need only prove

$$\frac{\partial}{\partial t}\sum_{i,j,k,l=1}^{3}(\delta_{ij}\delta_{kl}-\delta_{il}\delta_{jk})f_{ij}f_{kl}$$

$$-2\sum_{i,j,k,l=1}^{3}\frac{\partial}{\partial x_{j}}((\delta_{ij}\delta_{kl}-\delta_{il}\delta_{jk})v_{i}f_{kl})=0. \quad (5.216)$$

It is easy to verify the correctness of the above formula from system (5.181) and (5.205); here we omit the details, which are left to the reader.

Thus, from Theorem 2.1 of Chapter 2 we know that system (5.181), (5.182) of conservation laws is a first-order quasi-linear symmetric hyperbolic system with respect to the new unknown functions

$$Z=\frac{\partial \eta}{\partial U}=\left(\frac{\partial \eta}{\partial f_{11}},\ldots,\frac{\partial \eta}{\partial v_{3}}\right)^{T}.$$

From Theorem 2.1 of Chapter 2, we can prove that if the stored-energy function is strictly polyconvex (see Definition 5.9 in section 5.6.2 of this chapter for the concept of polyconvex), then the general system of nonlinear elastodynamics can be reduced to a first-order quasi-linear symmetric hyperbolic system of conservation laws (see [12]).

5.5.5 System of One-Dimensional Nonlinear Elastodynamics

In this subsection we discuss two cases which can be reduced to one-dimensional problems.

(1°) Pure axial deformation of isotropic materials.

Suppose that the material deforms only along the e_1 direction, and the deformation depends only on t and x_1, while it is independent of x_2 and x_3. This kind of pure axial deformation is given by the following formula:

$$y_1 = x_1 + u_1(t,x_1), \quad y_2 = x_2, \quad y_3 = x_3. \quad (5.217)$$

Then we obviously have

$$F=\begin{pmatrix} 1+\frac{\partial u_1}{\partial x_1} & 0 & 0 \\ 0 & 1 & 0 \\ 0 & 0 & 1 \end{pmatrix},$$

while the left Cauchy–Green strain tensor is

$$B=\begin{pmatrix} \left(1+\frac{\partial u_1}{\partial x_1}\right)^2 & 0 & 0 \\ 0 & 1 & 0 \\ 0 & 0 & 1 \end{pmatrix}.$$

Thus, from Theorem 5.3 we know that the Cauchy stress tensor T takes the following form:

$$T=\begin{pmatrix} t_{11}\left(\frac{\partial u_1}{\partial x_1}\right) & 0 & 0 \\ 0 & t_{22}\left(\frac{\partial u_1}{\partial x_1}\right) & 0 \\ 0 & 0 & t_{33}\left(\frac{\partial u_1}{\partial x_1}\right) \end{pmatrix},$$

5.5 System of Elastodynamics and Its Mathematical Structure

while the Piola stress tensor is

$$P = \begin{pmatrix} t_{11}\left(\frac{\partial u_1}{\partial x_1}\right) & 0 & 0 \\ 0 & \left(1+\frac{\partial u_1}{\partial x_1}\right)t_{22}\left(\frac{\partial u_1}{\partial x_1}\right) & 0 \\ 0 & 0 & \left(1+\frac{\partial u_1}{\partial x_1}\right)t_{33}\left(\frac{\partial u_1}{\partial x_1}\right) \end{pmatrix}.$$

From the above expression of P we can see that, for this kind of deformation, the components of the elastic tensor $A = (a_{ijkl})$ are zero, except a_{1111}, a_{2211}, and a_{3311}. And the strong ellipticity condition (5.177) implies that the above three components are all positive; in particular,

$$a_{1111} = \frac{\partial t_{11}}{\partial\left(\frac{\partial u_1}{\partial x_1}\right)} > 0. \tag{5.218}$$

Thus, the first equation in the conservation equation (5.74) of momentum is then reduced to

$$\rho_0 \frac{\partial^2 u_1}{\partial t^2} = \frac{\partial}{\partial x_1} t_{11}\left(\frac{\partial u_1}{\partial x_1}\right) + \rho_0 b_1. \tag{5.219}$$

Therefore, this kind of pure axial deformation is then reduced to a one-dimensional problem described by (5.219), while (5.218) means that (5.219) is a *one-dimensional quasi-linear wave equation*.

What needs to be explained is that the kind of deformation given by (5.217) is not easily realized in practice. For the real elastic rod, it cannot describe the deformation generated by the axial tension and compression. In addition, this kind of deformation is impossible for incompressible materials.

(2°) Pure shear deformation of isotropic materials.

This kind of deformation also happens only along the e_1 direction but depends only on t and x_2 (or x_3), i.e., takes the following form:

$$y_1 = x_1 + u_1(t, x_2), \quad y_2 = x_2, \quad y_3 = x_3. \tag{5.220}$$

Then we obviously have

$$F = \begin{pmatrix} 1 & \gamma & 0 \\ 0 & 1 & 0 \\ 0 & 0 & 1 \end{pmatrix},$$

where $\gamma = \frac{\partial u_1}{\partial x_2}$. The corresponding left Cauchy–Green strain tensor is

$$B = \begin{pmatrix} 1+\gamma^2 & \gamma & 0 \\ \gamma & 1 & 0 \\ 0 & 0 & 1 \end{pmatrix},$$

while the three principal invariants of B are, respectively,

$$I_1 = 3 + \gamma^2, \quad I_2 = 3 + \gamma^2, \quad I_3 = 1.$$

Noticing that
$$B^2 = \begin{pmatrix} (1+\gamma^2)^2+\gamma^2 & \gamma(1+\gamma^2)+\gamma & 0 \\ \gamma(1+\gamma^2)+\gamma & 1+\gamma^2 & 0 \\ 0 & 0 & 1 \end{pmatrix},$$

from Theorem 5.3 we know that
$$t_{13} = t_{23} = 0, \tag{5.221}$$
$$t_{12} = \gamma \mu(\gamma^2), \tag{5.222}$$

where $\mu(\gamma^2)$ is a function of γ^2, called the *generalized shear modulus*. Then it is easy to see that the Piola stress tensor is
$$P = \begin{pmatrix} t_{11} - \gamma t_{12} & t_{12} & 0 \\ t_{21} - \gamma t_{22} & t_{22} & 0 \\ 0 & 0 & t_{33} \end{pmatrix}.$$

Because the stress tensor P depends only on $\gamma = \frac{\partial u_1}{\partial x_2}$, the components of the elastic tensor $A = (a_{ijkl})$ are all zero, except $a_{1112}, a_{1212}, a_{2112}, a_{2212}$, and a_{3312}. Taking $\xi = (1,0,0)$ and $\eta = (0,1,0)$, from the strong ellipticity condition (5.177) we obtain
$$a_{1212} = \frac{dt_{12}(\gamma)}{d\gamma} > 0. \tag{5.223}$$

Under this kind of deformation, the first equation in the conservation equation (5.74) of momentum is reduced to
$$\rho_0 \frac{\partial^2 u_1}{\partial t^2} = \frac{\partial}{\partial x_2} t_{12}\left(\frac{\partial u_1}{\partial x_2}\right) + \rho_0 b_1. \tag{5.224}$$

Thus, the pure shear deformation is then reduced to a one-dimensional problem, and (5.223) means that (5.224) is a one-dimensional quasi-linear wave equation.

5.6 Well-Posed Problems of the System of Elastostatics

In this section we discuss the steady-state problem, i.e., the situation when the elastic body reaches equilibrium. At this time, the deformation of the elastic body no longer depends on the time t; that is, y and u depend only on x.

5.6.1 System of Linear Elastostatics

For the equilibrium of the linear case, the fundamental system (5.145) of elastic mechanics is reduced to
$$-\sum_{j,k,l=1}^{3} a_{ijkl} \frac{\partial^2 u_k}{\partial x_j \partial x_l} = \rho_0 b_i \quad (i=1,2,3), \tag{5.225}$$

where b is naturally a function of x. Assume that the elastic tensor $A = (a_{ijkl})$ satisfies the strong ellipticity condition (5.149); then system (5.225) is a *second-order linear elliptic*

5.6 Well-Posed Problems of the System of Elastostatics

system. Suppose that $\Omega \subset \mathbb{R}^3$ is a bounded domain; then a typical well-posed problem for system (5.225) is to find the solution u of this system in Ω such that it satisfies boundary conditions of the form (5.152) or (5.153) on the boundary $\partial\Omega$ of Ω. Certainly, on this occasion both h and σ do not depend on t.

For systems of elliptic type, even for the most common problems with boundary conditions (5.152), preserving the ellipticity of the system is generally not enough to ensure the uniqueness of solution. It is not difficult to see this by noticing the corresponding eigenvalue problem of the elliptic equation (system of equations). But from the mechanical point of view, if the elastic body is in its natural state before deformation, then under the conditions that there is no external force and the boundary points remain unchanged (the displacement is zero), the natural state should be the unique equilibrium. Therefore, the solution of the linear boundary value problem (5.225) and (5.152) should be unique, which is precisely guaranteed by the stability condition (5.170).

Suppose that $A = (a_{ijkl})$ satisfies the stability condition (5.170); then it is not hard to verify that

$$B(u,u) \stackrel{d}{=} \int_\Omega \sum_{i,j,k,l=1}^{3} a_{ijkl} \frac{\partial u_i}{\partial x_j} \frac{\partial u_k}{\partial x_l} dx$$

$$= \int_\Omega \sum_{i,j,k,l=1}^{3} a_{ijkl} e_{ij} e_{kl} dx$$

$$\geq \widetilde{\alpha} \int_\Omega |E|^2 dx. \tag{5.226}$$

Then we have the following theorem.

Theorem 5.7 (Korn inequality). *Suppose that $\Omega \subset \mathbb{R}^3$ is a bounded domain; then there exists a constant $C_0 > 0$ such that for any given $u \in (H_0^1(\Omega))^3$, we have*

$$\int_\Omega |E|^2 dx \geq C_0 \|u\|^2_{(H^1(\Omega))^3}, \tag{5.227}$$

where $H_0^1(\Omega)$ and $H^1(\Omega)$ are Sobolev spaces (see [4]).

Readers who are not familiar with Sobolev spaces can regard the u in Theorem 5.7 as a continuously differentiable vector function vanishing on the boundary $\partial\Omega$. In (5.227),

$$\|u\|^2_{(H^1(\Omega))^3} = \|u\|^2_{(L^2(\Omega))^3} + \sum_{i,j=1}^{3} \left\| \frac{\partial u_i}{\partial x_j} \right\|^2_{L^2(\Omega)}.$$

Proof of Theorem 5.7. We need only prove (5.227) for the sufficiently smooth vector function u vanishing on the boundary $\partial\Omega$. Denote

$$\frac{\partial u_i}{\partial x_j} = e_{ij} + r_{ij}, \tag{5.228}$$

where
$$r_{ij} = \frac{1}{2}\left(\frac{\partial u_i}{\partial x_j} - \frac{\partial u_j}{\partial x_i}\right).$$

Then we have
$$\frac{\partial u_i}{\partial x_j} \cdot \frac{\partial u_j}{\partial x_i} = e_{ij}^2 - r_{ij}^2. \tag{5.229}$$

Using Green's formula and noting that $u|_{\partial\Omega} = \mathbf{0}$, we have
$$\int_\Omega \frac{\partial u_i}{\partial x_j} \cdot \frac{\partial u_j}{\partial x_i} dx = \int_\Omega \frac{\partial u_i}{\partial x_i} \cdot \frac{\partial u_j}{\partial x_j} dx. \tag{5.230}$$

Thus, from (5.229) and (5.230) we get
$$\sum_{i,j=1}^{3} \int_\Omega r_{ij}^2 dx = \int_\Omega |\boldsymbol{E}|^2 dx - \int_\Omega (\operatorname{div} \boldsymbol{u})^2 dx$$
$$\leq \int_\Omega |\boldsymbol{E}|^2 dx. \tag{5.231}$$

From (5.228) we have
$$\left(\frac{\partial u_i}{\partial x_j}\right)^2 \leq 2(e_{ij}^2 + r_{ij}^2). \tag{5.232}$$

It follows immediately from the above two formulas that
$$\sum_{i,j=1}^{3} \int_\Omega \left(\frac{\partial u_i}{\partial x_j}\right)^2 dx \leq 4 \int_\Omega |\boldsymbol{E}|^2 dx. \tag{5.233}$$

Equation (5.233) together with the well-known *Friedrichs inequality* (see [1])
$$\int_\Omega |\boldsymbol{u}|^2 dx \leq C_1 \sum_{i,j=1}^{3} \int_\Omega \left(\frac{\partial u_i}{\partial x_j}\right)^2 dx$$

yields (5.227). C_1 in the above formula is a positive constant depending only on the domain Ω. Theorem 5.7 is proved. $\square$

Remark 5.2. In general, the Korn inequality takes the following form:
$$\int_\Omega |\boldsymbol{E}|^2 dx + \|\boldsymbol{u}\|_{(L^2(\Omega))^3}^2 \geq C_0 \|\boldsymbol{u}\|_{(H^1(\Omega))^3}^2, \quad \forall \boldsymbol{u} \in (H^1(\Omega))^3 \tag{5.234}$$

(see [4]).

From (5.226) and Korn inequality (5.227), we have

$$B(\boldsymbol{u},\boldsymbol{u}) \geq \alpha_0 \|\boldsymbol{u}\|^2_{(H^1(\Omega))^3}, \quad \forall \boldsymbol{u} \in (H^1_0(\Omega))^3, \tag{5.235}$$

where $\alpha_0 > 0$ is a positive constant.

The uniqueness of solution to the boundary value problem (5.225) and (5.152) follows immediately from inequality (5.235). For this, we need only prove that the corresponding homogeneous problem (i.e., the case with $\boldsymbol{b} \equiv \boldsymbol{0}$ and $\boldsymbol{h} \equiv \boldsymbol{0}$) admits only the trivial solution $\boldsymbol{u} \equiv \boldsymbol{0}$. Multiplying (5.225) by u_i, summing for i from 1 to 3, and then integrating over Ω, we get

$$-\sum_{i,j,k,l=1}^{3} \int_{\Omega} a_{ijkl} \frac{\partial^2 u_k}{\partial x_j \partial x_l} u_i \, dx = 0.$$

Applying Green's formula to the left-hand side of the above formula, we obtain

$$B(\boldsymbol{u},\boldsymbol{u}) = 0.$$

Thus, along with (5.235), we immediately have $\boldsymbol{u} \equiv \boldsymbol{0}$. This proves the uniqueness of solution to the boundary value problem (5.225) and (5.152).

Equation (5.235) is actually a stronger form of the Gårding inequality for elliptic well-posed problems. With the aid of this inequality, and using the Lax–Milgram representation theorem in functional analysis, we can prove the existence of a weak solution to the above well-posed problem. Then we can obtain the classical solution by proving the regularity of solution. The details, which we omit here, can be found in many textbooks about modern partial differential equations.

5.6.2 System of Nonlinear Elastostatics

For the equilibrium of the elastic body under finite deformation, system (5.173) is reduced to

$$-\sum_{j,k,l=1}^{3} a_{ijkl}(\nabla y) \frac{\partial^2 y_k}{\partial x_j \partial x_l} = \rho_0 b_i \quad (i=1,2,3), \tag{5.236}$$

where $\boldsymbol{b}$ does not depend on t. Under the assumption that the elastic tensor $\boldsymbol{A} = (a_{ijkl})$ satisfies the strong ellipticity condition (5.177), system (5.236) is a *second-order quasilinear elliptic system*. Its well-posed problem can be formulated as follows: Find the solution of system (5.236) in a bounded domain $\Omega \subset \mathbb{R}^3$ such that it satisfies the boundary conditions of the form (5.178), (5.179), or (5.180) on the boundary $\partial \Omega$ of Ω. Certainly, in the case of steady state, both $\boldsymbol{h}$ and $\boldsymbol{\sigma}$ are independent of the time t, while $\boldsymbol{\sigma}$ may depend on the unknown function—and even its tangential partial derivatives.

Example 5.1. Suppose that the boundary of the elastic body is under pressure which changes with the spatial position of the boundary points (for instance, the hydrostatic pressure). Denote this pressure as $\pi(\boldsymbol{y})$. Under this circumstances, the following boundary condition should be satisfied on the boundary of the elastic body:

$$\boldsymbol{Tv} = -\pi(\boldsymbol{y})\boldsymbol{v} \quad \text{on } \partial \Omega_1, \tag{5.237}$$

where $\partial\Omega_1$ represents the boundary of domain Ω_1 spatially occupied by the elastic body after deformation. Since we should discuss the problem in the domain Ω, we have to convert the boundary condition (5.237) into the condition on the boundary $\partial\Omega$ of the domain Ω. From Lemma 5.3 and (5.69) we can see that (5.237) can be reduced to

$$Pn = -J\pi(y)F^{-T}n \quad \text{on } \partial\Omega. \tag{5.238}$$

It is not hard to verify that $JF^{-T}n$ on $\partial\Omega$ depends only on the tangential partial derivatives of y. In fact, $JF^{-T} = \text{cof } F$ is the cofactor matrix of F, so the first component of $JF^{-T}n$ is

$$\begin{vmatrix} n_1 & n_2 & n_3 \\ \dfrac{\partial y_2}{\partial x_1} & \dfrac{\partial y_2}{\partial x_2} & \dfrac{\partial y_2}{\partial x_3} \\ \dfrac{\partial y_3}{\partial x_1} & \dfrac{\partial y_3}{\partial x_2} & \dfrac{\partial y_3}{\partial x_3} \end{vmatrix}.$$

After calculation we can see that it involves only the tangential derivatives with respect to y_2 and y_3. It can be verified similarly for the second and third components. Thus, boundary condition (5.238) can be written into the following form:

$$Pn = \sigma(y, \nabla_\tau y) \quad \text{on } \partial\Omega, \tag{5.239}$$

where ∇_τ represents the tangential gradient. ∎

Now we explore the existence and uniqueness of solutions to the well-posed problems for the system of nonlinear elastostatics. Physically intuitive examples show (see Example 5.2) that the uniqueness does not hold for the problem of three-dimensional elastostatics. That is to say, a qualified mathematical model has to admit several even infinitely many solutions.

Example 5.2. Consider the following pure displacement problem: Suppose that the reference configuration Ω is the spherical shell enclosed by two homocentric spheres, and the following boundary condition is given on the boundary $\partial\Omega$ of Ω:

$$y(x) = x \quad \text{on } \partial\Omega. \tag{5.240}$$

When the volume force $b \equiv 0$, obviously $y = x$ is a solution of system (5.236) satisfying the above boundary condition. Keeping the outer sphere unchanged, while rotating the inner sphere by an angle of integer multiples of 2π around an axis through the center of sphere, the produced deformation is also a solution of the above problem (see Figure 5.7). Thus we obtain infinitely many solutions to the given problem. ∎

As to the study of the existence of solutions, it is also quite difficult. In contrast, it is relatively easy to discuss the local problem (i.e., the deformation happens in the neighborhood of a known deformation). But what is more significant is the investigation of the nonlocal problem with large deformation. Generally speaking, the approaches to studying this kind of problem include the method of convexity and monotone operators, the extension method, the topology method and topological degree theory, and the variational

5.6 Well-Posed Problems of the System of Elastostatics

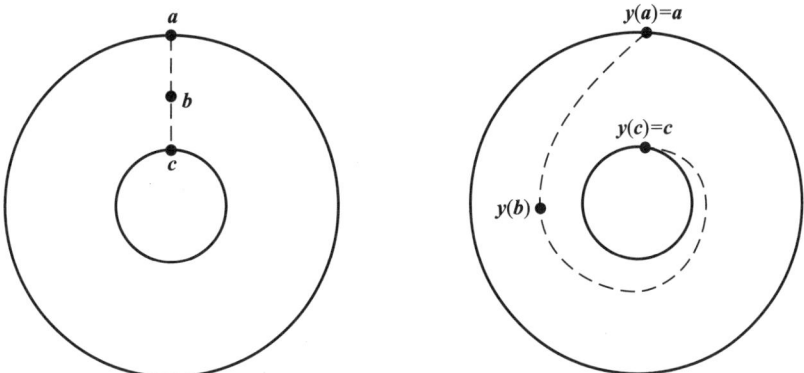

Figure 5.7.

method, etc. Among them, the variational method is more conspicuous. Now we briefly introduce the difficulties encountered in solving by the variational method as well as the resulting concept of polyconvexity of stored-energy function.

We take the given problem of system (5.236) under the mixed boundary condition (5.180) as an example for illustration. Assume that both h and σ are functions of only x, and the material is hyperelastic. Define the following potential-energy functional:

$$\Phi(y) = \int_\Omega \hat{W}(\nabla y) dx - \int_\Omega \rho_0 b \cdot y dx - \int_{\Gamma_2} \sigma \cdot y dS_0, \qquad (5.241)$$

where $\hat{W}$ is the stored-energy function. The first term on the right-hand side is the deformation energy of the elastic body, the second term is the potential energy of volume force, the third term is the potential energy of surface force acting on Γ_2, and therefore $\Phi(y)$ is the total potential energy of the elastic body. Now we first show that solving the boundary value problem (5.236) and (5.180) can be reduced to solving the minimum problem of the above functional $\Phi(y)$.

To specify the problem, we introduce the set

$$D = \left\{ y \in (W^{1,p}(\Omega))^3, \det\left(\frac{\partial y_i}{\partial x_j}\right) > 0, y|_{\Gamma_1} = h \right\},$$

where $W^{1,p}(\Omega)$ is a Sobolev space, and $1 < p < \infty$.

Suppose that $y^0 \in D$ is such that $\Phi(y)$ achieves its minimum on D, i.e.,

$$\Phi(y^0) = \inf_{y \in D} \Phi(y). \qquad (5.242)$$

Then for any given vector function $\phi \in (C^1(\overline{\Omega}))^3$ satisfying condition

$$\phi|_{\Gamma_1} = \mathbf{0}, \qquad (5.243)$$

the real function

$$\tau \mapsto \Phi(y^0 + \tau\phi)$$

achieves it minimum at $\tau = 0$, i.e.,

$$\frac{d}{d\tau}\Phi(\mathbf{y}^0 + \tau\boldsymbol{\phi})\bigg|_{\tau=0} = \mathbf{0}. \tag{5.244}$$

But from the expression (5.241) of functional $\Phi(\mathbf{y})$ we have

$$\begin{aligned}\frac{d}{d\tau}\Phi(\mathbf{y}^0 + \tau\boldsymbol{\phi}) \\ = \int_\Omega \left(\sum_{i,j=1}^3 \frac{\partial \hat{W}}{\partial f_{ij}}(\mathbf{F}^0 + \tau\nabla\boldsymbol{\phi})\frac{\partial \phi_i}{\partial x_j} - \rho_0 \mathbf{b} \cdot \boldsymbol{\phi}\right) dx \\ - \int_{\Gamma_2} \boldsymbol{\sigma} \cdot \boldsymbol{\phi}\, dS_0,\end{aligned} \tag{5.245}$$

where

$$\mathbf{F}^0 = \left(\frac{\partial y_i^0}{\partial x_j}\right).$$

The above two formulas lead to

$$\int_\Omega \left(\sum_{i,j=1}^3 p_{ij}(\mathbf{F}^0)\frac{\partial \phi_i}{\partial x_j} - \rho_0 \mathbf{b} \cdot \boldsymbol{\phi}\right) dx - \int_{\Gamma_2} \boldsymbol{\sigma} \cdot \boldsymbol{\phi}\, dS_0 = 0. \tag{5.246}$$

If there is a vector function $\mathbf{y}^0 \in (W^{1,p}(\Omega))^3$ satisfying the boundary condition $\mathbf{y}^0|_{\Gamma_1} = \mathbf{h}$ such that (5.246) is satisfied for any above-mentioned $\boldsymbol{\phi}$, then we say that $\mathbf{y}^0$ is a *weak solution* of the boundary value problem (5.236) and (5.180). If a weak solution $\mathbf{y}^0 \in (C^2(\Omega))^3 \cap (C^1(\overline{\Omega}))^3$, then it is easy to verify that $\mathbf{y}^0$ is a classical solution of this boundary value problem. What we need to explain is that functions in the function class D over which we find the minimum of functional $\Phi(\mathbf{y})$ are not required to satisfy boundary conditions on Γ_2. Boundary conditions on Γ_2 are naturally satisfied by the minimum function, which is called the *natural boundary condition*.

The usual steps for solving the minimum problem of functional $\Phi(\mathbf{y})$ are as follows:

(a) Prove that $\Phi(\mathbf{y})$ is bounded from below in D. This is usually not difficult to verify under certain assumptions.

(b) Find a *minimizing sequence* $\{\mathbf{y}^n\}$ of $\Phi(\mathbf{y})$ in D, i.e., the sequence $\{\mathbf{y}^n\}$ satisfying the following conditions:

$$\lim_{n\to\infty} \Phi(\mathbf{y}^n) = \inf_{\mathbf{y}\in D} \Phi(\mathbf{y}). \tag{5.247}$$

Then find a subsequence in $\{\mathbf{y}^n\}$ (for simplicity, still denoted by $\{\mathbf{y}^n\}$) such that when $n \to \infty$,

$$\mathbf{y}^n \rightharpoonup \mathbf{y}^0 \quad \text{weakly in } (W^{1,p}(\Omega))^3. \tag{5.248}$$

From (a) we know that $\Phi(\mathbf{y})$ has infimum in D, so there must exist a minimizing sequence. Finding the weakly convergent subsequence of the minimizing sequence usually depends

5.6 Well-Posed Problems of the System of Elastostatics

on some boundedness estimates on this sequence. This can also be done under certain assumptions.

(c) For the above weakly convergent subsequence, prove

$$\Phi(y^0) \leq \liminf_{n\to\infty} \Phi(y^n). \qquad (5.249)$$

It is easy to see that y^0 is exactly the function such that $\Phi(y)$ achieves it minimum as long as the above formula is satisfied. Equation (5.249) is called the *sequential weak lower semicontinuity* of functional $\Phi(y)$. Thus, the problem is converted into the discussion of whether the potential-energy functional $\Phi(y)$ has the sequential weak lower semicontinuity for general elastic bodies.

Deeper discussions discover that the requirement of the sequential weak lower semicontinuity to $\Phi(y)$ results in the requirement of convexity of the stored-energy function $\hat{W}$ with respect to F (see [6]). Now, for general elastic bodies, is the stored-energy function $\hat{W}$ convex with respect to F? The answer is disappointing. Since the set $\{F \in \mathbb{R}^{3\times3}, \det F > 0\}$ is not a convex set, we can prove that the convexity of $\hat{W}$ with respect to F conflicts with the following property of the stored-energy function:

$$\hat{W} \to +\infty \quad \text{as } \det F \to 0 \qquad (5.250)$$

(see [4]). Thus, the requirement of convexity of the stored-energy function $\hat{W}$ will exclude a vast collection of successful models of the stored-energy function (see section 5.4.3 in this chapter). This implies that the convexity of $\hat{W}$ with respect to F does not match the actual situation. Now, solving the problem seems to be a dilemma. But here we ignore a crucial piece of information: the components of F are not independent but form a gradient of a certain vector function. Therefore, the components of F satisfy the following relation:

$$\frac{\partial f_{ij}}{\partial x_k} = \frac{\partial^2 y_i}{\partial x_j \partial x_k} = \frac{\partial f_{ik}}{\partial x_j}; \qquad (5.251)$$

i.e., (f_{ij}) satisfies a restriction of a group of differential relations. That is, if we obtain the weak convergence $F^n \rightharpoonup F^0$ in a certain space, then we actually obtain far more than these. In fact, we can prove that if $p > 3$ and

$$y^n \rightharpoonup y^0 \quad \text{weakly in } (W^{1,p}(\Omega))^3, \qquad (5.252)$$

then

$$M^n \rightharpoonup M^0 \quad \text{weakly in } L^{p/m}(\Omega), \qquad (5.253)$$

where M^n and M^0 represent the $m \times m$ ($1 \leq m \leq 3$) subdeterminants corresponding to F^n and F^0, respectively (see [3], [4]). Because of this, J. Ball introduced the concept of polyconvexity for the stored-energy function, which is weaker than convexity (see [3]).

Definition 5.9. *If the function $\hat{W}(F)$ of third-order square matrix F can be represented as the convex function of the subdeterminants of F, then it is called* polyconvex.

Assume that the stored-energy function $\hat{W}$ is polyconvex; then it certainly can be represented in the following form:

$$\hat{W}(F) = W^*(F, \text{cof } F, \det F), \qquad (5.254)$$

where cof F is the cofactor matrix of F, while the function

$$(F, G, d) \mapsto W^*(F, G, d),$$

which depends on 19 arguments, is convex.

Thus, if we can find a weakly convergent minimizing sequence $\{y^n\}$ of functional $\Phi(y)$ in space $(W^{1,p}(\Omega))^3$,

$$y^n \rightharpoonup y^0 \quad \text{weakly in } (W^{1,p}(\Omega))^3, \tag{5.255}$$

then from (5.253) we must have

$$\text{cof } F^n \rightharpoonup \text{cof } F^0 \text{ weakly in } (L^{p/2}(\Omega))^9, \tag{5.256}$$
$$\det F^n \rightharpoonup \det F^0 \text{ weakly in } L^{p/3}(\Omega). \tag{5.257}$$

Then we find a sequence $(F^n, \text{cof } F^n, \det F^n)$ in $(L^{p/3}(\Omega))^{19}$, which converges weakly to

$$(F^0, \text{cof } F^0, \det F^0).$$

From the polyconvexity of $\hat{W}(F)$, i.e., the convexity of $W^*(F, G, d)$ with respect to (F, G, d), we can also derive the sequential weak lower semicontinuity of the functional $\Phi(y)$ with respect to (F, G, d) in $(L^{p/3}(\Omega))^{19}$. Hence we finish the proof of the existence of solutions.

In the above proof of the existence of solutions, we did not use the strong ellipticity condition of system (5.236). In fact, the polyconvexity of $\hat{W}$ may infer that $A = (a_{ijkl})$ satisfies the strong ellipticity condition (5.177) (see [3]).

The question of whether or not the weak solution obtained above has enough regularity to become a classical solution is still open, except for the one-dimensional case. Besides, if y^0 is a solution to the boundary value problem (5.236) and (5.180), then from the physical point of view, $x \mapsto y^0(x)$ should be a one-to-one onto mapping. Whether or not the weak solution we now obtain satisfies this requirement needs to be further studied.

To show that most models given in section 5.4.3 of this chapter are polyconvex, we prove the following theorem.

Theorem 5.8. *Assume that* $\Gamma : [0, +\infty) \to \mathbb{R}$ *is convex; then the stored-energy function given by* (5.121),

$$\hat{W}(F) = \sum_{i=1}^{M} a_i(\mu_1^{\alpha_i} + \mu_2^{\alpha_i} + \mu_3^{\alpha_i} - 3)$$
$$+ \sum_{i=1}^{N} b_i((\mu_2\mu_3)^{\beta_i} + (\mu_3\mu_1)^{\beta_i} + (\mu_1\mu_2)^{\beta_i} - 3)$$
$$+ \Gamma(\mu_1\mu_2\mu_3), \tag{5.258}$$

is polyconvex, where μ_1, μ_2, μ_3 *are the principal values of* $U = C^{1/2}$, *and* $a_i > 0$, $b_i > 0$, $\alpha_i \geq 1$, *and* $\beta_i \geq 1$ *are all constants.*

5.6 Well-Posed Problems of the System of Elastostatics

Proof. We prove only the case for $M = N = 1$, $\alpha_1 = \beta_1 = 2$ (obviously, this includes the material given by (5.122)). Please refer to [4] for the proof of general cases.

We first prove the following lemma.

Lemma 5.6. *The function*

$$g(\boldsymbol{F}) = \mathrm{tr}(\boldsymbol{F}^{\mathrm{T}}\boldsymbol{F}), \qquad \forall \boldsymbol{F} \in \mathbb{R}^{3\times 3} \tag{5.259}$$

is strictly convex.

Proof. It is easy to see that

$$g(\boldsymbol{F}) = \sum_{i,j=1}^{3} f_{ij}^2,$$

whose second-order partial derivatives are

$$\frac{\partial^2 g(\boldsymbol{F})}{\partial f_{ij} \partial f_{kl}} = \begin{cases} 2 & \text{if } (i,j) = (k,l), \\ 0 & \text{if } (i,j) \neq (k,l). \end{cases}$$

Therefore, the 9×9 matrix

$$\left(\frac{\partial^2 g(\boldsymbol{F})}{\partial f_{ij} \partial f_{kl}} \right)$$

is positively definite. This implies that $g(\boldsymbol{F})$ is a strictly convex function. The proof of the lemma is finished. □

Proof of Theorem 5.8 *(continued)*. It is easy to verify directly that

$$\mathrm{cof}(\boldsymbol{F}^{\mathrm{T}}\boldsymbol{F}) = (\mathrm{cof}\,\boldsymbol{F})^{\mathrm{T}}\mathrm{cof}\,\boldsymbol{F}; \tag{5.260}$$

then, from the second formula of (5.123) we have

$$(\mu_2\mu_3)^2 + (\mu_3\mu_1)^2 + (\mu_1\mu_2)^2 = \mathrm{tr}((\mathrm{cof}\,\boldsymbol{F})^{\mathrm{T}}\mathrm{cof}\,\boldsymbol{F}). \tag{5.261}$$

Thus, noting (5.123), (5.258) can be written as

$$\hat{W}(\boldsymbol{F}) = a_1 \mathrm{tr}(\boldsymbol{F}^{\mathrm{T}}\boldsymbol{F}) + b_1 \mathrm{tr}((\mathrm{cof}\,\boldsymbol{F})^{\mathrm{T}}\mathrm{cof}\,\boldsymbol{F}) \\ + \Gamma(\det \boldsymbol{F}) - 3(a_1 + b_1). \tag{5.262}$$

From Lemma 5.6, the first and second terms of (5.262) are convex functions of $\boldsymbol{F}$ and $\mathrm{cof}\,\boldsymbol{F}$, respectively. Also from the assumption of the theorem, the third term is a convex function of $\det \boldsymbol{F}$. So $\hat{W}(\boldsymbol{F})$ is polyconvex. The proof is finished. □

Finally, we want to point out that the stored-energy function of the Saint–Venant–Kirchhoff material (see (5.112)) is not polyconvex (see [4]).

Exercises

1. Prove Lemma 5.1.
2. Verify (5.36), i.e., prove
$$\frac{dJ}{dt} = J \operatorname{div}_y \boldsymbol{v}.$$
3. Prove that under the material description, the conservation law of momentum moment is equivalent to the symmetry of the second Piola stress tensor.
4. Assuming that the material is hyperelastic and the reference configuration is the natural state, prove that the tensor
$$\boldsymbol{A} = (a_{ijkl}) = \left(2\frac{\partial \overline{p}_{ij}}{\partial c_{kl}}\right)$$
obtained from linearization has the following symmetry:
$$a_{ijkl} = a_{klij}.$$
5. Assuming that the stored-energy function $\hat{W}$ of a hyperelastic material satisfies (5.93), prove that the Cauchy stress tensor $\boldsymbol{T}$ determined by it satisfies the objectivity hypothesis (5.81).
6. Assuming that the stored-energy function $\hat{W}$ of a hyperelastic material satisfies (5.111), prove that the Cauchy stress tensor $\boldsymbol{T}$ determined by it satisfies the isotropy condition (5.95).
7. In the case of linear elasticity, prove that for the isotropic material, the strong ellipticity condition (5.149) is equivalent to the Lamè constants satisfying
$$\mu > 0, \qquad \lambda + 2\mu > 0.$$
8. In the case of linear elasticity, prove that for the isotropic material, the stability condition (5.170) is equivalent to the Lamè constants satisfying
$$\mu > 0, \qquad \lambda + \frac{2}{3}\mu > 0.$$
9. Assuming that the eigenvalues of a 3×3 matrix $\boldsymbol{A}$ are $\lambda_1, \lambda_2, \lambda_3$, prove that the eigenvalues of cof $\boldsymbol{A}$ are
$$\lambda_2\lambda_3, \quad \lambda_3\lambda_1, \quad \lambda_1\lambda_2.$$
10. Prove that the function
$$\hat{W}(\boldsymbol{F}) = \begin{cases} \dfrac{1}{\det \boldsymbol{F}} & \text{if } \det \boldsymbol{F} > 0, \\ +\infty & \text{if } \det \boldsymbol{F} \leq 0 \end{cases}$$
is polyconvex.

Bibliography

[1] Gu C., Li T. et al. *Equations of Mathematical Physics* (in Chinese). Shanghai: Shanghai Press of Science and Technology, 1987.

[2] Courant R., Hilbert D. *Methods of Mathematical Physics*, Volume II. New York: John Wiley & Sons, 1989.

[3] Ball J. *Convexity conditions and existence theorems in nonlinear elasticity*. Arch. Rational Mech. Anal., 63 (1977), 337–403.

[4] Ciarlet P. G. *Mathematical Elasticity, Vol I: 3-dimensional elasticity*. Amsterdam: North-Holland, 1988.

[5] Ciarlet P. G., Geymonat G. *Sur les lois de comportement en élasticité non-linéaire compressible*. C. R. Acad. Sci. Paris, Sér. II, 295 (1982), 423–426.

[6] Evans L. C. *Weak Convergence Methods for Nonlinear Partial Differential Equations*. CBMS, Regional Conference Series in Mathematics 74, Providence, RI: AMS, 1990.

[7] Gurtin M. E. *Topics in Finite Elasticity*. CBMS-NSF, Regional Conference Series in Applied Mathematics 35, Philadelphia: SIAM, 1981.

[8] John F. *Finite amplitude waves in a homogeneous isotropic elastic solid*. Comm. Pure Appl. Math., 30 (1977), 421–446.

[9] Pazy A. *Semigroups of Linear Operators and Applications to Partial Differential Equations*. New York: Springer-Verlag, 1983.

[10] Gu C., Li T., Shen W. *Applied Partial Differential Equations* (in Chinese). Beijing: Higher Education Press, 1993.

[11] Struwe M. *Variational Methods: Applications to Nonlinear Partial Differential Equations and Hamiltonian Systems*. Berlin, New York: Springer-Verlag, 1990.

[12] Qin T. *Symmetrizing nonlinear elastodynamic system*. J. Elasticity, 50 (1998), 245–252.

Appendix A
Cartesian Tensor

Tensors are widely used in theoretical physics and continuum mechanics, not only for writing concisely, but also, and more importantly, to prevent important physical quantities in physics and mechanics from depending on the selection of coordinate system. In this appendix, any given coordinate system under consideration is the Cartesian coordinate system, and the corresponding tensor is called the *Cartesian tensor*.

A.1 Definition of Tensor

Under a selected measurement unit, a quantity which can be characterized by a number independent of the selection of coordinate system is called a *scalar*. For instance, the constants π and e used in mathematics, and the mass, density, temperature, and energy in physics are all scalars.

Under a selected measurement unit, a quantity which can be characterized by a number independent of the selection of coordinate system and a direction (namely, size and direction) is called a *vector*. For instance, the directed line segment, the displacement, velocity, force, and momentum in mechanics are all vectors.

We know that a vector can also be described by and perform calculations on its components. At this time, an appropriate coordinate system should be selected. The same vector has different components under different coordinate systems, but since these components under different coordinate systems describe the same vector, they should obey the corresponding coordinate transformation law. And this transformation law reflects that this objectively existing vector is independent of the selection of coordinate system. Therefore, a vector can be depicted by its components under a selected coordinate system along with the transformation law of its components under the coordinate transformation. To extend this basic concept, we use the concept of tensor.

Under the Cartesian coordinate system, we restrict ourselves only to the Cartesian coordinate transformation, i.e., translation, rotation and reflection. If both the new and old coordinate systems are assumed to be right-handed, then there are only translation and rotation.

Supposed that $O'x_1'x_2'x_3'$ and $Ox_1x_2x_3$ are the new and old right-handed coordinate systems, respectively. e_1', e_2', e_3' and e_1, e_2, e_3 are the unit coordinate vectors of the new and old coordinate systems, respectively. Assume

$$e'_i = \sum_{j=1}^{3} a_{ij} e_j \quad (i=1,2,3). \tag{A.1}$$

Noting that

$$e'_i \cdot e'_j = \delta_{ij}, \quad e_i \cdot e_j = \delta_{ij}, \tag{A.2}$$

where δ_{ij} is the Kronecker symbol, we have

$$e'_i \cdot e'_j = \sum_{k=1}^{3} a_{ik} e_k \cdot \sum_{l=1}^{3} a_{jl} e_l$$
$$= \sum_{k,l=1}^{3} a_{ik} a_{jl} \delta_{kl} = \sum_{k=1}^{3} a_{ik} a_{jk} = \delta_{ij}.$$

This shows that $A = (a_{ij})$ is an orthogonal matrix. Since we are restricted only to the coordinate transformation from one right-handed system to another, we should have $\det A = 1$. Since the inverse matrix of the orthogonal matrix A is $A^{\mathrm{T}} = (a_{ji})$, from (A.1) we have

$$e_i = \sum_{j=1}^{3} a_{ji} e'_j \quad (i=1,2,3). \tag{A.3}$$

Now let us look at the transformation law of the vector components under different Cartesian coordinate systems. Suppose that a is a vector whose components under the new and old coordinate systems are (a'_1, a'_2, a'_3) and (a_1, a_2, a_3), respectively. Then

$$a = \sum_{i=1}^{3} a_i e_i = \sum_{j=1}^{3} a'_j e'_j. \tag{A.4}$$

From (A.3), (A.4), we get

$$\sum_{i,j=1}^{3} a_i a_{ji} e'_j = \sum_{j=1}^{3} a'_j e'_j.$$

This implies that

$$a'_j = \sum_{i=1}^{3} a_{ji} a_i \quad (j=1,2,3),$$

which can also be written as

$$a'_i = \sum_{j=1}^{3} a_{ij} a_j \quad (i=1,2,3). \tag{A.5}$$

Since $A = (a_{ij})$ is an orthogonal matrix, it is easy to see from (A.5) that

$$a_i = \sum_{j=1}^{3} a_{ji} a'_j \quad (i=1,2,3). \tag{A.6}$$

A.2 Operations of Tensor

Equations (A.5) and (A.6) are consistent with the transformation expressions (A.1) and (A.3).

Thus another definition of vector is available: **Suppose that a quantity is represented by three components (numbers) a_1, a_2, a_3 under each Cartesian coordinate system,** which are transformed according to (A.5) under the coordinate transformation (A.1). Then this quantity is called a *vector* or a *first-order tensor*. We can similarly define second-order, third-order, and even higher order tensors by increasing the numbers of the component subscript.

Definition of second-order tensor. *If a quantity P is represented by nine components (numbers) p_{ij} ($i,j = 1,2,3$) under each Cartesian coordinate system $Ox_1x_2x_3$, and under the coordinate transformation (A.1), then these components are transformed into the nine components p'_{ij} ($i,j = 1,2,3$) under another coordinate system $O'x'_1x'_2x'_3$ according to the following transformation law:*

$$p'_{ij} = \sum_{k,l=1}^{3} a_{ik} a_{jl} p_{kl} \quad (i,j = 1,2,3). \tag{A.7}$$

Then the quantity P is called a second-order tensor, *denoted by*

$$P = \{p_{ij}\} = \begin{pmatrix} p_{11} & p_{12} & p_{13} \\ p_{21} & p_{22} & p_{23} \\ p_{31} & p_{32} & p_{33} \end{pmatrix},$$

where p_{ij} ($i,j = 1,2,3$) are called the components of this second-order tensor.

It is not hard to generalize the above definition to the *nth-order tensor* $P = \{p_{i_1 i_2 \cdots i_n}\}$. Under each Cartesian coordinate system $Ox_1x_2x_3$, the nth-order tensor P has 3^n components $p_{i_1 i_2 \cdots i_n}$ ($i_1, i_2, \ldots, i_n = 1,2,3$); under the coordinate transformation (A.1), the transformation relation between its components is given by

$$p'_{i_1 i_2 \cdots i_n} = \sum_{j_1, j_2, \ldots, j_n = 1}^{3} a_{i_1 j_1} a_{i_2 j_2} \cdots a_{i_n j_n} p_{j_1 j_2 \cdots j_n}$$
$$(i_1, i_2, \ldots, i_n = 1,2,3). \tag{A.8}$$

A.2 Operations of Tensor

(1°) Addition and substraction of tensor, and multiplication of tensor by a number.

We take the second-order tensor as an example. Suppose that $P = \{p_{ij}\}$ and $Q = \{q_{ij}\}$ are both second-order tensors; then $\{p_{ij} \pm q_{ij}\}$ is still a second-order tensor, denoted by $P \pm Q$, i.e.,

$$P \pm Q = \{p_{ij} \pm q_{ij}\}.$$

Suppose that α is a scalar; obviously $\{\alpha p_{ij}\}$ is still a second-order tensor, denoted by αP, i.e.,

$$\alpha P = \{\alpha p_{ij}\}.$$

(2°) Product of tensor.

Suppose that $P = \{p_{i_1 i_2 \cdots i_m}\}$ and $Q = \{q_{j_1 j_2 \cdots j_n}\}$ are mth-order and nth-order tensors, respectively. Set

$$r_{i_1 \cdots i_m j_1 \cdots j_n} = p_{i_1 \cdots i_m} q_{j_1 \cdots j_n}.$$

It is easy to verify that, $R = \{r_{i_1 \cdots i_m j_1 \cdots j_n}\}$ is an $(m+n)$th-order tensor, called the *tensor product* of P and Q, denoted by $R = P \otimes Q$.

Example A.1. Suppose that $a = \{a_i\}$ and $b = \{b_i\}$ are both vectors; then

$$C = a \otimes b = \{a_i b_j\} = \begin{pmatrix} a_1 b_1 & a_1 b_2 & a_1 b_3 \\ a_2 b_1 & a_2 b_2 & a_2 b_3 \\ a_3 b_1 & a_3 b_2 & a_3 b_3 \end{pmatrix}$$

is a second-order tensor. Physically, it is usually called the *dyad* (*dyadic vector*) of a and b. ∎

(3°) Contraction of tensor.

Suppose that $P = \{p_{i_1 i_2 \cdots i_n}\}$ is an nth-order tensor. If two subscripts of its component $p_{i_1 i_2 \cdots i_n}$ are taken to be the same and summed over from 1 to 3, it is not hard to verify that the resulting quantity is a component of an $(n-2)$th-order tensor, which is called the *contraction* of tensor P.

Now let us explain it from the case when the last two subscripts of the components of P are contracted. Suppose that

$$q_{i_1 i_2 \cdots i_{n-2}} = \sum_{k=1}^{3} p_{i_1 i_2 \cdots i_{n-2} k k};$$

then $Q = \{q_{i_1 i_2 \cdots i_{n-2}}\}$ is an $(n-2)$th-order tensor. In fact, noting that $A = (a_{ij})$ is an orthogonal matrix, we have

$$q'_{i_1 i_2 \cdots i_{n-2}} = \sum_{k=1}^{3} p'_{i_1 i_2 \cdots i_{n-2} k k}$$

$$= \sum_{k, j_1, \ldots, j_{n-2}, r, s = 1}^{3} a_{i_1 j_1} \cdots a_{i_{n-2} j_{n-2}} a_{kr} a_{ks} p_{j_1 \cdots j_{n-2} r s}$$

$$= \sum_{j_1, \ldots, j_{n-2}, r, s = 1}^{3} a_{i_1 j_1} \cdots a_{i_{n-2} j_{n-2}} \delta_{rs} p_{j_1 \cdots j_{n-2} r s}$$

$$= \sum_{j_1, \ldots, j_{n-2}, r = 1}^{3} a_{i_1 j_1} \cdots a_{i_{n-2} j_{n-2}} p_{j_1 \cdots j_{n-2} r r}$$

$$= \sum_{j_1, \ldots, j_{n-2} = 1}^{3} a_{i_1 j_1} \cdots a_{i_{n-2} j_{n-2}} q_{j_1 \cdots j_{n-2}}.$$

So Q is an $(n-2)$th-order tensor.

A.2 Operations of Tensor

Example A.2. The contraction $\sum_{i=1}^{3} a_i b_i$ of the dyad $\boldsymbol{a} \otimes \boldsymbol{b}$ is exactly the inner product of vectors $\boldsymbol{a}$ and $\boldsymbol{b}$. ∎

Example A.3. Suppose that $\{b_{ijkl}\}$ is a fourth-order tensor, $\{\varepsilon_{ij}\}$ is a second-order tensor; then

$$\{t_{ij}\} = \left\{ \sum_{k,l=1}^{3} b_{ijkl} \varepsilon_{kl} \right\}$$

is a second-order tensor. This can be easily shown by using the tensor product and contraction. ∎

Example A.4. Suppose that, for any given second-order tensor $\{\varepsilon_{ij}\}$,

$$\{t_{ij}\} = \left\{ \sum_{k,l=1}^{3} b_{ijkl} \varepsilon_{kl} \right\}$$

is always a second-order tensor, then $\{b_{ijkl}\}$ must be a fourth-order tensor. ∎

Now we prove the above conclusion. Under the new coordinate system $O' x_1' x_2' x_3'$,

$$t_{ij}' = \sum_{k,l=1}^{3} b_{ijkl}' \varepsilon_{kl}'. \tag{A.9}$$

But from the definition of tensor,

$$\begin{aligned}
t_{ij}' &= \sum_{p,q=1}^{3} a_{ip} a_{jq} t_{pq} \\
&= \sum_{p,q,r,s=1}^{3} a_{ip} a_{jq} b_{pqrs} \varepsilon_{rs} \\
&= \sum_{p,q,r,s,k,l=1}^{3} a_{ip} a_{jq} a_{kr} a_{ls} b_{pqrs} \varepsilon_{kl}'.
\end{aligned} \tag{A.10}$$

Since (A.9) and (A.10) hold for any given second-order tensor $\{\varepsilon_{kl}'\}$, we then have

$$b_{ijkl}' = \sum_{p,q,r,s=1}^{3} a_{ip} a_{jq} a_{kr} a_{ls} b_{pqrs},$$

i.e., $\{b_{ijkl}\}$ is a fourth-order tensor.

The conclusion of Example A.4 is usually called the *tensor identification theorem*. It is not hard to prove that this theorem is satisfied for tensors of any order.

A.3 Invariants of the Second-Order Symmetric Tensor

The invariant of second-order tensor P is a scalar determined by this tensor and is independent of the selection of coordinate system.

For the second-order tensor $P = \{p_{ij}\}$, if there exists a nonzero vector $a = (a_1, a_2, a_3)$ such that

$$Pa = \lambda a, \tag{A.11}$$

where λ is a scalar and

$$Pa = \left\{ \sum_{j=1}^{3} p_{ij} a_j \right\},$$

then the direction of a is called the *principal axis direction (principal direction)* of tensor P, and λ is called the *principal value* of P.

The principal value λ of tensor P is its invariant; i.e., under the coordinate transformation, it holds that

$$P'a' = \lambda a'. \tag{A.12}$$

In fact,

$$(P'a')_i = \sum_{j=1}^{3} p'_{ij} a'_j$$

$$= \sum_{j,k,l,m=1}^{3} a_{ik} a_{jl} p_{kl} a_{jm} a_m$$

$$= \sum_{k,l,m=1}^{3} a_{ik} \delta_{lm} p_{kl} a_m$$

$$= \sum_{k,l=1}^{3} a_{ik} p_{kl} a_l$$

$$= \sum_{k=1}^{3} a_{ik} \lambda a_k = \lambda a'_i.$$

This is exactly (A.12).

λ being the principal value of P is equivalent to λ satisfying

$$\begin{vmatrix} p_{11} - \lambda & p_{12} & p_{13} \\ p_{21} & p_{22} - \lambda & p_{23} \\ p_{31} & p_{32} & p_{33} - \lambda \end{vmatrix} = 0,$$

i.e.,

$$\lambda^3 - \lambda^2(p_{11} + p_{22} + p_{33})$$
$$+ \lambda \left(\begin{vmatrix} p_{22} & p_{23} \\ p_{32} & p_{33} \end{vmatrix} + \begin{vmatrix} p_{11} & p_{13} \\ p_{31} & p_{33} \end{vmatrix} + \begin{vmatrix} p_{11} & p_{12} \\ p_{21} & p_{22} \end{vmatrix} \right)$$
$$- \begin{vmatrix} p_{11} & p_{12} & p_{13} \\ p_{21} & p_{22} & p_{23} \\ p_{31} & p_{32} & p_{33} \end{vmatrix} = 0. \tag{A.13}$$

This is a cubic algebraic equation with respect to λ. From the relation between roots and coefficients, we get

$$I_1 \stackrel{d}{=} p_{11} + p_{22} + p_{33} = \lambda_1 + \lambda_2 + \lambda_3,$$

$$I_2 \stackrel{d}{=} \begin{vmatrix} p_{22} & p_{23} \\ p_{32} & p_{33} \end{vmatrix} + \begin{vmatrix} p_{11} & p_{13} \\ p_{31} & p_{33} \end{vmatrix} + \begin{vmatrix} p_{11} & p_{12} \\ p_{21} & p_{22} \end{vmatrix}$$
$$= \lambda_1\lambda_2 + \lambda_2\lambda_3 + \lambda_3\lambda_1,$$

$$I_3 \stackrel{d}{=} \begin{vmatrix} p_{11} & p_{12} & p_{13} \\ p_{21} & p_{22} & p_{23} \\ p_{31} & p_{32} & p_{33} \end{vmatrix} = \lambda_1\lambda_2\lambda_3,$$

where $\lambda_1, \lambda_2, \lambda_3$ are three roots of (A.13). Since λ_1, λ_2, and λ_3 are all invariants, the combinations I_1, I_2, and I_3 of components of tensor P given by the above three formulas are all invariants, called *the first, second, the third principal invariants* of P, respectively.

In particular, if

$$p_{ij} = p_{ji} \quad (i, j = 1, 2, 3), \tag{A.14}$$

then $P = \{p_{ij}\}$ is called a *second-order symmetric tensor*. It is easy to verify that this definition is independent of the selection of coordinate system. For second-order symmetric tensor P, there must exist three real principal values $\lambda_1, \lambda_2, \lambda_3$ and three principal axes perpendicular to each other. Taking the principal axes to be the new coordinate axes, the symmetric tensor $P = \{p_{ij}\}$ can be written in the following form:

$$\begin{pmatrix} \lambda_1 & 0 & 0 \\ 0 & \lambda_2 & 0 \\ 0 & 0 & \lambda_3 \end{pmatrix}.$$

This shows that a second-order symmetric tensor can be completely characterized by its three principal values λ_1, λ_2, and λ_3, and therefore all of its invariants can be represented by λ_1, λ_2, and λ_3. In other words, any possible invariants are essentially given by λ_1, λ_2, and λ_3. But the principal values of P are uniquely determined by its first, second, and third principal invariants, and therefore any invariants of the second-order symmetric tensor are given by its first, second, and third principal invariants I_1, I_2, and I_3.

A.4 Isotropic Tensor

For most tensors, their components will change values after the rotation transformation of coordinates. This kind of tensor is called the *anisotropic tensor*. But there are also some tensors whose components remain unchanged under the rotation transformation of coordinates; this kind of tensor is called the *isotropic tensor*. Strictly speaking, if each component of the nth tensor $\boldsymbol{H} = \{h_{i_1 i_2 \cdots i_n}\}$ is an invariant under the rotation transformation of coordinates, i.e., if it holds under the rotation transformation of coordinates that

$$h'_{i_1 i_2 \cdots i_n} = h_{i_1 i_2 \cdots i_n} \quad (i_1, i_2, \ldots, i_n = 1, 2, 3), \tag{A.15}$$

then $\boldsymbol{H}$ is called the isotropic nth order tensor. For example, the scalar (zero-order tensor) and $\{\delta_{ij}\}$ are both isotropic. The term *isotropy* comes from physics. For instance, the component p_{1111} of the fourth-order tensor $\{p_{ijkl}\}$ represents the tension elasticity coefficient of a certain material along the x_1 direction. When rotated to the new coordinate system $Ox'_1 x'_2 x'_3$, the tensor is $\{p'_{ijkl}\}$. At this moment, the component p'_{1111} represents the stretching elasticity coefficient of this material along the x'_1 direction. If the material is isotropic, the tension elasticity coefficients along different directions should be the same, and then we should have $p_{1111} = p'_{1111}$.

Now we give some specific expressions of isotropic tensors.

Theorem. (1) *The general form of the second-order isotropic tensor* $\boldsymbol{H} = \{h_{ij}\}$ *is*

$$h_{ij} = \lambda \delta_{ij} \quad (i, j = 1, 2, 3), \tag{A.16}$$

where λ is a scalar, and δ_{ij} is the Kronecker symbol.

(2) *The general form of the fourth-order isotropic tensor* $\boldsymbol{H} = \{h_{ijkl}\}$ *is*

$$h_{ijkl} = \lambda \delta_{ij}\delta_{kl} + \alpha \delta_{ik}\delta_{jl} + \beta \delta_{il}\delta_{jk}$$
$$(i, j, k, l = 1, 2, 3), \tag{A.17}$$

where λ, α, and β are all scalars.

Proof. We only prove (2). First, it is easy to verify that the fourth-order tensor given by (A.17) is isotropic. Second, we prove that if $\boldsymbol{H}$ is isotropic, then we have

$$h_{ijkl} = \begin{cases} \lambda + \alpha + \beta & \text{if } i = j = k = l, & (\text{A.18}_1) \\ \lambda & \text{if } i = j \neq k = l, & (\text{A.18}_2) \\ \alpha & \text{if } i = k \neq j = l, & (\text{A.18}_3) \\ \beta & \text{if } i = l \neq j = k, & (\text{A.18}_4) \\ 0 & \text{otherwise.} & (\text{A.18}_5) \end{cases}$$

As long as the above formulas are proved, (A.17) is satisfied accordingly.

Since $\boldsymbol{H}$ is isotropic, from the definition we know that

$$h_{ijkl} = h'_{ijkl} = \sum_{p,q,r,s=1}^{3} a_{ip} a_{jq} a_{kr} a_{ls} h_{pqrs}. \tag{A.19}$$

Now we finish our proof in four steps.

(1°) Take the coordinate transformation

$$e'_1 = e_2, \quad e'_2 = e_3, \quad e'_3 = e_1.$$

A.4 Isotropic Tensor

For this coordinate transformation,

$$(a_{ij}) = \begin{pmatrix} 0 & 1 & 0 \\ 0 & 0 & 1 \\ 1 & 0 & 0 \end{pmatrix}.$$

Applying this matrix to (A.19), we can see that for h_{ijkl}, when changing its subscripts $1 \to 2, 2 \to 3, 3 \to 1$, the components remain unchanged.

(2°) Rotate the coordinate system around the x_3 axis by an angle π to a new coordinate system. Then

$$e'_1 = -e_1, \quad e'_2 = -e_2, \quad e'_3 = e_3,$$

and correspondingly,

$$(a_{ij}) = \begin{pmatrix} -1 & 0 & 0 \\ 0 & -1 & 0 \\ 0 & 0 & 1 \end{pmatrix}.$$

Applying this matrix to (A.19), we can see that in the sum on the right-hand side of that formula, only the terms with $p = i, q = j, r = k$, and $s = l$ are not zero, and

$$h_{ijkl} = \pm h_{ijkl}. \tag{A.20}$$

The "$\pm$" in (A.20) are chosen according to the following rule: It takes "$-$" if and only if there is odd number 3 in i, j, k, and l; otherwise, it takes "$+$". When (A.20) takes "$-$" on the right-hand side, we have

$$h_{ijkl} = 0. \tag{A.21}$$

From (1°) we know that, when there is an odd number of 1 or 2 in i, j, k, and l, (A.21) is also satisfied. This proves (A.18$_5$).

(3°) Rotate the coordinate system around the x_3 axis by an angle $\pi/2$ to a new coordinate system. Then

$$e'_1 = e_2, \quad e'_2 = -e_1, \quad e'_3 = e_3,$$

and correspondingly,

$$(a_{ij}) = \begin{pmatrix} 0 & 1 & 0 \\ -1 & 0 & 0 \\ 0 & 0 & 1 \end{pmatrix}.$$

Applying this matrix to (A.19), we get

$$h_{1122} = a_{12}a_{12}a_{21}a_{21}h_{2211} = h_{2211}. \tag{A.22}$$

From (1°) we obtain

$$h_{1122} = h_{2233} = h_{3311}, \tag{A.23}$$
$$h_{2211} = h_{3322} = h_{1133}. \tag{A.24}$$

It is easy to prove from (A.22)–(A.24) that (A.18$_2$) is satisfied. Similarly we can prove (A.18$_3$) and (A.18$_4$).

(4°) Rotate the coordinate system around the x_3 axis by an angle $\pi/4$ to a new coordinate system. Then we have

$$e'_1 = \frac{\sqrt{2}}{2}e_1 + \frac{\sqrt{2}}{2}e_2,$$
$$e'_2 = -\frac{\sqrt{2}}{2}e_1 + \frac{\sqrt{2}}{2}e_2,$$
$$e'_3 = e_3,$$

and correspondingly,

$$(a_{ij}) = \begin{pmatrix} \frac{\sqrt{2}}{2} & \frac{\sqrt{2}}{2} & 0 \\ -\frac{\sqrt{2}}{2} & \frac{\sqrt{2}}{2} & 0 \\ 0 & 0 & 1 \end{pmatrix}. \tag{A.25}$$

From (A.19) we have

$$h_{1111} = \sum_{p,q,r,s=1}^{3} a_{1p}a_{1q}a_{1r}a_{1s}h_{pqrs}.$$

Plugging matrix (A.25) into the above formula, it is easy to see that in the sum on the right-hand side of the above formula, only the terms with p, q, r, and s taking 1 or 2 might not be zero. And then, from (A.18$_5$), we know that only the terms when there is an even number of 1 or 2 in p, q, r, and s might not be zero. So

$$h_{1111} = a_{11}^4 h_{1111} + a_{12}^4 h_{2222} + a_{11}^2 a_{12}^2 (h_{1122}$$
$$+ h_{2211} + h_{1212} + h_{2121} + h_{1221} + h_{2112})$$
$$= \frac{1}{4}(h_{1111} + h_{2222} + h_{1122} + h_{2211} + h_{1212}$$
$$+ h_{2121} + h_{1221} + h_{2112}). \tag{A.26}$$

From (A.18$_2$)–(A.18$_4$) we have

$$h_{1122} = h_{2211} = \lambda,$$
$$h_{1212} = h_{2121} = \alpha,$$
$$h_{1221} = h_{2112} = \beta.$$

Again from (1°) we get

$$h_{1111} = h_{2222} = h_{3333}.$$

Thus, from (A.26) we obtain

$$h_{1111} = \lambda + \alpha + \beta.$$

Equation (A.18$_1$) is then proved. The proof of the theorem is finished. □

A.5 Differentiation of Tensor

The above discussed tensor may be a constant tensor, i.e., each component is a constant under any given coordinate system, as well as a tensor function, i.e., its components are functions (depending on some parameters or the coordinates of spatial points). If the components of a tensor are only functions on some parameters, then any differential operations on the tensor can be taken as on its components, which is simpler. Here we emphasize the case when the components of a tensor are functions of the point coordinates. Suppose that

$$P = \{p_{i_1 i_2 \cdots i_n}(x)\}$$

is an nth-order tensor under the coordinate system $Ox_1x_2x_3$, i.e., under the new coordinate system

$$p'_{i_1 i_2 \cdots i_n}(x') = \sum_{j_1,\ldots,j_n=1}^{3} a_{i_1 j_1} \cdots a_{i_n j_n} p_{j_1 \cdots j_n}(x), \tag{A.27}$$

where $x = (x_1, x_2, x_3)$ and $x' = (x'_1, x'_2, x'_3)$. Now we define some differential operations of this tensor.

(1) Gradient.

The gradient ∇P (or grad P) of P is defined by

$$\nabla P = \{q_{i_1 \cdots i_n k}\}, \tag{A.28}$$

where

$$q_{i_1 \cdots i_n k} = \frac{\partial p_{i_1 \cdots i_n}}{\partial x_k}. \tag{A.29}$$

Now we show that the above defined $\{q_{i_1 \cdots i_n k}\}$ is an $(n+1)$th-order tensor. In fact,

$$q'_{i_1 \cdots i_n k} = \frac{\partial p'_{i_1 \cdots i_n}(x')}{\partial x'_k}$$

$$= \sum_{j_1,\ldots,j_n=1}^{3} \frac{\partial}{\partial x'_k}(a_{i_1 j_1} \cdots a_{i_n j_n} p_{j_1 \cdots j_n}(x))$$

$$= \sum_{j_1,\ldots,j_n,l=1}^{3} a_{i_1 j_1} \cdots a_{i_n j_n} \frac{\partial p_{j_1 \cdots j_n}(x)}{\partial x_l} \cdot \frac{\partial x_l}{\partial x'_k}$$

$$= \sum_{j_1,\ldots,j_n,l=1}^{3} a_{i_1 j_1} \cdots a_{i_n j_n} a_{kl} q_{j_1 \cdots j_n l}.$$

Example A.5. Suppose that $P = \{p\}$ is a zero-order tensor (scalar); then the gradient

$$\nabla P = \left\{\frac{\partial p}{\partial x_k}\right\}$$

of the scalar field is a vector, i.e., a first-order tensor. ∎

Example A.6. Suppose that $P = \{p_i\}$ is a first-order tensor (vector); then its gradient

$$\nabla P = \left\{ \frac{\partial p_i}{\partial x_j} \right\}$$

is a second-order tensor. ∎

(2) Divergence.
The divergence $\nabla \cdot P$ (or div P) of P is defined by

$$\nabla \cdot P = \left\{ \sum_{k=1}^{3} \frac{\partial p_{i_1 \cdots i_{n-1} k}}{\partial x_k} \right\}. \qquad (A.30)$$

Since the divergence $\nabla \cdot P$ of P is obtained by the contraction of its gradient ∇P, it is then an $(n-1)$th-order tensor.

Example A.7. Suppose that $P = \{p_i\}$ is a first-order tensor (vector); then its divergence

$$\nabla \cdot P = \left\{ \sum_{k=1}^{3} \frac{\partial p_k}{\partial x_k} \right\}$$

is the regular divergence of a vector field. This is a scalar, i.e., a zero-order tensor. ∎

Example A.8. Suppose that $P = \{p_{ij}\}$ is a second-order tensor; then it divergence

$$\nabla \cdot P = \left\{ \sum_{k=1}^{3} \frac{\partial p_{ik}}{\partial x_k} \right\}$$

is a first-order tensor, i.e., a vector. ∎

Appendix B
Overview of Thermodynamics

B.1 Objective of the Study of Thermodynamics

The objective of the study of thermodynamics is the physical system made up of a significant number of material particles (molecules, atoms, electrons, etc.). This is a dynamical system with a very large degree of freedom.

If a material particle system (hereafter referred to as *system*) has no interaction with the outside world, then this system is called an *isolated system*. If it has no material exchange with the outside world but may have energy exchange, then it is called a *closed system*.

Classical thermodynamics is aimed at the study of the system in an *equilibrium state*. The so-called equilibrium refers to a system in which the state near each point, from a macroscopic point of view, is the same, despite the complex motion of the material particles which compose the system, and accordingly, for the whole system, the state under consideration can be described by the corresponding state variables (e.g., the absolute temperature, etc.). In addition to the equilibrium state, the *thermodynamic state* is also discussed in classical thermodynamics. If a system as a whole is not in equilibrium, but it can be divided into a finite number of parts and each part is in equilibrium, then we say that the system is in the thermodynamic state. For instance, if a system is composed of two parts A and B which are both in equilibrium, then this system is in the thermodynamic state.

In thermodynamics, the discussion on the change of system state is restricted only to the case when the initial state and final state are both thermodynamic states, while the state at every moment of the process of change may or may not be a thermodynamic state. Generally speaking, these intermediate states are very complex. Only when the processing speed is very slow, compared with the speed at which system tends to internal equilibrium, can the intermediate state of the system be regarded as the thermodynamic state. An important idealized process of change is the *quasi-static process*. If a system is in equilibrium at each moment of the process of change, then this kind of process is called a quasi-static process. This is an idealized process which does not actually exist. But when the initial state and final state are both in equilibrium states and the process of change goes very slowly, this process can be approximately regarded as a quasi-static process.

B.2 The First Law of Thermodynamics; Internal Energy

The basic idea of the first law of thermodynamics is the conservation of energy. Precisely speaking, what is conserved is energy instead of heat.

The first law of thermodynamics. *When a closed system is changing from the initial state α_1 to the final state α_2, the sum of the work ΔW and the heat ΔQ of the system, obtained from the outside, depends only on the initial and final states of the system and has nothing to do with the intermediate process. Therefore, there exists a state variable U depending only on the system such that*

$$\Delta U \stackrel{d}{=} U(\alpha_2) - U(\alpha_1) = \Delta W + \Delta Q. \tag{B.1}$$

This state variable U is called the internal energy *of the system.*

The internal energy of the system is the sum of the kinetic energy caused by the irregular motion of the material particles (usually referred to as the molecules) in the system and the potential energy caused by their interaction.

B.3 The Second Law of Thermodynamics; Entropy

The first law of thermodynamics illustrates the energy relation which must be satisfied by the thermodynamic process, but does not point out the trend which the process changes. For example, suppose that a thermodynamic system is composed of two equilibria A and B, and the temperatures of the subsystems A and B are θ_1 and θ_2, respectively. The first law of thermodynamics tells us that if there is a heat exchange between A and B, then the heat gained by one subsystem must be equal to the heat lost by another subsystem, but the direction of the heat transfer is not indicated. However, experience tell us that the heat must be transferred from the higher-temperature subsystem to the lower-temperature subsystem. Revealing the existence of this asymmetry in nature, that is, the irreversibility of natural phenomena, depends on the second law of thermodynamics.

The second law of thermodynamics can be expressed in many forms. Here we take the form of Clausius inequality. This form of expression may easily derive the important state variable of entropy, although it is not very intuitive.

The process of change of a system with the same initial and final states is called a *cycle*.

The second law of thermodynamics. *Suppose that a system with initial equilibrium completes a cycle when staying in touch with a heat source. In this cycle, if the system absorbs heat $-dQ$ from the heat source with absolute temperature θ^e, then*

$$\oint \frac{-dQ}{\theta^e} \leq 0. \tag{B.2}$$

Equation (B.2) is called the Clausius inequality. *The above closed contour integral is taken along the cycle.*

Now we discuss the reversibility of the process of change. Suppose that the system under consideration changes from the state α_1 to the state α_2, while the external

B.3 The Second Law of Thermodynamics; Entropy

environment changes from the state β_1 to the state β_2. If there is a way that the system may change from the state α_2 back to the state α_1, while correspondingly the external environment changes from the state β_2 back to the state β_1, then this process is called a *reversible process*. Naturally occurring thermodynamic processes are all irreversible.

The quasi-static process, as a kind of idealized process, is reversible. This is because the quasi-static process is an equilibrium at every moment. This kind of process can be described by a curve in the phase space. For instance, for the fluid, only two state variables are independent. We may take these two independent variables as (V, θ), where V is the volume of the gas and θ is the absolute temperature. Then the quasi-static process with initial state α_1 and final state α_2 can be depicted by a curve C on the phase plan (V, θ) (see Figure B.1). This kind of process is certainly reversible. To this end we need only change the state α_2 back to the state α_1 along the original curve C in the opposite direction. From now on, the reversible process is referred to as the quasi-static process.

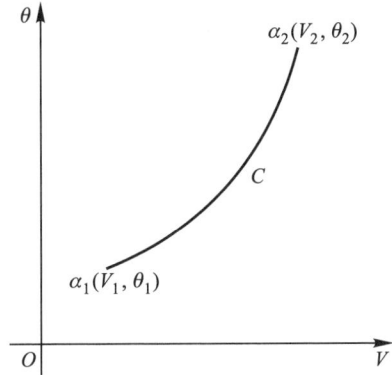

Figure B.1.

Now we investigate the cycle formed by the quasi-static process. Since the quasi-static process in reversible, from (B.2) we have

$$\oint \frac{-dQ}{\theta} = 0, \tag{B.3}$$

where θ is the absolute temperature of the system. Since (B.3) holds for any given closed curve in the phase space, $\frac{-dQ}{\theta}$ is then a total differential. That is, there exists a state variable S such that

$$dS = \frac{-dQ}{\theta}. \tag{B.4}$$

S is called the *entropy* of the system. For the case of the above-mentioned fluid, we have

$$-dQ = dU + pdV, \tag{B.5}$$

where U is the internal energy and p is the pressure; i.e.,

$$-dQ = \left(\frac{\partial U}{\partial V} + p\right) dV + \frac{\partial U}{\partial \theta} d\theta. \tag{B.6}$$

Equation (B.4) shows that the right-hand side of (B.6) is generally not a total differential but has the integrating factor $\frac{1}{\theta}$.

In the above we have defined only the entropy to the equilibrium of the system. But the entropy has the additivity, and the entropy of the system in the thermodynamic state is defined as the sum of the entropy of each subsystem in equilibrium.

Suppose that a thermodynamic system changes from the equilibrium α_1 to the equilibrium α_2 through a certain process L. Then we must have

$$\int_{\alpha_1 L \alpha_2} \frac{-\mathrm{d}Q}{\theta^e} \leq S(\alpha_2) - S(\alpha_1). \tag{B.7}$$

In fact, let R be the quasi-static process changing from α_1 to α_2 (see Figure B.2). From the Clausius inequality (B.2), we have

$$\int_{\alpha_1 L \alpha_2} \frac{-\mathrm{d}Q}{\theta^e} + \int_{\alpha_2 R \alpha_1} \frac{-\mathrm{d}Q}{\theta} \leq 0.$$

But (B.4) gives

$$\int_{\alpha_2 R \alpha_1} \frac{-\mathrm{d}Q}{\theta} = S(\alpha_1) - S(\alpha_2),$$

so (B.7) holds. Equation (B.7) can also be written as

$$\frac{-\mathrm{d}Q}{\theta^e} \leq \mathrm{d}S. \tag{B.8}$$

Hence, if the system is isolated or adiabatic, then we have

$$\mathrm{d}S \geq 0. \tag{B.9}$$

It can be shown that this conclusion still holds for the isolated or adiabatic system with the initial and final states, which are both thermodynamic states. For an isolated or adiabatic thermodynamic system, (B.9) is a condition such that the process may be carried out.

It is not hard to obtain the following from (B.9).

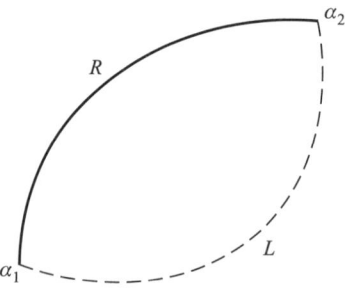

Figure B.2.

Another expression of the second law of thermodynamics. *It is impossible to transfer the heat from a lower-temperature system to a higher-temperature system without causing other changes.*

B.4 Legendre Transform

As we mentioned before, we can take two different thermodynamic state variables as the arguments and obtain the corresponding equation of state to establish the fundamental relations of thermodynamics. To show how to transform between arguments generally, we first introduce the Legendre transform.

We know that the surface in the (x,y,u) space can be regarded as the trajectory of points as well as the envelope of its tangent planes. If the equation of the surface is given by

$$u = u(x,y), \qquad (B.10)$$

then the normal vector across the point $(x,y,u(x,y))$ on it is $(u_x, u_y, -1)$, and thus the tangent plane across this point is

$$\bar{u} - u - (\bar{x} - x)u_x - (\bar{y} - y)u_y = 0,$$

i.e.,

$$u_x \bar{x} + u_y \bar{y} - \bar{u} = xu_x + yu_y - u, \qquad (B.11)$$

where $(\bar{x}, \bar{y}, \bar{u})$ is the variable coordinate on the tangent plane, and $u = u(x,y)$. Thus we obtain a family of tangent planes depending on two parameters (x,y). Denoting

$$\xi = u_x, \quad \eta = u_y, \quad \omega = xu_x + yu_y - u, \qquad (B.12)$$

this tangent plane can be reduced to

$$\xi \bar{x} + \eta \bar{y} - \bar{u} = \omega. \qquad (B.13)$$

(ξ, η, ω) is called the *plane coordinate* of this tangent plane. After having set it, this tangent plane can be determined according to the above formula. Thus, the plane coordinate of the family (B.11) of the tangent planes of the surface (B.10) is given by (B.12). In principle, we can solve x and y as functions of (ξ, η) from $\xi = u_x(x,y)$, $\eta = u_y(x,y)$ and plug them into the last formula of (B.12) to obtain

$$\omega = \omega(\xi, \eta). \qquad (B.14)$$

This formula is called the *tangent plane coordinate equation* of the surface.

These two expressions of the surface can be exchanged with each other. From the above discussion we know that if the point coordinate equation (B.10) of the surface is given, then in principle the tangent plane coordinate equation (B.14) can be obtained accordingly. In addition, from (B.12) we have

$$\omega_\xi = x + \xi \frac{\partial x}{\partial \xi} + \eta \frac{\partial y}{\partial \xi} - u_x \frac{\partial x}{\partial \xi} - u_y \frac{\partial y}{\partial \xi} = x. \qquad (B.15)$$

Similarly,
$$\omega_\eta = y. \tag{B.16}$$

Combining (B.12), (B.15), and (B.16), we conclude that the point coordinate equation $u = u(x, y)$ and the tangent plane coordinate equation $\omega = \omega(\xi, \eta)$ of the surface satisfy the following relations:
$$\begin{cases} \omega = x\xi + y\eta - u, \\ \xi = u_x, \quad \eta = u_y, \\ x = \omega_\xi, \quad y = \omega_\eta. \end{cases} \tag{B.17}$$

On the contrary, if the tangent plane coordinate equation (B.14) is given, namely, a two-parameter family of planes
$$\xi x + \eta y - u = \omega(\xi, \eta) \tag{B.18}$$

is given, where (x, y, u) is the variable coordinate on this plane, then the envelope of this family of planes determines a surface. According to the method of finding the envelope surface, we differentiate (B.18) once with respect to parameters ξ and η and get
$$x = \omega_\xi, \quad y = \omega_\eta. \tag{B.19}$$

Solving ξ and η as functions of (x, y) and plugging them into
$$u = \xi x + \eta y - \omega, \tag{B.20}$$

we get the surface $u = u(x, y)$. This is exactly the point coordinate equation of the envelope surface to be sought. From (B.19)–(B.20), we have
$$u_x = \xi + x\frac{\partial \xi}{\partial x} + y\frac{\partial \eta}{\partial x} - \omega_\xi \frac{\partial \xi}{\partial x} - \omega_\eta \frac{\partial \eta}{\partial x} = \xi. \tag{B.21}$$

Similarly,
$$u_y = \eta. \tag{B.22}$$

Then we obtain that the equations are completely the same as (B.17).

From relation (B.17) we can see that for $\omega = \omega(\xi, \eta)$ determined by $u = u(x, y)$, the point coordinate equation determined by it must still be the original $u = u(x, y)$. Therefore, the point coordinate equation and the tangent plane coordinate equation of the surface are not only equivalent to each other but also related to each other by a group of symmetric formulas in (B.17), and there exists a duality between them.

The transform which turns the point coordinate equation of the surface into the tangent plane coordinate equation of the surface is called the *Legendre transform*. This transform is different from ordinary variable transformations, since the transform itself depends on the function $u = u(x, y)$ under discussion. Because of this, a number of nonlinear problems can be greatly simplified, and even reduced to linear problems.

Now we discuss the conditions under which the Legendre transform can be carried out. It is easy to see that, as long as we can solve x and y as functions of (ξ, η) from
$$\xi = u_x, \quad \eta = u_y, \tag{B.23}$$

B.4 Legendre Transform

the Legendre transform can be carried out. From the implicit function theorem, at the point under consideration, if the Jacobian determinant

$$\begin{vmatrix} u_{xx} & u_{xy} \\ u_{xy} & u_{yy} \end{vmatrix} \neq 0, \tag{B.24}$$

then there exists at least a neighborhood of this point such that x and y can be solved uniquely from (B.23) in this neighborhood. For the surface $u = u(x,y)$ satisfying

$$u_{xx}u_{yy} - u_{xy}^2 = 0, \tag{B.25}$$

the Legendre transform fails. The surface satisfying this relation is a developable surface, that is, the envelope of a one-parameter family of tangent planes. At this time, each tangent plane is tangential to the surface along a straight line, and certainly it is impossible to establish the one-to-one correspondence between the point on the surface and the tangent plane.

Now we prove that if

$$\begin{vmatrix} u_{xx} & u_{xy} \\ u_{xy} & u_{yy} \end{vmatrix} = a \neq 0, \tag{B.26}$$

then the Jacobian determinant of the corresponding inverse transform is

$$\begin{vmatrix} u_{\xi\xi} & u_{\xi\eta} \\ u_{\xi\eta} & u_{\eta\eta} \end{vmatrix} = \frac{1}{a}. \tag{B.27}$$

This shows that the Legendre transform is feasible conversely. In fact, differentiating (B.23) once with respect to ξ and η, we get

$$u_{xx}\omega_{\xi\xi} + u_{xy}\omega_{\xi\eta} = 1,$$
$$u_{xx}\omega_{\xi\eta} + u_{xy}\omega_{\eta\eta} = 0,$$
$$u_{xy}\omega_{\xi\xi} + u_{yy}\omega_{\xi\eta} = 0,$$
$$u_{xy}\omega_{\xi\eta} + u_{yy}\omega_{\eta\eta} = 1,$$

i.e.,

$$\begin{pmatrix} u_{xx} & u_{xy} \\ u_{xy} & u_{yy} \end{pmatrix} \begin{pmatrix} \omega_{\xi\xi} & \omega_{\xi\eta} \\ \omega_{\xi\eta} & \omega_{\eta\eta} \end{pmatrix} = \begin{pmatrix} 1 & 0 \\ 0 & 1 \end{pmatrix}.$$

Equation (B.27) follows immediately from this.

The above method of introducing the Legendre transform for the function of two variables can be similarly applied to the general function $u = u(x_1, \ldots, x_n)$ of n variables (≥ 1), whose rule of Legendre transform is

$$\begin{cases} u + \omega = x_1\xi_1 + \cdots + x_n\xi_n, \\ u_{x_1} = \xi_1, \ldots, u_{x_n} = \xi_n, \\ \omega_{\xi_1} = x_1, \ldots, \omega_{\xi_n} = x_n. \end{cases} \tag{B.28}$$

The condition under which this Legendre transform can be carried on is similar to the case of two variables, we do not go into detail here.

B.5 Thermodynamic Functions

As we mentioned before, for the fluid system in equilibrium, only two are independent among its state variables. By the Legendre transform, taking different independent state variables, we can obtain various kinds of thermodynamic functions.

From (B.4), (B.5) we have

$$dU = \theta dS - pdV. \tag{B.29}$$

If we denote by e and τ the internal energy per unit mass and the volume per unit mass (i.e., the specific volume), while still denoting by S the entropy per unit mass, then (B.29) can be written as

$$de = \theta dS - pd\tau. \tag{B.30}$$

Thus, we can take $\tau \ (= \frac{1}{\rho}; \rho$ is the density) and S as the independent variables, and $e = e(\tau, S)$ is the corresponding equation of state. Then we have

$$\theta = \frac{\partial e}{\partial S}, \quad p = -\frac{\partial e}{\partial \tau}. \tag{B.31}$$

We assume that (these assumptions can usually be satisfied)

$$\frac{\partial e}{\partial S} > 0, \quad \frac{\partial e}{\partial \tau} < 0, \tag{B.32}$$

and the matrix

$$\begin{pmatrix} \dfrac{\partial^2 e}{\partial S^2} & \dfrac{\partial^2 e}{\partial S \partial \tau} \\ \dfrac{\partial^2 e}{\partial S \partial \tau} & \dfrac{\partial^2 e}{\partial \tau^2} \end{pmatrix} \tag{B.33}$$

is positive definite.

Now we utilize the Legendre transform to obtain other independent variables and thus obtain new thermodynamic functions.

Regarding S as a parameter, we take the Legendre transform with respect to τ and $e(\tau, S)$. From the positive definiteness of (B.33), this is possible. From the relation (B.28) as $n = 1$, we have

$$\begin{cases} e + \omega = \tau \xi, \\ \dfrac{\partial e}{\partial \tau} = \xi, \quad \dfrac{\partial \omega}{\partial \xi} = \tau. \end{cases} \tag{B.34}$$

From (B.31) and (B.34) we know that $p = -\xi$. Denoting $i = -\omega$, the first formula of (B.34) can be written as

$$i = e + p\tau. \tag{B.35}$$

i is a function of p and S: $i = i(p, S)$, called the *enthalpy* (per unit mass). From the last formula of (B.34) we have

$$\frac{\partial i}{\partial p} = \tau. \tag{B.36}$$

B.5 Thermodynamic Functions

Using (B.35) and noting (B.31), it is not hard to verify that

$$\frac{\partial i}{\partial S} = \frac{\partial e(\tau, S)}{\partial S} = \theta. \tag{B.37}$$

Thus, when taking p and S as independent variables, τ and θ, as functions of p and S, can be given by (B.36) and (B.37), and we also have

$$\mathrm{d}i = \tau \mathrm{d}p + \theta \mathrm{d}S. \tag{B.38}$$

For the isobaric process, we have $\mathrm{d}i = \theta \mathrm{d}S = -\mathrm{d}Q$. That is, in the isobaric process, the increment of enthalpy is equal to that of heat, so sometimes the enthalpy is also called the *heat function*. We point out that the enthalpy is the state variable, while the heat is not.

Similarly, regarding τ as a parameter, we take the Legendre transform with respect to S and $e(\tau, S)$. The transform relation (B.28) leads to

$$\begin{cases} e + \omega = S\xi, \\ \dfrac{\partial e}{\partial S} = \xi, \quad \dfrac{\partial \omega}{\partial \xi} = S. \end{cases} \tag{B.39}$$

We know from (B.31) that $\theta = \xi$. Denoting $F = -\omega$, the first formula of (B.39) can be written as

$$F = e - \theta S. \tag{B.40}$$

F is a function of τ and θ: $F = F(\tau, \theta)$, called the *Helmholtz free energy* (per unit mass). From the last formula of (B.39) we have

$$\frac{\partial F}{\partial \theta} = -S. \tag{B.41}$$

Similarly to (B.37), it is easy to verify that

$$\frac{\partial F}{\partial \tau} = \frac{\partial e(\tau, S)}{\partial \tau} = -p. \tag{B.42}$$

Thus, when taking τ and θ as independent variables, S and p, as functions of τ and θ, can be given by (B.41) and (B.42), and we also have

$$\mathrm{d}F = -p\mathrm{d}\tau - S\mathrm{d}\theta. \tag{B.43}$$

Now we explain the physical meaning of the Helmholtz free energy. We know that the energy released by a thermodynamic system cannot all be used to do work, some of which must be transferred into the form of heat. What we are interested in is the part that can be used to do work. For the isothermal process, from (B.40) we have

$$\mathrm{d}F = \mathrm{d}e - \theta \mathrm{d}S = \mathrm{d}e - -\mathrm{d}q, \tag{B.44}$$

where $-\mathrm{d}q$ stands for the heat transferred to the system (per unit mass) from the outside, and $\mathrm{d}w = \mathrm{d}e - -\mathrm{d}q$ is the work done by the outside to the system (per unit mass). So, $-\mathrm{d}F$ is exactly the work that the system is able to do to the outside in the isothermal process.

Now we take the Legerndre transform with respect to τ, S, and $e(\tau, S)$. The transform relation (B.28) gives

$$\begin{cases} e + \omega = \tau \xi + S\eta, \\ \dfrac{\partial e}{\partial \tau} = \xi, \quad \dfrac{\partial e}{\partial S} = \eta, \\ \dfrac{\partial \omega}{\partial \xi} = \tau, \quad \dfrac{\partial \omega}{\partial \eta} = S. \end{cases} \quad (B.45)$$

From (B.31) we know that

$$\xi = -p, \quad \eta = \theta. \quad (B.46)$$

Denoting $G = -\omega$, we then have

$$G = e + p\tau - \theta S = i - \theta S. \quad (B.47)$$

G is a function of p and θ: $G = G(p, \theta)$, which is called the *Gibbs free energy* (per unit mass) and also called the *thermodynamic potential*. From the last two formulas of (B.45) we get

$$\frac{\partial G}{\partial p} = \tau, \quad \frac{\partial G}{\partial \theta} = -S. \quad (B.48)$$

Thus, when taking p and θ as independent variables, τ and S, as functions of p and θ, can be given by (B.48), and we also have

$$dG = \tau \, dp - S \, d\theta. \quad (B.49)$$

B.6 Expressions of Internal Energy and Entropy

Now we discuss, when taking τ and θ as independent variables, how to utilize the equation of state $p = p(\tau, \theta)$ and the specific heat at constant volume c_V to express the internal energy and entropy of the fluid. The so-called specific heat at constant volume is the heat absorbed by the fluid per unit mass to raise the temperature by one degree when keeping the volume constant. Obviously,

$$c_V = \frac{\partial e(\tau, \theta)}{\partial \theta}. \quad (B.50)$$

From (B.30) we have

$$\theta \, dS = de + p \, d\tau; \quad (B.51)$$

then

$$\theta \frac{\partial S}{\partial \theta} = c_V \quad (B.52)$$

B.6 Expressions of Internal Energy and Entropy

and

$$\theta \frac{\partial S}{\partial \tau} = \frac{\partial e}{\partial \tau} + p. \tag{B.53}$$

But from (B.43) we have

$$\frac{\partial S}{\partial \tau} = \frac{\partial p}{\partial \theta}. \tag{B.54}$$

Thus, (B.53) gives

$$\frac{\partial e}{\partial \tau} = \theta \frac{\partial p}{\partial \theta} - p. \tag{B.55}$$

Form (B.50) and (B.55) we have

$$\begin{aligned} de &= \frac{\partial e}{\partial \tau} d\tau + \frac{\partial e}{\partial \theta} d\theta \\ &= \left(\theta \frac{\partial p}{\partial \theta} - p\right) d\tau + c_V d\theta, \end{aligned} \tag{B.56}$$

while from (B.52) and (B.54) we have

$$\begin{aligned} dS &= \frac{\partial S}{\partial \tau} d\tau + \frac{\partial S}{\partial \theta} d\theta \\ &= \frac{\partial p}{\partial \theta} d\tau + \frac{c_V}{\theta} d\theta. \end{aligned} \tag{B.57}$$

The equation of state $p = p(\tau, \theta)$ and c_V can usually be obtained by experiments. Thus, the expressions of e and S can be obtained by (B.56) and (B.57).

For the *ideal gas*, the equation of state is

$$p = R \frac{\theta}{\tau}, \tag{B.58}$$

where R is a positive constant. From (B.55) we now have

$$\frac{\partial e}{\partial \tau} = 0. \tag{B.59}$$

This implies that the internal energy e of the ideal gas is a function of only θ: $e = e(\theta)$, and so is the specific heat at constant volume c_V: $c_V = c_V(\theta)$. Thus, from (B.50) we get

$$e = \int c_V(\theta) d\theta, \tag{B.60}$$

while from (B.57), the expression of entropy is given by

$$\begin{aligned} S &= \int \frac{R}{\tau} d\tau + c_V(\theta) \frac{1}{\theta} d\theta \\ &= \int R d\ln \tau + c_V(\theta) d\ln \theta. \end{aligned} \tag{B.61}$$

When c_V is a constant (when the temperature is not too high, the ideal gas can be regarded as satisfying this condition), we say that the gas is a *polytropic gas*. At this moment, from (B.60) and (B.61), and noticing that the internal energy should be 0 when the absolute temperature is 0, we have

$$e = c_V \theta, \tag{B.62}$$

$$S = c_V \int \mathrm{d}\ln\left(\theta \tau^{\frac{R}{c_V}}\right)$$
$$= c_V \ln(\theta \tau^{\gamma-1}) + \text{constant}, \tag{B.63}$$

where $\gamma = \frac{R}{c_V} + 1$ is called the *adiabatic exponent*. Using the equation of state (B.58), the above formula can also be written as

$$S = c_V \ln\left(\frac{p\tau^\gamma}{R}\right) + \text{constant}$$
$$= c_V \ln(p\tau^\gamma) + \text{constant}. \tag{B.64}$$

Form (B.64) we also get

$$p = (\gamma - 1)\tau^{-\gamma} \exp\left(\frac{S - S_0}{c_V}\right),$$

where S_0 is a constant. Thus, from (B.62), (B.58), and the definition of γ, the internal energy can be expressed by the following formula:

$$e = \tau^{-(\gamma-1)} \exp\left(\frac{S - S_0}{c_V}\right). \tag{B.65}$$

Using this formula, it is easy to verify the assumption (B.32) and the positive definiteness of the matrix (B.33) for the polytropic gas.

Index

absolute temperature, 70
adiabatic exponent, 76, 260
admissible boundary condition, 27
Alfvén characteristic speed, 155
Alfvén theorem, 147
Ampère theorem, 9
Ampère–Biot–Savart law, 8
anisotropic tensor, 243

Beltrami equations, 179
bound charge, 41
boundary value problem with equivalued surface, 53

Cartesian tensor, 237
Cauchy elastic, 194
Cauchy problem, 23, 78, 104–106, 208
Cauchy strain tensor, 180
Cauchy stress tensor, 188
centered rarefaction wave, 115
characteristic curve, 88
Ciarlet–Geymonat material, 202
Clausius inequality, 250
closed system, 249
compressible Ogden material, 202
compressive shock, 116
conduction current, 42
conservation and transformation law of (electromagnetic) momentum, 19
conservation and transformation law of electromagnetic energy, 18
conservation law of magnetic field, 147
constitutive equation, 97
constitutive relation, 175, 194
contact discontinuity, 111
continuity equation, 71, 136, 183
continuity equation of electric current, 7
contraction of tensor, 240

conversation law of electric charges, 7
Coulomb gauge, 36
Coulomb's law, 2
current density, 7
cycle, 250

Darwin model, 58, 61
deflagration, 161
deformation gradient tensor, 178
detonation, 161
dielectric constant, 2, 41
differential form of Ohm's law, 51
diffusion equation, 142
dipole, 36
Dirichlet boundary condition, 50
Dirichlet integral, 50
displacement current, 43
displacement gradient tensor, 180
displacement vector, 179
dyadic vector, 19, 240

elastic body, 175
elastic material, 194
elastic tensor, 213
electric conductivity, 51, 67
electric dipole moment, 36
electric displacement vector, 41
electric field, 2
electric field intensity, 1, 3
electric field line, 3
electric flux, 3, 4
electric flux density, 41
electric neutrality, 132
electric polarizability, 41
electromagnetic energy, 18
electromagnetic energy density, 21, 47
electromagnetic energy flux density vector, 21, 47

electromagnetic field, 1
electromagnetic induction, 11
electromagnetic momentum, 19
electromagnetic momentum density vector, 21, 47
electromagnetic momentum flux density tensor, 21, 47
electromagnetic wave, 15, 27
electrostatic field, 2, 48
electrostatic field potential, 6
energy density, 71
energy flux vector, 71
enthalpy, 256
entropy, 75, 79, 218, 251
entropy condition, 116
entropy flux, 218
entropy flux vector, 79, 84
entropy function, 84
entropy inequality, 116, 218
equation of state, 70
equations in Eulerian representation, 120
equations in Lagrangian representation, 120, 158, 171
equilibrium state, 249
equivalued surface boundary condition, 50, 52
Euler equation, 73, 98, 138
Eulerian coordinates, 120, 176
Eulerian representation, 177
expansive coefficient of viscosity, 97

"frozen" principle of magnetic field lines, 147
Faraday's law of electromagnetic induction, 12
fast characteristic speed, 155
field with source, 5
finite elasticity theory, 176
first coefficient of viscosity, 97
first law of thermodynamics, 250
first-order quasi-linear hyperbolic system, 87, 157, 172
first-order quasi-linear symmetric hyperbolic system, 76, 78, 151, 168, 219, 221
first-order quasi-linear system of partial differential equations, 75

first-order symmetric hyperbolic system of partial differential equations, 23
first-order system of partial differential equations, 22
first-order tensor, 239
Fourier's experimental law, 92
free electromagnetic wave, 28
frictional shear stress, 90
Friedrichs inequality, 226

Gauge invariance, 34
Gauge transform, 34
Gauss's law, 4
generalized Hooke's law, 204
generalized Newton's law, 97
generalized shear modulus, 224
geometric nonlinearity, 175
Gibbs free energy, 258

harmonic equation, 56
heat conductivity, 92
heat flux density vector, 92
heat function, 257
Helmholtz free energy, 257
homogeneous material, 194
Hugoniot equation, 112
Hugoniot function, 112
hyperbolic system, 88, 215
hyperelastic, 197

ideal fluid, 69
ideal gas, 70, 259
ideal magnetofluid, 149
ideal magnetohydrodynamics system, 149
incompressible fluid, 105
infinitesimal strain tensor, 180
initial-boundary value problem, 23, 25, 105, 106, 208, 214
interface condition, 44
internal energy, 70, 250
irrotational field, 6
isentropic flow, 89
isolated system, 249
isotropic, 198
isotropic tensor, 244

Index

Joule heat, 141
Joule–Lenz law, 141
jump condition, 111

k-shock, 116
kinetic coefficient of viscosity, 97
Korn inequality, 225, 226

Lagrangian coordinates, 120, 176
Lagrangian representation, 176
Lamé constant, 200
Laplace operator, 10
largest nonnegative subspace, 27
left Cauchy–Green strain tensor, 179
leftward shock wave, 116
Legendre transform, 82, 85, 254
Lenz's law, 12
linear elastic body, 175
linear elasticity theory, 175
local speed of sound, 77
longitudinal field, 31
longitudinal wave, 30
longitudinal wave condition, 31
Lorentz condition, 35
Lorentz force, 16
Lorentz force formula, 1, 16
Lorentz gauge, 35

magnetic conductivity, 8, 42
magnetic dipole moment, 66
magnetic field, 8
magnetic field intensity, 42
magnetic induction intensity, 1, 8
magnetic susceptibility, 42
magnetization, 42
magnetohydrodynamics, 133
magnetohydrodynamics system, 134, 141
magnetostatic field, 9, 53
mass density, 69
mass flux vector, 70
material coordinates, 120, 176
material derivative, 177
material nonlinearity, 175
material representation, 176
Maxwell's equations, 1, 13
Maxwell's equations in a vacuum, 15
method of energy integral, 23
minimizing sequence, 230

momentum density vector, 70
momentum flux tensor, 70
Mooney–Rivlin material, 202

nth order tensor, 239
natural boundary condition, 230
Navier–Stokes equations, 106
Neo–Hookean material, 202
Neumann boundary condition, 50
Newtonian fluid, 90
non-Newtonian fluid, 90
nonlinear elasticity theory, 176
normal stress, 188

objectivity hypothesis, 195, 196
Ogden material, 202
one-dimensional quasi-linear wave equation, 223, 224

p-system, 124
Piola stress tensor, 193
Piola stress vector, 193
plane coordinate, 253
plane electromagnetic wave solution, 28
plane wave, 28
plasma, 131
Poisson equation, 11
Poisson's ratio, 205
polar decomposition, 178
polarization, 41
polyconvex, 231
polytropic gas, 70, 260
poynting vector, 18, 47
Prandtl's relation, 117
pressure, 70
principal axis direction (principal direction) of tensor, 242
principal invariant of tensor, 243
principal value of tensor, 242

quasi-electrostatic model, 58
quasi-electrostatic state, 57
quasi-harmonic equation, 48, 52, 56
quasi-linear symmetric hyperbolic-parabolic coupled system, 103, 149, 168
quasi-static process, 249

radiation region, 40
Rankine–Hugoniot conditions, 111
reference configuration, 176
relative dielectric constant, 41
relative magnetic permeability, 42
resistivity, 51
response function, 194
reversible process, 251
right Cauchy–Green strain tensor, 179
rightward shock wave, 116
rotational field, 9

scalar, 237
scalar potential, 31, 34
scalar potential method, 54
scalar potential of magnetostatic field, 55
second coefficient of viscosity, 97
second law of thermodynamics, 250, 253
second Piola stress tensor, 193
second-order hyperbolic system, 208
second-order linear elliptic system, 225
second-order quasi-linear elliptic system, 227
second-order quasi-linear hyperbolic system, 214
second-order symmetric tensor, 243
second-order tensor, 239
sequential weak lower semicontinuity, 231
shear modulus, 206
shear stress, 188
shock wave, 111
skin effect, 49
slow characteristic speed, 155
Sobolev space, 108, 225, 229
solenoidal field (source-free field), 11
spatial coordinates, 176
spatial representation, 177
specific heat at constant volume, 70
specific volume, 75
St. Venant–Kirchhoff material, 201
stability condition, 212
static field region, 39
steady-like field equation, 58
stored-energy function, 197
strain energy function, 197
strain–stress relation, 205
stress, 90, 183

stress tensor, 93
stress vector, 93, 184
stress–strain relation, 205
strictly hyperbolic system, 88, 157
strong ellipticity condition, 208, 214
surface force, 183
symmetric parabolic system, 103
symmetric parabolic system in the sense of Petrovsky, 103
system of conservation laws, 78
system of reacting fluid dynamics, 164

tangent plane coordinate equation, 253
tensor identification theorem, 93, 241
tensor product, 19, 240
thermodynamic potential, 258
thermodynamic state, 249
thermodynamical shock condition, 112
three-dimensional Navier–Stokes equations, 106
total energy, 162
total flux boundary condition, 50
total flux boundary value problem, 53
transverse field, 30, 53
transverse wave, 30
transverse wave condition, 30
two-dimensional Navier–Stokes equations, 108

vector, 237, 239
vector potential, 31, 34
vector potential method, 53
vector potential of magnetostatic field, 53
velocity vector, 69
viscosity coefficient, 90
volume elastic modulus, 207
volume force, 183
volume force density, 72

wave equation, 27, 210
wave field region, 40
wave vector, 29
weak solution, 230
weakly space-like surface, 23

young modulus, 205